Eco-friendly Agro-biological Techniques for Enhancing Crop Productivity

Rakesh Singh Sengar • Ashu Singh
Editors

Eco-friendly Agro-biological Techniques for Enhancing Crop Productivity

Springer

Editors
Rakesh Singh Sengar
Department of Agricultural Biotechnology
College of Agriculture
S.V.P University of Agriculture
& Technology
Meerut, Uttar Pradesh, India

Ashu Singh
Department of Agricultural Biotechnology
College of Agriculture
S.V.P University of Agriculture
& Technology
Meerut, Uttar Pradesh, India

ISBN 978-981-10-6933-8 ISBN 978-981-10-6934-5 (eBook)
https://doi.org/10.1007/978-981-10-6934-5

Library of Congress Control Number: 2018934387

Printed on acid-free paper

This Springer imprint is published by the registered company Springer Nature Singapore Pte Ltd. part of Springer Nature.
The registered company address is: 152 Beach Road, #21-01/04 Gateway East, Singapore 189721, Singapore

Contents

About the Editors

R. S. Sengar the author of the book is Professor and Head, Department of Agriculture Biotechnology, College of Agriculture in Sardar Vallabhbhai Patel University of Agriculture and Technology, Meerut. Dr. Sengar has 24 years of teaching, research, and extension experience to his credit. He remained associated with the teaching for a period of about 10 years at G.B. Pant University of Agriculture and Technology, Pantnagar, U.S. Nagar (Uttarakhand). Prof. Sengar possesses a career of leadership in improving agrarian livelihood, food, nutrition, and ecological security in cutting-edge research and technology development, in higher education and human resources development, and in policy and program formulation, execution, and appraisal, each pursued in national and global capacities. He has supervised 11 Ph.D. students and published more than 98 research papers in Indian and foreign journals of repute. Dr. Sengar contributed more than 115 abstracts/papers to different conferences/symposiums/workshop organized at national and international forums. He also participated in several international programs and seminar/conferences as chairperson of technical sessions, group leader, invited speaker, and delegate. He has published 885 popular articles in Hindi and English languages in important journals and magazines. He has also published 11 books for scientific community and farmers. Dr. Sengar has delivered more than 205 talks at All India Radio and Television at Delhi, Lucknow, Rampur, and Bareilly centers. Dr. Sengar is life member of several professional societies. He is a member of editorial boards and review committees of few journals. He has been the principal and co-investigator of 10 projects financed by various government departments, while he has honor of mentorship of two national projects under Department of Science and Technology, Government of India, and has a wide exposure to international programs and also in handling international projects. He has widely traveled and has visited United Kingdom, Nepal, and Dubai under various programs. Dr. Sengar is the recipient of the "Best Writer" Award from Vishwa Agro Marketing and Communication, Kota, Rajasthan; "Kunwar Saxena Bahadur SRDA" Award from Society for Recent Development in Agriculture, Meerut; "Man of the Year" and "Research Board of Advisors" Awards from American Biographical Institute, Inc. USA; and "Aryabhat" Award from Vigyan Bharti, New Delhi. Dr. Sengar has also received Gold Medal 2011 from Hi-tech Horticulture Society, Meerut; Dr. J.C. Edward Medal 2012 from Bioved Research Society, Allahabad; "Scientist of the Year – 2013" from The

Academy of Environmental Biology, Lucknow; Life Time Achievement Award – 2014 from Dr. Babasaheb Ambedkar National Institute of Social Sciences, Dr. Ambedkar Nagar (Mhow), Madhya Pradesh; Dr. R.B. Lal Vishisht Sikshak Puruskar – 2015 from Uttar Pradesh Academy of Agricultural Sciences, Lucknow; Farm and Food Award – 2016 from Farm and Food, New Delhi; and Young Achiever Award – 2016 from Society of Plant Research, Bareilly. He is fellow of the Society of Plant Research, Bareilly; Society for Recent Development in Agriculture, Meerut; Gramin Vikas Avam Siksha Prasar Samiti, Agra; and Technical Society for Basic and Allied Sciences, New Delhi. Dr. Sengar has made significant contributions in the fields of plant physiology, biotechnology, and crop production and management systems besides providing leadership in shaping up national level research institutions and agricultural universities. Dr. Sengar has assisted several developing countries in enhancing their human resource, research and technology development capacities, and in augmenting productivity, sustainability to enhance farmer's income and alleviate hunger and poverty crises.

Ashu Singh was born in Allahabad, India. She received the B.Tech. degree in Genetic Engineering from the Allahabad Agricultural Deemed University, Allahabad, India, in 2008, and the M.Tech. degree in Biotechnology Engineering from the Sardar Vallabhbhai Patel University of Agriculture and Technology (State Agriculture University) Modipuram, Meerut, in 2011. In 2013, she joined the Department of Agriculture Biotechnology, Sardar Vallabhbhai Patel University of Agriculture and Technology (State Agriculture University) Modipuram, Meerut, as a Women Scientist (WOS-A). Since April 2013, she has been with the Department of Agriculture Biotechnology, College of Agriculture, where she is working on varietal improvement of aromatic rice. Her current research interests include molecular characterization and gene mapping of salinity and drought resistance of aroma rice (*Oryza sativa* L.). Miss Singh is a National Woman Scientist Fellow of the Department of Science and Technology New Delhi, India. She has a long experience of eight trainings of more than 3 months at different reputed institutes and research laboratories with technical expertise in molecular breeding and also has made significant contribution in the vast arena of stress physiology, biotechnology, and basmati rice production while exploring the applicability of research in enhancing productivity. She also holds Diploma in "Industry Programme in Bioinformatics" (BII, Noida). She has submitted research study on "Importance of Protein Folding for Protein Structure." Apart from this she is also an Associate Editor at *Biotech Today* journal published under the Society of Green World for Sustainable Environment along with being an active reviewer of many reputed journals. Miss Singh contributed 12 research papers in reputed indexed journals worldwide and attended several conferences/symposiums/workshops organized at national and international forums.

Organic Agriculture: Techniques to Improve Crop Production

Somasundaram Eagan and Udhaya Nandhini Dhandayuthapani

Abstract

Global awareness of health and environmental issues is increasing in recent years, and there is a growing demand for organically grown food products worldwide. We are also convinced that modern agriculture may not be able to meet the requirements of the ever-increasing population in the future. So, we are searching for alternative technologies. Several alternatives have been proposed such as low external input agriculture, sustainable agriculture, organic farming, biodynamic farming, etc. However, they require, sometimes, little or considerable external inputs whose availability may be uncertain in the future. Hence for the developing countries, other alternatives, viz. traditional methods, have special advantages over modern agricultural techniques. For centuries, farmers have planned agricultural production and conserved natural resources by adopting indigenous knowledge. Indigenous technical knowledge (ITK) has immense potential for innovation, especially at the grassroots level. India is a country populated by a number of indigenous communities, most of which have their own set of unique traditional knowledge and technology base. Also the capital and technological skill requirements in the use of traditional technologies are generally low, and their adoption often requires little restructure of the traditional societies. The traditional technologies are nothing but indigenous technical knowledge. By adopting such indigenous knowledge, our ancestors did not face any problem of large-scale pest outbreak or economic crisis unlike today's farmers. Energy crisis, higher fertilizer cost, sustainability in agriproduction system and ecological stability are the important issues which renewed the interest of farmers and research workers to opt for non-chemical sources of plant nutrients like bio-fertilizers, *Azolla* and organic manures, viz. farmyard manure, vermicompost, poultry manure and green manure.

S. Eagan (✉) · U. N. Dhandayuthapani
Department of Sustainable Organic Agriculture, Tamil Nadu Agricultural University, Coimbatore, Tamil Nadu, India

R. S. Sengar, A. Singh (eds.), *Eco-friendly Agro-biological Techniques for Enhancing Crop Productivity*, https://doi.org/10.1007/978-981-10-6934-5_1

Keywords
Environmental issues · Modern agriculture · Indigenous technical knowledge · Economic crisis and Organic farming

1 Introduction

Recently, pollution has drawn lot of attention at local and international level as well, and it was well documented that chemicals used in agriculture are the main sources of soil and water pollution. Presently the farming situation urges need to develop farming techniques, which are sustainable from environmental, production and socio-economic points of view. Modern agricultural production throughout the world does not appear to be sustainable in the long run. Sustainable agricultural development refers to a range of strategies for addressing many problems that affect agriculture. A sustainable agriculture is a system of agriculture that is committed to maintain and preserve the agriculture base of soil, water and atmosphere ensuring future generations the capacity to feed themselves with an adequate supply of safe and wholesome food. Such sustainable development in the agriculture, forestry and fishery sectors conserves land, water, plant and animal genetic resources and is environmentally non-degrading, technically appropriate, economically viable and socially acceptable. Such concerns imparted a way to organic farming. It is the need of the day to understand the prospects and problems of organic farming to launch a successful and flawless organic production programme in the farm environment.

India over several millenniums has been the treasure of biological wealth, intellectual knowledge and spiritual wisdom. During both pre- and post-Vedic periods, arts, science and technology had flourished well in various fields. Indian agriculture to some extent is much traditional and dates back to about 10,000 years. The traditional practices or grassroots technologies are low cost, easy to adopt as they depend on local resources and often not polluting the environment and food chain system.

Through the long path from primitive agriculture to the modern farming, people have developed a number of farming techniques through their own age-old experiments by trial and error in their attempt to overcome numerous problems faced during the farming operations. This knowledge is based on many generations of insight gained through close interaction within the natural and physical micro-environments (Kolawole 2001). Derived from experience gained over the centuries and adopted to the local culture and environment, traditional knowledge is transmitted orally from generation to generation. It is derived in the form of stories, songs, folklores, proverbs, cultural values, beliefs, rituals, community laws, local language and agricultural practices, including the development of plant species and animal breeds. It is the basis for local-level decision in agriculture, health care, food preparation, education, natural resource management and a host of their activities in rural communities (Warren 1993). Indigenous knowledge is the information base for a society, which facilitates communication and decision-making. Indigenous information systems are dynamic and are continually influenced by internal creativity and experimentation as well as by contact with external systems.

Though the indigenous technical knowledge (ITK) is region specific, it can be applicable to similar agroclimatic conditions because most of the indigenous agricultural technologies have got scientific rationale (Sabarathinam 1997). Now the need has come to re-examine and then gradually reintroduce the effective traditional technologies of crop production. However, one may doubt about the productivity of the indigenous practices and varieties. These fears could be dispelled by the reports of Rathnam (1966) that the paddy productivity in the Chengalpattu area of Tamil Nadu during the 1780s was more than 2500 lbs/acre. Comparatively in the 1960s, the productivity was only 1680 lbs/acre. Similar data exist for North Arcot, South Arcot, Tanjore and Coimbatore districts of Tamil Nadu, for the late eighteenth century to early nineteenth century.

2 Rationale and Principles of ITKs

The ecologically sound designs or elements of indigenous practices, which have been losing importance in high technology production system, must be saved and synthesized appropriately to attain sustainable farming. The total eco-design involves combining local elements with innovation-based traditional methods with newer elements to form a balanced whole. In order to encourage and consistently apply this indigenous knowledge, which is degenerating, it needs scientific backup. Though the ITKs are region specific, once the advantages are ascertained, the principle underlying the practices may be extrapolated to similar agroclimatic conditions, by carrying out the experiments using the materials locally available in the farm holdings. An agricultural technology based on Indigenous knowledge may bring moderate to high levels of productivity using local resources. Given favourable political, social and ecological conditions, such agrotechnologies may be sustainable at a low cost for a long period (Quiroz 1996). Also, in our National Agricultural Policy, concerted efforts were made to pool, distil and evaluate traditional practices, knowledge and wisdom on organic farming and to harness them for sustainable agricultural growth. Efforts are being made as programmes for utilization of domestic and agricultural waste for organic matter repletion and pollution control.

All ITKs go by the principle of 'permanence'. It is not so with modern technologies with synthetic inputs. All the ills in crop production (soil mining, degradation, pollution, etc.) are due to decreasing attention to ITK. Indigenous practices in agriculture are organic in nature. They do not cause any damage to the air, water and soil, safe to the human beings and are free from causing environmental pollution. Therefore, now the need has come to re-examine and then reintroduce the effective traditional methods of crop production and protection using organic sources, because there is considerable demand and scope for development of organic technologies either individually or as a package, without necessarily aiming at full adoption of organic system (Sharma 2002).

Indigenous agricultural practices can play a key role in the design of sustainable and eco-friendly agricultural system, increasing the likelihood that the rural populations will accept, develop and maintain innovations and interventions. In the

Declaration Statement of the 88th session of Indian Science Congress held in New Delhi, 2001, it is opined that if the modern techniques are integrated with the traditional and indigenous practices, that will alleviate the poverty and results in the prosperity of the country. Many indigenous practices documented by Sundaramari and Ranganathan (2003) illustrate how well the farmers in the tropics learned to manipulate and derive technologies from local resources and natural processes, applying the principles of agroecology without knowing that this term exists.

Some of the agroecological practices are discussed below for the benefit of farmers.

3 Scope and Importance

In Tamil Nadu, more than 65% of the population is engaged in agriculture, and state economy largely depends on agriculture sector. The importance of indigenous technology and practice to sustainability is being brought through pooling of traditional knowledge, shortlisting and evaluating them in the context of modern scientific and technological environment and harnessing it for sustainable agriculture growth. For achieving sustainability in agriculture, indigenous technical knowledge (ITK) is essential, and it should be identified, classified and documented for refining supported with scientific reasons and mixing with scientific technology through various modes for its immediate acceptance and adoption of the technology in the user system without any time lag. Since, time lag is the prime factor in the diffusion and adoption process of any technology. Use of indigenous knowledge in the development projects generally facilitates the participation of the beneficiaries. A blend of indigenous knowledge and modern technology may be most appropriate for sustainable development (Radhakrishnan et al. 2009).

Transferring of indigenous knowledge from generation to generation has ended because now it has been replaced by modern scientific knowledge. Today in India, as in many other developing countries with a rich agricultural tradition of their own, the words 'improved agriculture' and 'progressive agriculture' have become synonymous with the spread of high-yielding varieties grown with ever-increasing doses of chemical fertilizers and pesticides. Wherever the new crop varieties have spread, time-honoured crop rotations, intercropping patterns and other important features of traditional agriculture have been harshly uprooted. It is very much essential to concretize this experience into a system and develop it as methodology.

Characteristics of ITK

- ITK is dynamic but not static.
- Exogenous knowledge and endogenous creativity bring change to ITK.
- ITK is intuitive in its mode of thinking.
- ITK is mainly qualitative in nature.
- ITK study needs a holistic approach.
- ITK, if properly tapped, can provide valuable insights into resources, processes, possibilities and problems in particular area.

- ITK is recorded and transferred through oral tradition.
- ITK is learned through observation and hands-on experience.
- ITK forms an information base for variety.
- ITK reflects local tradition.

4 Advantages of Indigenous Technical Knowledge

- Two powerful advantages:
  - Low cost and non-monitoring
  - Readily available
- ITK system: socially desirable, economically affordable, sustainable and involves minimum risk to rural farmers and producers – believed to conserve resources.
- ITK: problem-solving strategies for local communities especially small and marginal farmers:
  - ITK is the basis for self-sufficiency and self-determination.
  - People are familiar with indigenous practices and techniques.
  - ITK draws on local resources.
- ITK provides effective alternatives to western know-how.
- Indigenous technologies and practices are often cheaper than modern agriculture.
- Efficiency:
  - ITK is cost-effective.
  - ITK uses appropriate technology.
- Effectiveness:
  - ITK improves chances of adoption.
  - ITK is integral to local communities.
- Sustainability:
  - ITK facilitates mutual adaptation and learning.
  - ITK empowers local communities.

5 Indigenous Techniques for Crop Production (Somasundaram et al. 2015)

5.1 Agricultural Practices

5.1.1 Crop Rotation

Crop rotation involves alternating different crops in a systematic sequence on the same land. To stimulate the microbial population that is present in the root zone of plants (the rhizosphere), crop rotation is a very important technique that has to be followed by the farmers. Different species of plants are raised to reduce pest and disease attack, maintain the fertility status of soil and ultimately increase the yield. There are four fundamental reasons for including crop rotation in organic farming are as follows: (a) not all plants have the same nutritive needs, (b) soil structure is improved through the action of roots, (c) pest buildup is avoided and (d) it helps against the buildup of weeds. Component crops are so chosen so that soil health is not impaired, e.g. cotton-gram and sugarcane-wheat. Or it means growing a set of crop in a regular succession on a piece of land in a specific period of time, with an object to get maximum profit least investment without impairing soil fertility.

5.1.2 Companion Crops/Mixed Crops

Companion crops/mixed crops are the crops grown generally to avoid diseases and pest attacking the main crop. The companion crop attracts the pests that attack the main crop in turn maintaining the quality and quantity of yield. This helps in reducing the manpower and extra costs incurred on pesticides or insecticides.

For example, brinjals and chillies are usually attacked by fruit borer. If flower crop like chrysanthemum flower is grown as a companion crop along with the main crop, then the pest attack can be avoided.

5.1.3 Mulching

Mulching is suitable for long duration and wide distant plants. It is mainly undertaken during summer season to reduce evapotranspiration and irrigation.

Mulching is a way of avoiding direct sun rays falling on the crops by covering the plants with waste materials like banana leaves, coconut leaves and grass.

Collect the raw materials (dry paddy grass, dried banana leaves, dried coconut leaves and other waste materials) used for mulching, and heap in one place. Cover the banana, vegetables, fruits and mulberry plants completely with mulching materials in the summer months. The raw materials are either spread on the ground leaving a space for passage or mulched around individual plants. Before mulching is undertaken, care should be taken to see that there are no weeds in the plot.

Uses

- Maintains moisture and reduces repeated irrigation.
- Controls the weeds and reduces labour cost on weeding.
- Materials get converted to good farm manure.
- Maintains soil structure around the plants and in turn helps in in situ vermicomposting. High-quality vermicompost is formed in the mulched area.

5.1.4 Green Manuring

Addition of leaves, stems and other parts of the plant to the soil when they are still green to improve the fertility of the soil is called green leaf manure.

1. *In Situ Green Leaf Manuring*

Growing various types of leguminous plants like sunhemp, dhaincha, horse gram, niger, cowpea, jute, etc. in the field before 15 days of sowing/transplanting and finally incorporating the same in the soil during ploughing.

2. *Ex Situ Green Leaf Manuring*

Getting the green leaves, stems and branches from the trees grown outside like *Gliricidia*, *Cassia semia*, jackfruit, *Pongamia*, niger, etc. and incorporating the same within the soil just 15 days before sowing or transplantation of the crop in the field.

Uses

The green leaf manure rots and provides more nitrogen to the plants ultimately saving organic manure application. Addition of various types of green leaves to the soil controls pests and diseases. Two to three tons of green leaf manure can be added to one acre of land.

5.1.5 Tank Silt Application

Silt is a combination of sand and clay particles collected from tanks or lakes in the villages. It is mainly done to improve soil and moisture conservation as well as increase the aeration and porosity of the soil.

The main field before sowing of the crop must be prepared well by ploughing and bunding along the margins to avoid runoff. Silt is collected in tractors from the tanks/lakes and spread evenly on the surface of the soil. Since silt has adhesive property as it contains kaolinite and montmorillonite, it gets mixed in the main field on the onset of monsoon which in turn helps in retention of moisture and nutrients in the soil making it available to plants. Twenty to 25 tractor loads of silt are needed for 10 acres of land. Silt can be applied to the soil once in 3 years.

Uses of Silt Application

Application of silt in the main land helps in retention of nutrients in the soil and increases the fertility of soil which in turn increases crop yield. Application of silt increases the water holding capacity of the soil providing moisture throughout the growing season of the crop. Silt application can convert barren land into fertile soil. Silt application can be applied to any type of soil but best suited for soils having sandy/sandy loam property.

Silt application helps to increase the water table of the soil.

Before application of silt, farmers have to get their soil tested and complete knowledge of silt content from the soil testing laboratory and take necessary steps as per the lab recommendations.

5.1.6 Intercropping

Cultivating companion crops in between main crops. Space in between main crops is utilized judiciously. The intercrops should be selected such that they should be fast growing than main crop. Intercrops should not compete with main crop. Also they should fix atmospheric nitrogen.

5.1.7 Cover Crop

Cover crop means plants or a green manure crop grown for seasonal soil protection or soil improvement. Cover crops help control soil movement and protect the soil surface between crops. Cover crop reduces wind erosion by shielding the soil with vegetation and anchoring the soil with roots. In India, green manure crops like sunhemp, cowpea, dhaincha, lupins, etc. are more commonly used. Legume cover cropping in grape, mango, guava and other fruit crops is becoming a common practice in the management of orchards. Cowpea and French beans grow well under guava and sapota tree. In some places to prevent soil erosion, certain permanent cover crops like *Calapogonium muconoides*, *Centrosema pubescens* and *Peuraria phaseoloides* are raised in the alley spaces. They are leguminous crops, are established in a short period and dry up during summer to conserve moisture. With summer showers they come up again because of their profuse seeding habit and spread themselves as a vegetative mat by the time the heavy monsoon starts pouring in. Such permanent cover cropping is a common feature in rubber plantations of Kerala and Kanyakumari district.

5.2 Plant Growth Promoters

5.2.1 Panchagavya

- Fresh cow dung 4 kg
- Cow urine 3 l
- Cow milk 2 l
- Curd 2 l
- Cow desi ghee 1 kg

Mix all the ingredients thoroughly and ferment for 15 days with twice stirring per day. Dilute 3 l of panchagavya in 100 l water and spray over soil. Twenty litre of panchagavya is needed per acre for soil application along with irrigation water. Panchagavya can also be used for seed treatment. Soak seeds for 20 min before sowing (Table 1).

5.2.2 Fish Amino Acid/Meat Gunabasalam

This input is mostly used in organic farming in Japan, Korea, etc. 1 kg fish not suitable for consumption along with 1 kg of indigenous sugar kept inside an air tight glass contained with lid. Water should not be added during the process and kept for 21 days for anaerobic fermentation. Filter on the 22nd day to get approximately 300–400 g honey like extract spray at 1% (1 ml/l of water) which produces greening effect on crops.

5.2.3 Vermiwash

Take a plastic drum with 250 kg capacity and make a small hole at the bottom of the container. Fill the container at the bottom with small stones/pebbles to a height of

Table 1 Panchagavya spray schedule for various crops

Crops	:	Time schedule
Rice	:	10, 15, 30 and 50th DAT
Sunflower	:	30, 45 and 60 DAS
Black gram	:	Rainfed: first flowering and 15 DAS
		Irrigated: 15, 25 and 40 DAS
Green gram	:	15, 25, 30, 40 and 50 DAS
Castor	:	30 and 45 DAS
Groundnut	:	25 and 30th DAS
Bhendi	:	30, 45, 60 and 75 DAS
Moringa	:	Before flowering and during pod formation
Tomato	:	Nursery and 40 DAT, seed treatment with 1% for 12 h
Onion	:	30, 45 and 40 DAT
Rose	:	At the time of pruning and budding
Jasmine	:	Bud initiation and setting
Vanilla	:	Dipping setts before planting
Fruit trees		
First time	:	One month before flowering
Second time	:	15 DAS
Third time	:	When the fruit is in pea size
Fourth time	:	After harvest once
Leafy crops		
Greens, curry leaf, tea, coffee	:	2% weekly once

Note: *DAS* days after sowing, *DAT* days after transplanting

1½ inches. Add water to test if the water flows out through the hole. Above the stones, fill the container with sand (1 inch) and waste materials like manure, fertile soil and vegetable peels (30–40 cm) as the food for earthworms. Add 1 kg of fresh cow dung into the drum and release 200–300 earthworms in the drum. Take a 5 l capacity bucket and make a small hole at the bottom. Fill the bucket with water and place it on the drum so that water drips into the drum. Fill this bucket with water as it turns empty. Collect the vermiwash in a bottle that trickles slowly from the bottom of the drum.

5.3 Nerium Leaf Extract Spray

- 1 kg of *Nerium* seeds was powdered well and mixed with 5 l of curd and 5 l of milk in an earthen pot and left for 3 days for fermentation.
- After the fermentation period, the solution is filtered using cotton cloth, and 1 l of the filtrate solution is mixed with 9 l of water and used as a spray solution for 1 acre of the cropped field.

Uses

- Mix 1 l of vermiwash with 5 l of water and use it for crops. It can be sprayed directly to the crop or applied along with irrigation.
- The vegetable plants become very healthy and strong. Controls deflowering and increases the yield of a crop.
- Vermiwash mixed with cow's urine and water in the ratio of 1:1:6 is sprayed to crops for effective control of caterpillars and aphids in vegetable crops.

5.3.1 Farmer's Effective Micro-organism

- Pumpkin 3 kg
- Banana 1 kg
- Papaya 3 kg
- Jaggery 3 kg
- Egg 5 numbers
- Non-chlorinated water 10 l

Preparation

Cut the vegetables into small pieces. Transfer these pieces into a clean plastic container. Mix jaggery in 10 l of non-chlorinated water till it dissolves well and add the eggs to it. Mix all the contents. Close the container with airtight lid. Open lid after 10 days to release the air. Mix well again. Keep the setup closed for 45 days. After 45 days, there will be three layers in the container. The upper thin layer is in white colour, which indicates successive fermentation. The middle layer will be pure brown-coloured liquid, and the lower layer will be the semi-solid formed by the dissolved vegetables. Open the tap fixed at the bottom of the container to collect the semi-solid portion in one container. The upper and middle portions are collected in another container. Two to five percent concentration in water can be used as foliar spray on any crop.

5.4 Weed Management

5.4.1 Hand Weeding

Hand weeding is a slow but very useful method. Here as weeds are pulled up, then weeds loose anchorage from the soil, fall dry desiccate and subsequently die. Thus weeds do not compete with the crop. This cannot be recommended for a larger area but can be adopted successfully as a last resort to remove weeds, which have become big and problematic in specific situations.

5.4.2 Hand Hoeing

- The surface soil is to be digged for shallow depth with the help of hand hoes, by which the weeds are uprooted and removed.
- After hand hoeing, the field should not be irrigated and is subjected to drying to avoid re-establishing of uprooted weeds.

- This method can be practised in irrigated upland crops like finger millet, pearl millet, onion, etc.

5.4.3 Digging

Digging is useful for patch or spot control of obnoxious/perennial weeds. Digging is very useful in the case of perennial weeds to remove the underground propagating parts of weeds from the deeper layer of the soil. They can be eliminated by digging with crowbar, pickaxe, etc. *Cynodon dactylon* can be effectively controlled by this method.

5.4.4 Mowing

Mowing is the cutting of weeds to the ground level. It is useful more in non-cropped areas than cropped areas. Mowing improves aesthetic value of an area. It is effective against erect and herbaceous weeds. In fields where it is not practicable to plough, such as a new meadow or new seeding of alfalfa or clover, many weeds can be controlled by mowing them about the time the blossoms appear or at least before seeds are matured. This cutting can be done with a scythe or mowing machine. A single mowing will often not prevent seed production. The new stems produced below the initial cut will flower and form seeds. Two or three mowings will be needed to ensure prevention of seed formation.

5.4.5 Cutting

- Weeds are cut above the ground surface leaving stubble. It is the most common practice against brush and trees.
- Cutting is done with the help of axes and saws.

5.4.6 Dredging and Chaining

- Dredging and chaining methods are used to control aquatic weeds growing in shallow ditches.
- Removing of weeds along with their roots and rhizomes with the help of mechanical force is called dredging.
- The floating aquatic weeds are removed by chaining.
- A very heavy chain is pulled over the water bodies to collect the weeds.

5.4.7 Planting in Narrow Rows

Keeping all other factors constant, narrow-spaced crops smother weeds more efficiently than wide-spaced crop. Wide-spaced crop allows more light to reach weeds. Hence, weed growth is more.

5.4.8 Planting Direction

Crops sown in north-south direction suppress weed growth better than east-west direction. East-west crop allows sunlight to reach weeds throughout the day, whereas north-south sown crop utilizes more light, shades weeds and does not allow light to reach weeds.

5.4.9 Plant Density

In the crop-weed competitive situation, a higher number of plants gain comparative edge, more plant result in broader canopy cover and thus higher weed suppression capacity.

5.4.10 Mulching

Mulches stop weed growth by restricting the penetration of sunlight to the soil surface and in the case of surface mulches of cover crops have the potential to release inhibitory (allelopathic) chemicals into the soil environment that inhibit weed seedling growth. Many weed seeds require light to stimulate germination, so mulches reduce the germination of such seeds. Straw, manures, crops, biodegradable plastics and newspapers can also be used as mulch materials. Sometimes the soil as such will be used as dust mulch in crops.

5.4.11 Soil Solarization

Solarization in agriculture would include thermal, chemical and biological changes caused by solar radiation when covered by clear plastic films especially when the soil has high moisture content. The possible mechanisms of weed control by solarization:

- Killing of germinating seeds
- Killing of seeds stimulated to germinate in the moistened mulched soil
- Indirect microbial killing of seeds weakened by sublethal heating
- Direct killing of weeds due to heat

In general, solarization increases heat. This increased heat of temperature induces weed seeds to germinate and get killed.

5.5 Plant Protection Measures

5.5.1 Preventive Practices

1. *Selection of adapted and resistant varieties*
   - Choose varieties which are well adapted to the local environmental conditions (temperature, nutrient supply, pests and disease pressure), as it allows them to grow healthy and makes them stronger against infections of pests and diseases.
2. *Selection of clean seed and planting material*
   - Use safe seeds which have been inspected for pathogens and weeds at all stages of production.
   - Use planting material from safe sources.
3. *Use of suitable cropping systems*
   - Mixed cropping systems: can limit pest and disease pressure as the pest has less host plants to feed on and more beneficial insect life in a diverse system.

- Crop rotation: reduces the chances of soilborne diseases and increases soil fertility.
- Green manuring and cover crops: increase the biological activity in the soil and can enhance the presence of beneficial organisms (but also of pests; therefore a careful selection of the proper species is needed).

4. *Use of balanced nutrient management*
   - Moderate fertilization: steady growth makes a plant less vulnerable to infection. Too much fertilization may result in salt damage to roots, opening the way for secondary infections.
   - Balanced potassium supply contributes to the prevention of fungi and bacterial infections.
5. *Input of organic matter*
   - Increases micro-organism density and activity in the soil, thus decreasing population densities of pathogenic and soilborne fungi.
   - Stabilizes soil structure and thus improves aeration and infiltration of water.
   - Supplies substances which strengthen the plant's own protection mechanisms.
6. *Application of suitable soil cultivation methods*
   - Facilitates the decomposition of infected plant parts.
   - Regulates weeds which serve as hosts for pests and diseases.
   - Protects the micro-organisms which regulate soilborne diseases.
7. *Use of good water management*
   - No waterlogging: causes stress to the plant, which encourages pathogen infections.
   - Avoid water on the foliage, as waterborne disease spread with droplets and fungal disease germinates in water.
8. *Conservation and promotion of natural enemies*
   - Provide an ideal habitat for natural enemies to grow and reproduce.
   - Avoid using products which harm natural enemies.
9. *Selection of optimum planting time and spacing*
   - Most pests or diseases attack the plant only in a certain life stage; therefore it's crucial that this vulnerable life stage doesn't correspond with the period of high pest density and thus that the optimal planting time is chosen.
   - Sufficient distance between the plants reduces the spread of a disease.
   - Good aeration of the plants allows leaves to dry off faster, which hinders pathogen development and infection.
10. *Use of proper sanitation measures*
    - Remove infected plant parts (leaves, fruits) from the ground to prevent the disease from spreading.
    - Eliminate residues of infected plants after harvesting.

5.5.2 Habitat Diversification

Habitat diversification makes the agricultural environment unfavourable for growth, multiplication and establishment of insect pest populations. The following are some approaches by which the pest population can be brought down.

Table 2 Effect of intercropping system on pest levels

Sl. no.	Crop		Pest reduced
	Sole crop	Intercrop	
1.	Sorghum	Red gram	Earhead bug
2.	Sorghum	Cowpea	*Chilo partellus* stem borer
3.	Pigeon pea	Sorghum	Leafhopper *Empoasca kerri*
4.	Green gram	Sorghum	*E. kerri* leafhopper
5.	Ground nut	Sorghum	*E. kerri* leafhopper
6.	Pigeon pea	Sorghum	*H. armigera*
7.	Chickpea	Wheat, mustard or safflower	*H. armigera*
8.	Sugarcane	Green gram, black gram	Early shoot borer

Intercropping System

Intercropping system has been found favourable in reducing the population and damage caused by many insect pests due to one or more of the following reasons.

- Pest outbreak less in mixed stands due to crop diversity than in sole stands
- Availability of alternate prey
- Decreased colonization and reproduction in pests
- Chemical repellency, masking, feeding inhibition by odours from non-host plants
- Act as physical barrier to plants

The following table gives a few examples of intercropping system where reduction in damage level was noticed (Table 2).

Interplanting maize in cotton fields increased the population of Araneae, Coccinellidae and Chrysopidae by 62.8–115.7% compared with control fields. Maize also acted as a trap crop for *H. armigera* reducing the second-generation eggs and damage to cotton. Intercropping pulses in cotton reduced the population of leafhopper on cotton, and *Lablab* bean in sorghum reduced the sorghum stem borer incidence. Hence it is highly important that appropriate intercropping systems have to be evolved where reduction in pest level occurs.

Trap Cropping

Crops that are grown to attract insects or other organisms like nematodes to protect target crops from pest attack. This is achieved by:

- Preventing the pests from reaching the crop
- Concentrating them in a certain part of the field where they can be economically destroyed (Table 3)

Diamondback moth population in cabbage can be managed by growing paired mustard rows with every 25 cabbage rows. The first row of mustard should be sown 15 days before and second 25 days after cabbage planting. Growing castor along the border of cotton field and irrigation channels acts as indicator or trap crop for

Table 3 List of successful examples of trap crop

Sl. no.	Main crop	Trap crop	Pest
1.	Tobacco/cotton/groundnut	Castor	*Spodoptera litura*
2.	Maize	Sorghum	Shoot fly, stem borer
3.	Cotton	Onion/garlic	*Thrips tabaci*

Table 4 Role of planting dates on pest population and damage

Sl. no.	Host plant	Insect	Response
1.	Rice	Leaf folder	Early-planted rice (up to the third week of June) suppressed population
2.		BPH	Planting in the end of July in kharif and early in rabi escapes attack in AP
3.		Gall midge	Lowest incidence if planted in Aug or Oct
4.	Sorghum	Shoot fly	Advancing sowing date (Sept–Oct) decreased incidence
5.	Cotton	Leafhopper	Higher incidence in late sown crop
6.	Chickpea	*H. armigera*	For every 10-day delay in sowing, 4.02% increase in pod damage
7.	Tomato	Whitefly (*B. tabaci*)	Incidence less if planted within Jul–Nov
8.	Chillies	Thrips	Late-planted crop severely affected by thrips and leaf curl virus

Table 5 Effect of plant density on pest population

No.	Crop	Spacing/density	Insect	Response
1.	Rice	Dense planting	Leaf folder, BPH	High incidence
2.	Chickpea	Dense plant population	*H. armigera*	High incidence
3.		Less dense population	*Aphis craccivora*	High incidence
4.	Sugarcane	Dense seed rate	Topshoot borer	Low incidence
			Early shoot borer	High incidence

Spodoptera litura. Planting 40-day-old African tall marigold and 25-day-old tomato seedlings (1:16 rows) simultaneously reduces *Helicoverpa* damage.

Planting Dates and Crop Duration

Planting dates should be so adjusted that the susceptible stage of crop synchronizes with the most inactive period or lowest pest population. The plantings should be also based on information on pest monitoring, as the data varies with location. Crop maturity also plays an important role in pest avoidance. The following table shows the importance of planting dates on pest population and damage (Table 4).

Planting Density

Plant nutrient status, interplant spacing, canopy structure, etc., affect insect behaviour in searching food, shelter and oviposition site. It also affects natural enemy population. The effect of plant density on pest population is shown in Table 5.

Table 6 Alternate hosts to be removed to reduce damage by pests

Sl. no.	Crop	Pest	Alternate host to be removed
1.	Groundnut	Thrips (*Caliothrips indicus*)	*Achyranthus aspera*
2.	Rice	Gall midge	Wild rice (*O. nivara*)
3.		GLH	*Leersia hexandra*
			Echinochloa colonum
			E. crus-galli
			C. dactylon
4.		WBPH	*Chloris barbata*
5.	Sorghum	Earhead midge	Grassy weeds

Table 7 Effect of irrigation on pest population/damage

Sl. no	Crop	Insect	Response
1.	Rice	Mealy bug	Continuous ponding of 5 cm water reduced incidence
2.	Rice	Caseworm and BPH	Draining of water to field capacity reduces incidence
3.	Fruit tree nursery	Termite	Copious irrigation reduces incidence
4.	Groundnut	Aphids	Copious irrigation increased incidence

Destruction of Alternate Host Plants

Many insects use a wide range of cultivated plants especially weeds as alternate hosts for off-season carry-over of population. Weeds around the crop can alter the proportion of harmful and beneficial insects that are present and increase or decrease crop damage (Table 6).

Destruction of off types and volunteer plants, thinning and topping, pruning and defoliation and summer ploughing are other cultural methods which can reduce pest load in field.

Water Management

Availability of water in requisite amount at the appropriate time is crucial for proper growth of crop. Hence, water affects the associated insects by many ways such as nutritional quality and quantity, partitioning of nutrients between vegetative growth and reproduction, etc. The following table shows the effect of irrigation on pest population/damage (Table 7).

Crop Rotation

Sustainable systems of agricultural production are seen in areas where proper mixtures of crops and varieties are adopted in a given agroecosystem. Monocultures and overlapping crop seasons are more prone to severe outbreak of pests and diseases. For example, growing rice after groundnut in garden land in puddled condition eliminates white grub.

Organic Manure

Application of press mud in groundnut at 12.5 t/ha had a better influence on leaf miner with lower leaflet damage at 38.84% and 2.48 larval numbers per plant during summer 1991. It was 34.93% and 2.72 numbers during kharif, 1991. Farmyard manure, *Azospirillum* and phosphobacteria have no significant influence on the control of leafhopper and fruit borer in bhendi. The incidence of paddy plant and leafhopper was low in *Azospirillum* combined with farmyard manure. Application of organic manure lowered the rice gall midge incidence (5.28%)

5.5.3 Herbal Pesticides

Neem Leaf Extract

- 1 kg of green neem leaf is soaked overnight in 5 l of water and crushed, and the extract is filtered. Add 10 ml of emulsifier (neutral pH adjuvant).
- This is beneficial against leaf eating caterpillars, grubs, locusts and grasshoppers.

Neem Cake Extract

- 100 g of neem cake is soaked overnight in 1 l of water in a muslin pouch and crushed, and the extract is filtered. Add 1 ml of emulsifier (neutral pH adjuvant) per litre of water

Neem Oil Spray

- 15–30 ml neem oil is added to 1 l of water and stirred well. To this emulsifier is added (1 ml/1 l).

Fermented Curd Water

In some parts of central India, fermented curd water (butter milk or *Chaach*) is also being used for the management of whitefly, jassids, aphids, etc.

Take the distasteful butter milk and add equal amount of water in it. Keep it for 2 days in a semi-shaded place. Now take it and add 40 l of water in 10 l of the extract to form 50 l of the solution. Spray this solution in 1 ha of field in such a way that all plants get bath in the fogging spray in the early morning time.

5.5.4 Herbal Pesticide Formulation

- 500 g neem seeds, 1000 g tobacco, 100 g *Acorus calamus*, 250 g asafoetida and 50 g *Sapindus emarginata* seeds are ground, and the extract is sprayed for one acre cotton to control pests.

5.5.5 Neem-Cow Urine Extract

- 5 kg of neem leaves, 5 l of cow urine, 2 kg of cow dung, 100 l of water.
- Crush all ingredients and ferment for 24 h with intermittent stirring, filter and squeeze the extract and dilute to 100 l of water.
- Use this extract to fill in the spray machine and spray it over one acre of the crop.

5.5.6 Chilli-Garlic-Ginger Extract

Crush green chillies (0.5 kg), ginger (0.5 kg) and garlic (1 kg) separately with required quantity of water. Soak tobacco leaves (0.5 kg) in 1 l hot water overnight and take the solution for use. Mix 200 ml neem oil with one packet of shampoo (use of shampoo is to mix thoroughly neem oil and water). Then mix chilli, garlic and ginger extract and tobacco leaf extract and neem oil in a mud pot or plastic bucket, and close the lid. Keep this solution for 1 or 2 days and apply to the crops. This solution should be used within 3 days of preparation. 10 ml of the solution is diluted in 1 l of water and sprayed to crops for effective control of pests. Spraying is done regularly at an interval of 10–12 days during flowering and fruiting stage (Table 8).

5.6 Soil and Water Management

Table 8 Important ITKs practised for various soil and water management

No	Categories	Name of ITK
1.	Agronomic measures	1. Cover cropping
		2. Criss-cross ploughing
		3. Hoeing with local hoes
		4. Set furrow cultivation
		5. Application of manure (FYM)
		6. Green manuring
2.	Tillage	1. Summer ploughing/off-season tillage
		2. Repeated tillage during monsoon season
3.	Bunding and terracing (mechanical and vegetative barrier)	1. Vegetative barrier
		2.Compartmental bunding
		3. Peripheral bunding/field bunding
		4. Conservation bench terrace
		5. Strengthening bunds by growing grasses
		6.Growing of *Saccharum*
4.	Land configuration	1. Use of indigenous plough for formation of broad bed and furrows
		2. Levelling the plots by local leveller
		3. Opening up set furrow
		4. Conservation furrow: Gurr
5.	Soil amendment/ mulching	1. Retention of sunflower stalks
		2. Mulching of sal leaf in turmeric
		3. Crop residue application in the field
6.	Water harvesting, seepage control and groundwater recharge	1. Dug wells
		2. Haveli/Bharel system
		3. Bandh system of cultivation
		4. Earthen check dams
		5. Field water harvesting
		6. Nadi farming system
		7. Collection of sub-surface runoff water and recycling in Diara land
		8. Rainwater harvesting in kund/tanka

(continued)

Table 8 (continued)

No	Categories	Name of ITK
7.	Furrow opening in standing crops	After intercultivation, a small furrow is opened with the help of blade hoe after every two crops, in standing crops, viz. cotton sorghum having row spacing of at least 45 cm. It is done for rainwater conservation
8.	Nadi farming system	To collect runoff during kharif for lifesaving irrigation during drought spell or pre-sowing irrigation (Palewa) for rabi crops
9.	Application of white soil as lining material in farm pond	To work as a sealant material for lining dugout farm pond
10.	Wider row spacing in pearl millet	Rainwater conservation and weed control
11.	Rainwater harvesting in kund/tanka	The harvested water in kund/tanka is used for drinking and establishment of tree
12.	Mulching in turmeric	To conserve rainwater
13.	*Agave* sp. as vegetative barrier	To reduce runoff velocity and to increase infiltration opportunity time
14.	Broad bed and furrow practice	To harvest rainwater and dispose of runoff
15.	Set-row cultivation	For harvesting rainwater and maintaining soil structure
16.	Summer/pre-monsoon tillage	Conservation tillage to harvest early showers, facilitate timely seeding and weed control
17.	Ridge and furrow planting for modulation of overland flow	Conservation of rainwater, modulating excess water, control soil loss and boosting productivity
18.	Application of tank silt	To increase the fertility and water holding capacity of soil

6 General Indigenous Practices for Agriculture (Somasundaram et al. 2015)

6.1 Soil and Water Management

- In Tuticorin district, to improve the 'theri' lands, 200 tonnes to tank silt is applied per acre followed by 50 tonnes per year for the next few years.
- Crop residues and tree branches are burnt on the soil surface at the end of summer to improve the structure of clayey soil and to make ploughing easy (Farmers of Konkan Region in Maharashtra).
- Grassed waterways are adopted, or vetiver (*Vetiveria zizanoides*) is planted along the bunds in red soils to check the soil erosion (Farmers in Karnataka, India).
- Terracing is a traditional conservation method employed to prevent soil erosion particularly in steep areas (Farmers in the Philippines).
- Leaves of *Butea monsperma* are spread in the field and burnt along with the dried wheat stalks to improve the soil fertility (Farmers in Gujarat).
- Wetland crops require more intensive care when compared to the garden land crops.
- Waterlogged dry lands are unsuitable for cultivation.

- Soil character decides the choice of crops for cultivation. Red soil is suitable for continuous cropping. Black soil has more water holding capacity than the red soil. Sandy soil is less suitable for the cultivation of many crops.
- Excessive application of farmyard manure (FYM) improves the soil texture.
- Tank silt is applied to increase the soil texture.
- Addition of red soil to black soil increases the fertility of the black soil and vice versa.
- Penning of sheep/cattle in the field during summer season to improve the soil fertility.
- To increase the soil fertility of rainfed areas, mixed cropping or intercropping of legumes can be practised.

Rainfall

- Rainwater received during the 'Magha' constellation is stored and used as a growth promoter on the standing crop in the future (Farmers in Gujarat).
- The large number of fireflies seen at night on the forest trees is a sign that the monsoon will start early (Farmers in Maharashtra).
- If there is rain, accompanied with lightning and mild thunder on the second day of Jyeshtha month (May–June), there will be no rain for the next 72 days (Farmers in Gujarat).
- If the velocity of wind is not high during the Mrigashirsha constellation and high heat is not experienced during the *Rohini* constellation, a drought can be expected to follow.
- Movement of clouds in one direction is a sign of rainy weather, while the opposite direction meant dry weather.
- A red sky at sunrise and sunset considered a warning for rainless days ahead.
- Cold nights with mist and dew considered signs of impending dry weather, while hot and warm feelings during the day signalled rainy evenings.
- Thunder in summer and lightening in rainy season bring heavy rain.
- Drizzling occurring in 'Aani' (June–July) assures rain in 'Avani' (Aug–Sept).
- Rainfall during 'Margali' (Dec–Jan) does not benefit the crop but affects the grains.

Preparatory Cultivation

- During the summer season itself, the soil is prepared well and applied powdered FYM (Farmers in Kerala).
- A piece of raw/fermented areca nut with a betel leaf and a pinch of lime are chewed together and spit on the ploughing spot. If the soil becomes blackish, it indicates that the soil is well tilled. Or if the soil turns reddish, then the test is negative, indicating the need for more tillage (Farmers in Assam).

Seeds and Sowing

- Seeds soaked in 'kumkum' (Vermillion) water or milk for a day before sowing germinate faster and grow in to healthy plants (Farmers in Karnataka).
- For increasing the productivity of seeds, it is to be soaked in the paste of ankola fruit (*Alangium salvifolium*) or in its oil or in the oil of 'Shelsh mantak' fruit (*Cordia latifolia*) (Farmers in Maharashtra).

Manures and Manuring

- In 'theri' lands of Tuticorin district, FYM is applied by all the farmers to improve the structure and fertility of soils (Farmers in Tamil Nadu).
- If leaves of *Agaves* spp. are cut and embedded in the soil for 45 days, the soil is enriched (Farmers in Karnataka).
- About 10 kg of neem cake is soaked in 10 l of cow urine along with ½ kg. of waste asafoetida and leaf overnight. In the next day, it is sprayed for 1 ac after dilution as liquid manure (Farmers in Tamil Nadu).
- About 20–25 kg of common salt is mixed with 10 cartloads of natural manure and applied during May to increase crop yields (Farmers in Gujarat).
- To make the soil soft and fertilized, gingelly plants in full bloom may be cut into pieces and mixed with the soil (Farmers in Maharashtra).

Weed Management

- To control *Cynodon dactylon*, harvested dried stalks of cumin crop are spread in the field. As the stalks decompose and mix with the soil, the weed is destroyed (Farmers in Gujarat).
- Farmers change the variety of paddy crop in each season to control the weeds like 'dhakura' (Farmers in Uttaranchal).
- Manure made with mango leaves is applied in the fields to control *Cyperus rotundus* (Farmers in Tamil Nadu).
- Use of multivariate seeds (MVS) for mixed sowing navathaniyam as intercrops.
- Repeated ploughing will reduce weed population.
- Crop yield will be less in the fields having 'Arugu' (*Cynodon dactylon*) weeds.
- To control 'Arugu (*Cynodon dactylon*) grass' in black soils, the field is kept fallow for 3 years.
- Cultivating rice once in 3 years in garden lands to control 'Arugu' (*Cynodon dactylon*) weeds.
- Raising and ploughing the green manure crops like dhaincha (*Sesbania* spp.) and kolingi (*Tephrosia purpurea*) in the field before their flowering to reduce weed population.
- Raising *Calotropis gigantea* as a green manure to check the growth of Aarai (*Marsilea quadrifolia*) weed.

- Growing horse gram to control nut grass (*Cyperus rotundus*).
- Growing cowpea as a green manure to control nut grass.
- Allowing swine in the fields to eradicate nut grass.
- Frequently ploughing the fields by wooden plough made up of neem trees and frequent application of neem cake in the soil to control nut grass.
- Dissolving 1 kg of salt and 100 g sarvodaya soap in 10 l water and spraying this solution to control all the weeds except nut grass.
- To control nut grass in the field, 50 kg neem cake is applied both at the time of ploughing and sowing.
- Dissolving 200 g salt in 1 l water and spraying to eradicate congress weed (*Parthenium hysterophorus*).
- Continuous submergence of field for some time controls the weeds.
- Keeping the irrigation channels free from weeds.

Rat Control

- Rats do not live in fields where sheep penning is being practised.
- Planting closely notchi (*Vitex negundo*) and erukku (*Calotropis gigantea*) around the fields as a fence helps to control rat problem.
- Putting the branches of Thangarali (*Tecoma stans*) around the fields to control rats.
- To control rats in paddy fields, Channampoo (*Cycas circinalis*) flowers are cut into pieces and placed in many places whose bad odour drives away the rats.
- Pieces of palmyra (*Borassus flabellifer*) leaves are tied on the poles fixed on the field. The sound produced by the leaves scares away the rats.
- Providing owl stands near the rat holes will help in reducing the rat damage.
- To reduce rat population, rat holes are dug and rats are killed after each harvest.
- To catch the rats, a trap made up of wire loops on bamboo pegs is being used.
- Big round-shaped earthen pots are buried on the field at ground level. Half of the pot is filled with mud slurry on which baiting material is put on a coconut shell. Attracted rats fall inside the pot and they cannot climb up and get killed.
- Use of soaked rice as bait attracts more rats.
- Putting fresh cow dung on both the fields and bunds to reduce rat problem.
- Papaya – 3 fruits/acre.

Pest and Disease Management

- To control pod borer (*Leucinodes orbonalis*) in vegetables, *Anethum sowa* plants are sown in rows along with the vegetables. The strong odour of the plant repels the insects (Farmers in Karnataka, India).
- Mint and marigold are planted in the farm to repel insects (Farmers in the Philippines).

- To project the plants from the white ants, *Vetiveria zizanoides*, *Euphorbia tirucalli* and *Calotropis gigantea* are planted at intervals in the field (Farmers in Maharashtra).
- A mixture is made with 1 l of neem oil, 3 kg of fine sand and 3 kg of cow dung and heaped in shade covering with a moist sack for 3 days. On the fourth day, the mixture is dissolved in 150 l of water and sprayed to control all sucking pests (Farmers in Tamil Nadu).
- About 10 kg of dried cow dung is ground into fine powder and mixed with ash (obtained from brick kiln) and dusted in the early morning, to control pests and diseases (Farmers in Tamil Nadu).
- Garlic acts on a wide spectrum of organisms in unrelated crop plants singly or in combination with neem products, chilli, asafoetida, etc. Besides garlic is effective against bacteria, fungi and nematodes (Farmers in Tamil Nadu).

Storage Pest Management

- Lime juice is mixed with grains and then sun-dried before storage to prevent insect pests (Farmers in Nigeria).
- Seeds are safely stored in earthen pots after mixing with the leaves of neem and *Vitex negundo* (Farmers in Gujarat, India) (Table 9).

Table 9 Usage of green plant parts and powders made of dried green parts and fruit

Method	Effect
Fresh or dried leaves of Annona added to the commodity in layers	Strong repulsive and insecticide effect during 3–4 months on bruchids and sorghum and millet pests
Mint (*Mentha* spp.) leaves added to the grain/seed at 0.5–2 weight %	Insecticide effect supposed to act on pests of cereals
Crushed *Lantana* parts – sandwich technique or as a top layer	Repulsive effect on bruchids of grain legumes acting up to 6 months
Dry or powdered neem or *Melia* leaves mixed to the grain/seed or applied in layers	Insecticide and repulsive inhibition of development. Acts mainly on stored product beetles up to 1 year
Ocimum canum (hoary basil) leaves entire or as powders applied in sandwich technique	Insecticide effect on beetles
Entire or powdered fruits of red pepper (*Capsicum* spp.)	Insecticide and repulsive effect
Entire or powdered fruits of black pepper (*Piper* spp.)	Comparable to red pepper effect lasting for 3 months
Neem kernel powder added at a rate of 0.5–4 vol. %	Strong repulsive and insecticide effect, but stronger
Annona grain powder added at a rate of 0.5–2 weight %	Strong repulsive and insecticide effect
Powder of dried rhizomes of *Acorus calamus* (added at a rate of 0.2–1 weight %)	Insecticide, repulsive effect and inhibition of development against many pests for more than 6 months

7 Conclusion

Though ITKs are sustainable, using them alone will not meet out the food requirement of exploding population. So, to keep in phase with the present level of food production and to sustain the higher levels of productivity, ITKs are to be integrated with modern technologies. To move towards interactive technology development from the conventional transfer of technology approach, it is feasible, efficient and cost-effective to learn from the village-level experts. Keeping ITK as a basis, during the process of developing technologies, results in a basket of sustainable options rather than fixed packages. Hence, ITK is the product of centuries of trial and error, natural selection and keen observations that can form a knowledge base on which researchers and extension workers can plan their research strategy and experimental procedure. Validating farmers' experiments creates an environment of respect for local people and village-level extension workers, thus leading to their increased participation and empowerment. The conservation and sustainable use of biological diversity will have lasting consequences in developing appropriate technologies.

References

Kolawole OD (2001) Local knowledge utilization and sustainable rural development in the 21st century. Indigenous Knowl Dev Monit 9(3):13–15

Quiroz C (1996) Local knowledge systems contribute to sustainable development. Indigenous Knowl Dev Monitor 4(1):3–5

Radhakrishnan T, Anandaraja M, Ramasubramanian M, Nirmala L, Israel Thomas M (2009) Traditional agricultural practices-applications and technical implements. New India Publishing Agency, New Delhi

Rathnam R (1966) Agricultural development in Madras state prior to 1900. New Century Book House, Madras

Sabarathinam VE (1997) Rationalization of indigenous technical knowledge in Indian agriculture. In: Shivashankar K (ed) 3rd IFOAM – ASIA science conference and general assembly "Food security in harmony with nature" held at UAS, Bangalore, 1st – 4th December 1997, p 97

Sharma SK (2002) A synoptic view of linkages of organic farming with productivity and sustainability of India. In: CAS training on organic agriculture – a paragon for sustainability held at JNKVV, Jabalpur, March 26th – April 15th, 2002, p 29

Somasundaram E, Asoka Raja N, Mohamed Amanullah M, Nandhini DU (2015) A guide to organic farming. University Press, Tamil Nadu Agricultural University, Coimbatore – 641003, India

Sundaramari M, Ranganathan TT (2003) Indigenous agricultural practices for sustainable farming. Agrobios, Jodhpur

Warren MD (1993) Using indigenous knowledge in agricultural development. The World Bank, Washington, DC

Check for updates

Recent Biotechnological Approaches to Produce Climate-Ready Crops and Enhancing Crop Productivity

Gourav Kumar, Gopal Kumar Prajapati, Amar Nath Mishra, and Dev Mani Pandey

Abstract

Living beings are dependent on green energy for their survival and crops are playing most potential role in it. Sudden changes, i.e., biotic and abiotic stresses, are always a big factor affecting the productivity of crops, and in this adverse environmental conditions, it is very much imperative to find the alternate strategies for increasing crop productivity and to feed the increasing worldwide populations. We have always the knowledge of obtaining the best cultivar as earlier the farmers used to select the healthy seedlings for agriculture. Gregor Johann Mendel's work Mendelian inheritance gives us a platform to use the conventional methods with scientific and validated models for better yields of crops. Therefore, initiatives need to be taken to tackle this problem. In recent years advanced biotechnological approaches, tools, and techniques have played important role. With the use of these advanced technologies researchers are able to do whole genome sequencing, annotation, prediction, and validation of gene expression, prediction of possible function as well as their involvement in various metabolic pathways, and identification of the presence of set of nucleic acid sequences (motif) in the promoter region as well as their protein-DNA interaction. Efforts has also been done to develop the genetically modified crop having important characters and to give good yield under adverse conditions.

G. Kumar · G. K. Prajapati · D. M. Pandey (✉)
Department of Bio-Engineering, Birla Institute of Technology,
Mesra, Ranchi, Jharkhand, India
e-mail: dmpandey@bitmesra.ac.in

A. N. Mishra
Department of Chemical Engineering, Birla Institute of Technology,
Mesra, Ranchi, Jharkhand, India

R. S. Sengar, A. Singh (eds.), *Eco-friendly Agro-biological Techniques for Enhancing Crop Productivity*, https://doi.org/10.1007/978-981-10-6934-5_2

Keywords
Bioinformatics · Molecular beacon · Molecular markers · Next-Generation sequencing · Plant breeding · Polymerase chain reaction · Plant tissue culture · Restriction enzymes

1 Introduction

India is an agricultural country with varied climate. Indian agriculture includes mixing of traditional and modern farming techniques, which vary from traditionally using cattle for plowing farms to sophisticated machineries. Traditional farming techniques have mixed consequences of low per capita productivity and diminishing farmer income in contrast with conservation of natural and environmental factors. Uses of chemicals to enhance the captivity of crop have many compromising effects including loss in soil fertility and repercussion of severe health problems. So use of natural resources, i.e., biotechnology and modern science technologies is playing a prime role nowadays. India as well as the world economy is in one way or the other dependent on agricultural (Fig. 1).

2 Plant Breeding and Biotechnology

Cultivation of prominent crop can be achieved through two methods manipulating genetic variability and plant evaluation. Previously, farmers used to seed only those plants that have given best in the field. Pollination of plants in controlled environment aimed for specific crosses resulting in a new potential generation.

Several plant species have been genetically modified in labs for better yield, using *Agrobacterium* and vector for transfer of genes or using other vector-independent options, viz., liposome, gene gun, or microinjection methods, although almost all cases require some type of tissue culture technology (Gupta et al. 2017).

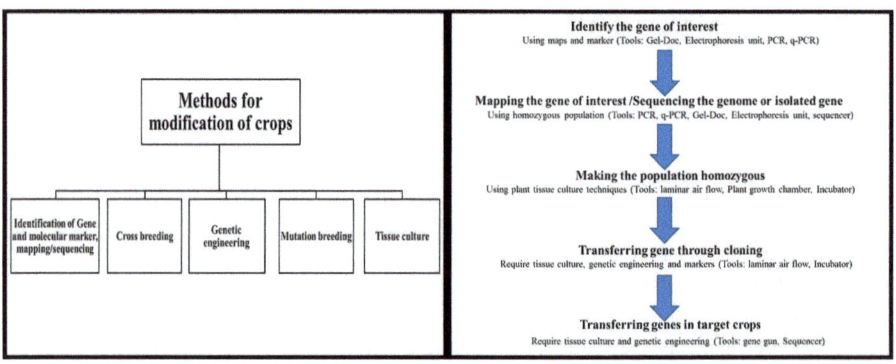

Fig. 1 (**a**) A Generalized method used for improvement of crop (**b**) Flowchart of making a genetically modified crop

2.1 Plant Tissue Culture

To fight with adverse conditions and in realm of crop improvement techniques of plant, tissue culture plays a very crucial role as it introduces an array of genetic variation in plant species for obtaining genetically modified crops which are tolerant for any adverse conditions. It is a set of in vitro techniques, using methods and strategies from subject of plant biotechnology/botany. Tissue culture along with molecular techniques is used for successful gene transfer into desired genome running fruitful and controlled biochemical pathways. Culture of anthers, embryos, microspores, ovules, and protoplasts are the most commonly used techniques to create modified and potential genetic variation in the breeding lines via haploid production. Somaclonal variation is obtained through making chromosomal rearrangements in somatic cells for both quantitative and qualitative advancement, and gametoclonal variants through arranging gametic cells are also obtained from plant tissue culture (Gupta et al. 2017).

2.2 Wide Hybridization

Fertilization in plants angiosperm requires fusion of two sperm nuclei with the egg. For achieving this pollen grains need to travel to the stigma of the mother plant then leading germination of the pollen tube which penetrates the stigma and style and reaches to the ovule. This whole process ends up with discharge of sperm within the female gametophyte resulting in embryo and nutritional endosperm, respectively (Chopra and Sharma 1991). Now this is the step where several factors can block this hybridization. These failures may be overcome through the:

(a) *In vitro fertilization*: Used to overcome barriers found prior to fertilization.
(b) *Embryo culture*: Used to overcome barriers found prior to fertilization.
(c) *Protoplast fusion*: Used to produce exceptional hybrid plants fusing protoplasts to create unique somatic hybrid cells that generated immense excitement in the 1960s. Simply, protoplasts are plant cells whose cell wall has been removed using enzymes such as cellulase to leave a spherical and quite fragile cell, surrounded by the plasma membrane. To fuse protoplasts, polyethylene glycol (PEG) (Kao et al. 1974) or electrical pulses (Zimmerman and Scheurich 1981) are normally used.

2.3 Micropropagation

Micropropagation, also known as clonal propagation, is a widely used technique for in vitro growing of multiple copies for isolated plant cells, tissue, or organ (Bhojwani and Dantu 2013). This technique is commonly practiced between gardeners for growing ornamental plants and can also be used for vegetable crops if it is cost-effective.

2.4 Synthetic Seed

Artificial seeds are encapsulated somatic embryos, shoot buds, or cell aggregates, made experimentally through a hydrogel or nacked or polyoxyethylene glycol. They have the ability to grow as normal plants either in vitro or ex vitro and retain the same potential as earlier even after storage (Attree and Fowke 1993). Two types of synthetic seeds are produced:

(a) *Desiccated synthetic seeds*: These seeds are produced either nacked or through polyoxyethylene glycol encapsulation.
(b) *Hydrated synthetic seeds*: These seeds are made using hydrogels like carrageenan, potassium alginate, sodium alginate, or sodium pectate as encapsulating material.

Propagation of hybrid and genetically modified plants, maintenance of genetic uniformity, propagation of endangered species, and preservation of elite genotypes are some of the major applications of artificial seeding (Fig. 2).

Fig. 2 Process of tissue culture in controlled environment

2.5 Molecular Farming

Molecular farming is a tool of biotechnology through which crop's genetic modification can induce leading to production of desired proteins and chemical products. Thus molecular farming opens a viable horizon for biotechnology, pharmaceuticals, and biopharmaceutical industries (Ahmad 2014); for example, in the case of any desired seeds, a protein accumulated in seed can be stored in the seed and in the future processed using standard and optimized methods, making pilot production easy and cost-effective.

2.6 Germplasm Preservation and Transport

Cell culture for germplasm preservation and transport is a convenient technique, used in varying degrees by many germplasm repositories, biotechnology companies, and laboratories involved in commercial propagation and academic research. The preservation of wild or diverse population germplasm or that of cultivated crops, especially those that are vegetatively propagated, is an important tool for crop improvement (Fang et al. 2016). This technique reduces the labor involved in maintaining plant lines and also minimizes loss due to disease. Plant material can be cold-stored so that there is only minimal growth; cryopreserved in liquid nitrogen in the form of shoot tips, somatic embryos, and callus; and then used as needed (Ferrari et al. 2016). This minimizes the need to maintain stock plants in the greenhouse or under normal in vitro conditions requiring frequent maintenance. This technique also facilitates international transport and exchange of germplasm through quarantine facilities as most in vitro plant material is free from microbial or insect contaminants and can even be certified as virus-free.

3 PCR

The structure of nucleic acid reveals number of opportunities for scientist working in the arena of molecular biology. Proteins, the body building material or building blocks for living organism, are the result of translation of genes/RNA. So we can say the life is based on cascade of genetic actions, and almost all the problems of living being are eradicated through study and alteration of gene and their regulation. Kary Mullis in the year 1980 developed PCR (polymerase chain reaction) which becomes a milestone in the history of molecular biology. In PCR the ability of DNA polymerase to synthesize new strand of DNA complementary of desired template is used to amplify the gene, and successful products are named as amplicon. DNA polymerase is able to add a nucleotide only to a pre-existing 3′ OH and hence requires a short nucleotide sequence called primer which fulfills this need, making it feasible to amplify required region in the template. PCR reaction ends with the accumulation of billions of specific sequence copies.

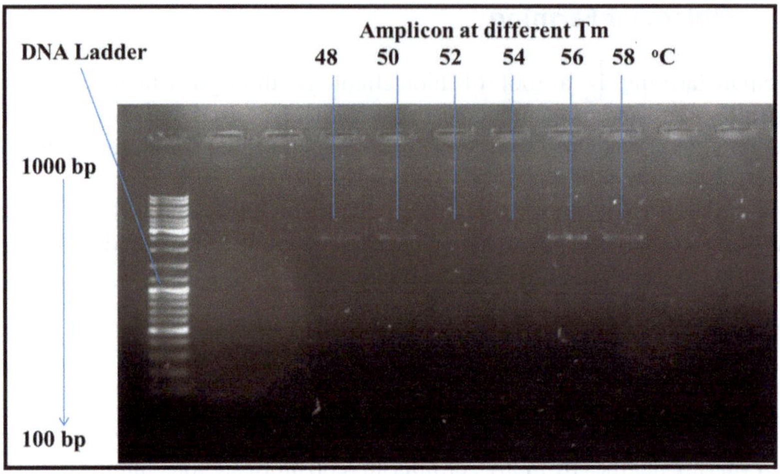

Fig. 3 Gel electrophoresis after gradient PCR amplification

For conducting PCR DNA template (i.e., sample DNA which contains the required target sequence that needs to amplify), DNA polymerase, primers (primers are short segments of complementary single-stranded DNA), and nucleotides (deoxynucleotide triphosphates that are single units of the purines and pyrimidine bases, i.e., adenine, thymine, guanine, and cytosine) and $MgCl_2$ are required. PCR results are visualized using electrophoresis (Fig. 3).

Application

Polymerase chain reaction (PCR) methods have revolutionized identification and diagnosis of plant pathogen and their effects. Progression of genomic studies makes DNA sequence data available for remedy of many plant disease and increasing crop yield. PCR is the most versatile and fairly simple technique that is employed in detecting plant pathogens. Initial reports for PCR-based assays for plant pathogen detection were reported in the early 1990s by Rasmussen and Wulff (1991) in whole plants and Prosen et al. (1991) in seeds.

There are certain genes which are resistant to plant pathogen, using those genes in production of genetically modified/engineered crop. Leister et al. (1996) used tobacco and *Arabidopsis thaliana* as a source for production of pathogenic resistance gene for potato. Through the genome mapping of amplified PCR potato fragments, they have concluded that these genes may be effective against nematode, fungi, viruses, and bacteria.

Limitations of PCR

The target sequence of the PCR reaction is amplified in the exponential phase which is where the starting amount of target of target sequence can be ascertained. PCR, though highly selective for the target sequence, is very sensitive toward inhibitors that may be present in the sample, and the "plateau effect" due to exhaustion of reagents leads to unreliable endpoint quantification of PCR target.

3.1 Real-Time PCR

Real-time PCR is also known as quantitative polymerase chain reaction (qPCR) that is a very much powerful and sensitive technique that has been extensively used for gene expression analysis and validation studies, SNP analysis, miRNA analysis, etc. This measures the amplification of nucleic acids on real-time scale.

3.2 Digital Real-Time PCR

Digital PCR has been a new approach for detection and quantification of nucleic acid by counting the number of target molecules directly and not depending on reference standards. Literatures about principle and application of digital PCR for absolute quantification give a strong idea (Pohl and Shih 2004).

4 Next-Generation Sequencing (NGS)

DNA or RNA as per discussed earlier is a prime factor in regulation or may be called the beginning of life. So every hidden fact can be revealed just by unfolding its structure or the sequence of nucleotides. Study of genomes and its structure briefs the structure of protein and its metabolic process; identification of new gene and associations leading to identification of diseases, associated pathogen, and phenotypes; and finally identification of potential drug targets. Sequencing of genome can be done through the numerous methods. Sanger and Coulson (1975) developed the plus and minus method, considering both strands of DNA as plus and minus. It is a fast and simple method of determining DNA sequence. If we considered both strands of DNA, it requires a comparison of both the "plus" and "minus" sequences to determine the actual sequence of DNA. The limitation is its ability to determine sequence of only single-stranded DNA, and through the development of this method, the genome of ΦX174 bacteriophage was sequenced.

4.1 Maxam-Gilbert Sequencing

Maxam and Gilbert DNA sequencing method was similar to the Sanger and Coulson method in using polyacrylamide gels to separate both strands terminating each base throughout the target sequence, only differing at generated products ending in a specific base. Their method started with a radiolabeled double-stranded DNA restriction fragment cleaved by base-specific chemical reactions. One reaction cleaves at both purines (the "A + G" reaction), one preferentially at A ("A > G"), one at pyrimidines ("C + T"), and one at cytosines only ("C"). Unlike the plus and minus method, the chemical method produced bands for every sequence position.

4.2 Chain Termination Methods

Frederick Sanger (1918–2013 two Nobel prizes in chemistry) is well known for his contribution in the form of invention and development of pioneer key methods for both protein and nucleic acid sequencing. While working as head of protein chemistry division of Medical Research Council's Laboratory of Molecular Biology (MRC-LMB), he began developing new methods for sequencing RNA finally leading into new approaches for sequencing DNA. The whole studies started with the use of "plus and minus" method followed by the development of the dideoxynucleotide chain termination aka Sanger sequencing method. He also extended his performed sequence assembly by joining overlapping DNA fragments to form larger polynucleotide sequences. This work resulted in the first fully sequenced DNA-based genome, that of φX174, a 5386 base-pair bacterial virus and earned his second Nobel Prize in Chemistry. The hazardous radioactive labels were replaced by less dangerous or safe fluorescent dyes mostly because of Sanger's work. During the subsequent 25 years, the Sanger sequencing method was used almost exclusively for DNA sequencing, including the completion of the worldwide effort to sequence the three billion base-pair human genome, a major portion of which was conducted at the aptly named Sanger Centre (currently the Sanger Institute) in Hinxton. Without any doubt, based on his quiet work and significant accomplishments, Fred Sanger is regarded as one of the fathers of both modern proteomics and genomics.

4.3 Pyrosequencing (454)

Commercial launch of the Roche 454 pyrosequencing was a milestone in genome sequencing (Raza and Ahmad 2016). The technique has obvious advantage over the traditional Sanger sequencing method in terms of average read length being close to 500 base pairs.

4.4 Sequencing by Synthesis (Illumina)

Sequencing or cluster amplification of gene is successfully achieved through a library preparation which is done using the following steps (Burke and Darling 2016).

Fragmentation: If the clusters are large in size (~1000 bases), their amplification would be increasingly inefficient and have lower yield. Hence, the Illumina sequencing libraries provide best results with fragment size of 200–500 bases, after accounting for adapters.

End repair: Random fragmentation generates dsDNA blunt-ended fragments resulting in both the presence and absence of 5′ phosphate moiety in these fragments. This necessitates generation of uniformity in the blunt-ended fragments by phosphorylation using a mixture of enzymes, which is basically the end repair of the gene.

A-tailing: Before ligating the adaptor, addition of a single adenine to the 3′ ends prevents concatemerization (formation of multiple DNA sequence copies arranged end to end in tandem; concatenate means to link together in a chain or in a series) of templates increasing the efficiency of adapter ligation.

Adapter ligation: Dissimilar adapter sequences must be added to template strands for successful cluster amplification and sequencing reactions.

Cleanup: Cleanup can be performed either by column or by SPRI (solid-phase reversible immobilization) beads. These beads are carboxyl-coated magnetic particles that have the ability to reversibly bind with DNA in the presence of PEG (polyethylene glycol) and salt. The latter process (AMPure XP SPRI beads) is preferred due to easy scalability.

Size selection: Removal of undesirable fragment size can be achieved at either the DNA shearing stage by modulating the conditions, by single AMPure XP treatment (for removing fragments with size typically below 150–200 bases), or by double treatment (for selecting upper or lower cutoff value for fragment size).

PCR: Properly ligated template strands (with adapters at both ends) are enriched by PCR amplification of the libraries so as to:

- Increase the quantity of library, so that there is enough DNA for accurate quantification.
- Add oligonucleotide sequence for hybridization to the flow cell surface.

Quantification: Accurate quantification is required as data yield is reduced both in case of low cluster density (leads to increased cost for sequencing) and high cluster density (results in cluster overlap.

Denaturation: Denaturation is done before the sample is loaded on the cBot or directly onto the sequencer. The subsequent steps are processed by the sequencer or on the cBot.

Hybridization and extension: After complete PCR reaction, template strands have an exact gene sequence matched with one of the forward flow cell primers at one end, while the reverse orientation to the other flow cell primer complimentary to the template at their opposite end is observed. Thus, only the complementary end of each template strand hybridizes to a flow cell primer.

Cluster amplification: The flow cell primer, hybridized to the tethered template strand, is extended using *Bst* polymerase, to generate a double-stranded product.

Linearization: The structure of the adapters results in the presence of approximately 12 nucleotides at the ligation junction. The sequencing primers must hybridize efficiently to the strands within a cluster, and each cluster must only be sequenced in a single direction; one of the flow cell primers is cleaved, making the clusters single stranded due to selective removal of one strand.

Sequencing by synthesis: At the end hybridized flow cells undergo repeated cycles of nucleotide incorporation, imaging, and cleavage.

4.5 SOLiD Sequencing

Identification and removal of inaccurate sequence reads can be done through re-sequencing of genome with Applied Biosystem (Life Technology) SOLiD system. It is tag-based applications such as gene expression and ChIP-seq (Edwards and Batley 2010).

5 Restriction Enzymes

Restriction enzymes are basically proteins which act as DNA-cutting enzymes found in bacteria (and harvested from them for use). Because they cut within the molecule, they are often called restriction endonucleases. Werner Arber, Hamilton O. Smith, and Daniel Nathans discovered and characterized restriction enzymes in the late 1960s and early 1970s. More than 2500 Type II restriction enzymes have been recognized from several bacterial species which are the source of more than 2500 recognized Type II restriction enzymes which identify roughly 200 discrete sequences which may range from 4 to 8 bases in size.

They cleave DNA at or near specific recognition nucleotide sequences known as restriction sites. Restriction enzymes are involved in protecting bacterial cell from infecting microbes by cleaving foreign DNA. This also gives them the ability to manipulate DNA fragments, which makes these enzymes immensely useful for recombinant DNA technology and genetic engineering.

Depending on the structure and if they cleave at the recognition site or away from it, restriction enzymes are categorized into four types. In order to cleave DNA, all restriction enzymes make two incisions, once through each sugar-phosphate backbone (i.e., each strand) of the DNA double helix. The enzyme "scans" the DNA molecule for a particular sequence, generally of four to six nucleotides, called as recognition site. Once the site is identified, enzyme digestion takes place and the DNA strands are cleaved. As we know DNA strands are antiparallel palindromic sequences; therefore the recognition sites are present on both the strands in opposite directions. Hence the enzyme cleaves both the strands. Restriction enzymes like HaeIII and AluI cut straight across the double helix producing "blunt" ends, whereas BamHI, HindIII, and EcoRI cut in offset fashion producing overhanging piece of single-stranded DNA (Heinrichs 2007; Arber and Linn 1969; Brownlee 2005; Konforti 2000) (Fig. 4).

Four categories of restriction enzymes are:

Type I enzymes combine restriction modification and cleave DNA far from recognition sequence. These enzymes have little to no practical use as their restriction fragments are indiscreet and produce indistinct pattern on electrophoresis.

Type II enzymes are the most commonly used enzymes used for gene cloning and related experiments in labs cut definite positions nearby or inside the recognition sequences within gene resulting in production of restriction fragments which have distinct gel banding patterns and are separate. Examples of Type II enzymes are those HhaI, HindIII, and NotI.

Type III enzymes are restriction, modification enzymes with numerous combinations which differ from other class of enzymes cleaving outside of their recognition sequences, usually not giving complete digests.

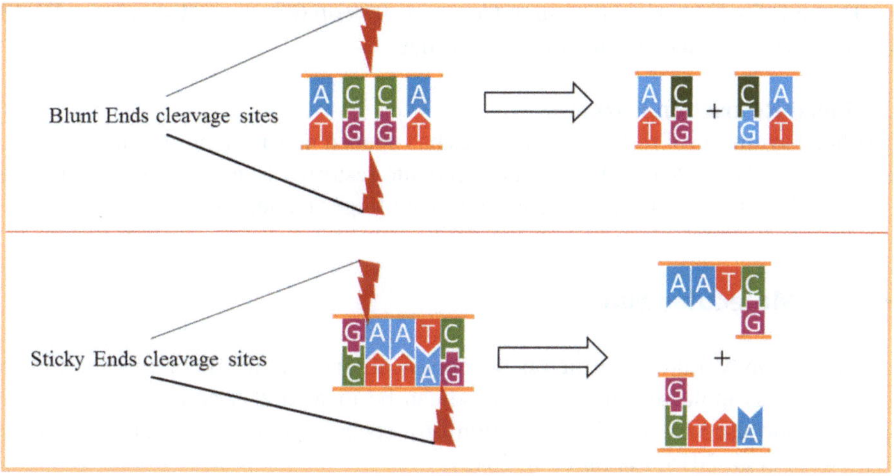

Fig. 4 Endonuclease activity of enzymes

Table 1 Types of restriction endonuclease enzyme and their activities

Types and activities of restriction enzymes	
Type I	Cleaves gene randomly, away from its target sequence
Type II	Cleaves gene at specific positions close or within its target
Type II G	Cleaves outside its target with both restriction and modification enzymatic activities in the same protein
Type II P	Cleaves symmetric targets and cleavage sites
Type II S	Ability to recognize asymmetric DNA
Type III	Cleaves outside its target in opposite orientation within the same DNA
Type IV	Cleaves modified (e.g., methylated) target DNA

Type IV enzymes only recognize and cleave most probably methylated DNA, and best examples are the Mrr and McrBC systems of *E. coli* (Table 1).

Type V restriction endonuclease may cleave variable ranges of length through utilizing the guide RNAs for targeting specific sequences although it doesn't read non-palindromic sequences (a sequence in which forward and reverse are not same like AAAATTTTAAAA). They are present in invading organisms. These characters make these enzymes candidature promising for future genetic engineering applications.

Nomenclature of Restriction Enzyme

All restriction enzymes are named on the bacterium name from which it was isolated using its genus, species, and strain for nomenclature.

Example: EcoRI restriction endonuclease isolated from *E. coli* was named as per the description below.

E stands for *Escherichia* (genus); **Co** stands for *coli* (species); **R** stands for **RY13** (strain); while **I** stands for identification order.

Artificial Restriction Enzymes
Artificial synthesis of restriction enzymes has opened new horizons giving liberty to researchers. They cleave DNA at a desired site resulting in more precise sequence for genetic engineering experiments, e.g., zinc-finger nucleases.

6 Molecular Markers

A specific DNA fragment which is phenotypically neutral inside of genome can be considered as molecular markers. These can be located at specific regions of the genome and used to "flag" the position of a specific gene or a specific character (Hayward et al. 2015; Zargar et al. 2015).

6.1 Restriction Fragment Length Polymorphism (RFLP)

RFLP is a molecular marker based on the differential hybridization of cloned DNA to DNA fragments in a sample of restriction enzyme digested DNAs. In RFLP, genomic DNA is digested with restriction enzymes and subsequently digested DNA fragments are size fractionated by gel electrophoresis, then transferred on membranes using Southern blotting technique and lastly hybridized with the labeled probe to visualize DNA polymorphisms. RFLP markers are identified by a specific restriction enzyme-probe combination.

Advantages of RFLP are high genomic abundance; no sequence information; co-dominant markers; can use filters numerous times, across species, and in plants reliably (well-tested), respectively; and crime detection.

Disadvantages are as follows: need of large amount of good-quality DNA, requirement of radioactive labeling, difficult to automate, and cloning and characterization of probe are required.

6.2 Random Amplification of Polymorphic DNA (RAPD)

It is a PCR-based method, employing single primers of random nucleotide sequence to amplify unknown PCR fragments from genomic template of targeted DNA. In RAPD analysis, the target sequences are unidentified. In RAPD, PCR is usually carried out with random primers. It can be used in measurements of genetic structure of populations, germplasm characterization, genetic diversity, development of markers linked to a trait of interest, genetic mapping, interspecific hybridization, in case of somaclonal variation identification of clones, and verification of genetic identity. Advantages of RAPD are less amount of DNA is required, high genomic abundance, ideal for automation, no radioactive labeling, large genome coverage, no prior knowledge of sequence information, and relatively faster.

6.3 Amplified Fragment Length Polymorphism (AFLP)

This technique is an amalgamation of the RFLP and PCR techniques used for quick screening of genetic diversity. DNA fragments with dissimilar nucleotide sequence of which huge number of copies have been amplified are AFLPs. Two different enzymes are used for restriction digestion of DNA, ligation of two adopters, and selective amplification resulting in separation on denaturing gel analysis of amplified fragments. The amplified products are normally separated on denaturing polyacrylamide gel and further are visualized using autoradiography. The technique requires more amount of DNA. It rapidly generates high numbers of significantly replicable markers from DNA of any organism and, hence, permits high-resolution genotyping. The time and cost-efficiency, replicability, and resolution of AFLPs are greater or equal to those of other markers. It is a better technique than RAPD and RFLP because of high genomic abundance, no need for sequence information, can be used across species, high polymorphism, and works with smaller RFLP fragments, while having these qualities makes it less satisfactory: very complicated due to changes in patterns with respect to materials used, cannot acquire consistent map, and required very good primers.

6.4 Simple Sequence Repeat Microsatellite Polymorphism

Microsatellite or simple sequence repeats (SSRs) offer fairly broad genomic coverage as they are flexible to automation. These markers have wide range for genetic analysis in crop improvement strategies. Because of their abundance, high polymorphism, no radioactive labeling having locus identity, being multi-allelic hypervariability, and suitability for high-throughput analysis, they are widely used in crops. Their limitations include inability to be used across species, need for sequence information, and they are still not well-tested.

6.5 Single-Nucleotide Polymorphism (SNP)

Current advances in DNA sequencing allow development of high-throughput assays; thus single-nucleotide polymorphisms turn out to be dominant molecular marker applications. SNP discovery comprises of finding differences between two sequences. Traditionally this has been accomplished through PCR amplification of genes of interest from multiple individuals trailed by sequencing of these amplicons. Sequences are then aligned for identification of any polymorphisms. Because of its expensive and time-consuming, in silico approaches of SNP and SSR discovery are now being accepted. A very large volume of next-generation sequencing data is produced at the expense of fewer amounts. Whole genome sequencing of complete genome provides better chance to recognize the variety of genetic diversity in a population and better understanding of relationship between the inherited genome and heritable traits.

6.6 Variable Number Tandem Repeat (VNTR)

A variable number tandem repeat (VNTR) as his name suggests is a short nucleotide sequence organized as a tandem repeat. High mutation rate associated with it has been exploited in producing highly discriminatory VNTR profiles. VNTR analysis has resolved very closely related strains of pathogen progenitor of many plant diseases.

6.7 Simple Sequence Length Polymorphism (SSLP)

Simple sequence length polymorphisms (SSLPs) are tandem repeat sequences employed as genetic markers with polymerase chain reaction (PCR). An SSLP is a form of polymorphism (the existence of two or more forms of individuals within the same species) that shows a variance in DNA sequence amid individuals. SSLPs are recurring sequences over changing base lengths in intergenic regions of DNA. Since the flanking sequences are the same in all individuals, PCR primers that will selectively amplify the repeat sequences can be used to detect differences in repeat number (length of the amplified sequence). Difference in the length of SSLPs can be used to understand genetic difference among two individuals in a certain species.

An example of the usage of SSLPs (microsatellites) is seen in a study by Rosenberg et al. in which Rosenberg and his team used SSLPs to cluster different continental races. Rosenberg studied 377 SSLPs in 1000 people in 52 different regions of the world. By using PCR and cluster analysis, Rosenberg was able to group individuals that had the same SSLPs. These SSLPs were exceptionally useful to the experiment because they do not disturb the phenotypes of the individuals, thus being unaffected by natural selection.

6.8 Short Tandem Repeat (STR)

Short tandem repeats (STRs), also known as simple sequence repeats (SSRs) or microsatellites, are repeating sequences of two to six base pairs of DNA. It is a type of variable number tandem repeat (VNTR). The eukaryotic genome is crowded with repeated DNA sequences. These repeated sequences come in several sizes and are classified conferring to the number of contiguous repeated units, the length of the core repeat units, and/or the overall length of the repeat region. STRs have become prevalent DNA markers because they are effortlessly amplified by polymerase chain reaction (PCR) without the problem of differential amplification; that is, the PCR products for STRs are normally comparable in amount, making analysis easier.

STRs are repeated many times in a head-tail manner, i.e., the 16 bp sequence of "gatagatagatagata" would signify four head-tail copies of tetramer "geta." The polymorphism in STRs is because of different number of copies of repeat elements that can arise in a population of individuals. This kind of DNA analysis works to examine distinct areas in DNA. The differences for particular specific DNA regions in one individual versus another can allow differentiating between individuals.

Laser-induced fluorescence detection of dye-labeled polymerase chain reaction (PCR) products is the most common form of STR used which is further followed by capillary electrophoresis (CE) size-based separation. This unit defines the techniques and marker systems most widely used around the world in building criminal DNA databases and conducting forensic casework.

Regarding the benefits of STRs, it allows "mistake-proof" information to be gathered even in poor conditions as well as helps in detection of genetic diseases.

7 Molecular Beacon

Now probe-based techniques like molecular beacon, minor groove-binding (MGB) assays, etc. are used to identify the specific sequences in the nucleic acids based on the real-time PCR. Molecular beacon probes have specifically increased the sensitivity and precision over the conventional PCR with respect to the detection and quantification of target genes and set of conserved sequences in the promoter region.

Molecular beacons were first developed by Tyagi and Kramer (1996). They can be employed to monitor real-time amplification during PCR, gene mutation, detection of pathogens, and genetic analysis and in various other biological contexts. Real-time detection of nucleic acid amplification both in vivo and in vitro became more convenient with the development of molecular beacons (Tyagi and Kramer 1996; Kostrikis et al. 1998; Tyagi et al. 1998). Real-time PCR technique is the qualified version of conventional PCR technique. On the basis of specificity, detection chemistries can be split into two groups—nonspecific detection using DNA-binding dyes and specific detection using target-specific probes.

7.1 Nonspecific Detection Using DNA-Binding Dyes

The product accumulation leads to increased fluorescence of the reporter dye that binds to DNA with each successive cycle of amplification, for example, SYBR Green (Andersen et al. 2006).

7.2 Specific Detection Using Target-Specific Probes

Specific detection of real-time PCR is carried out with some oligonucleotide probes labeled with both a fluorescent dye and an acceptor or quencher dye (Hiyoshi and Hosoi 1994; Chen et al. 1997). The amount of fluorescence is directly proportional to the amount of target DNA generated during the PCR process (Parashar et al. 2006), for example, TaqMan probe, molecular beacon, scorpion probe, etc.

7.3 Structure and Design of Molecular Beacon

MBs are representing a class of nucleic acid probes that have unique DNA hairpin structures that specifically target complementary DNA or RNA (Wu et al. 2012) based on the signal transduction mechanism of FRET (Förster or fluorescence resonance energy transfer) (Marras et al. 2002). It is reported that molecular beacon is defined as a single-stranded oligonucleotide probe with a special stem-loop structure. The structure of MBPs (15–30 nt long) consists of the four parts, complimentary stem, one loop, fluorescent dye at 5′, and a quencher at 3′ (Tyagi and Kramer 1996; Kostrikis et al. 1998; Ortiz et al. 1998). The stem was 5–7 nt long at both ends, complementary to each other. The loop has a probe sequence which varies from 15 to 30 nt where probe sequence is attached.

7.4 Application

In the late years, gene detection method has been built up rapidly. Molecular beacon probe-based detection was found to be more precise, sensitive, and less cumbersome based on the real-time PCR (Tyagi and Kramer 1996; Andersen et al. 2006; Kamboj et al. 2014; Göhring et al. 2014). Many scientists used MB technology in detecting liquid-phase hybridization (Poddar 1999), effectively reducing cross-contamination of nucleic acids (Lee et al. 2005), real-time nuclear acid detection (Pas et al. 2005; Ye et al. 2009; Elsayed et al. 2006; Lata et al. 2009), design of biosensors (Fang et al. 1999), protein-DNA interactions (Fang et al. 2000), disease-causing pathogen (Tomlinson et al. 2007), and gene expression study (Smith et al. 2014).

7.5 Nucleic Acid Detection

Molecular beacons are oligonucleotide probes that can report the presence of specific nucleic acids in homogeneous solutions (Tyagi and Kramer 1996). In the presence of target DNA, the MBs hybridize with the target DNA and form double-stranded DNA (dsDNA), the fluorophore FAM is detached from the quencher BHQ-1, and thus the fluorophore produces fluorescence. Many researcher groups have been using MB probe for the detection of the specific nucleic acid sequences (Tyagi and Kramer 1996; Marras et al. 1999; Xiang et al. 2014, Prajapati et al. 2013; Prajapati and Pandey 2015, Wang et al. 2013).

7.6 Molecular Beacon-Based SNP Genotyping Methods

Several methods are used for the detection of mutation, but MBs have higher specificity. Even a single-base mismatch of target sequences or insertion mutations can be detected. Molecular beacons are used for detecting single-nucleotide polymorphisms (SNPs) and allele discrimination because of their excellent selectivity in differentiating single-nucleotide mutation (You et al. 2013).

7.7 DNA-Protein Interaction

Using molecular beacons (MBs) for DNA-binding proteins is a well-developed technology (Li et al. 2000; Fang et al. 2000). Molecular beacon assemblies can detect sequence-specific DNA-binding proteins (Dummit and Chang 2006; Krusiński et al. 2008). MB probe was employed to identify the presence of *E. coli* single-stranded DNA-binding protein (SSB). The binding of SSB to the MB was expected to disrupt the hairpin conformation of the oligonucleotide, resulting in the release of the fluorophore and the restoration of fluorescence (Lohman and Ferrari 1994). Molecular beacon aptamer (MBA) (Tan et al. 2004), a new class of protein probe, has been established by coalescing of florescent MB probe and the protein-binding specificity of non-florescent aptamer. Aptamer is an innovative and new class of short DNA/RNA sequence that rivals antibody in protein recognition (Ellington and Szostak 1990; Robertson and Joyce 1990). Now, aptamers hold many advantages over antibodies: easier synthesis, easier storage, more flexible labeling, faster tissue penetration, better reproducibility, shorter blood residence, etc.

8 Computational Approaches

There is remarkable potential to analyze the genetic basis of crop performance. Even though this analysis is likely to be iterative and integrated, it is very much beneficial. Computational tools can be used to study interactions (eg. protein-DNA promoter motifs, Fig. 5), genomics and proteomics of plant to study intensively

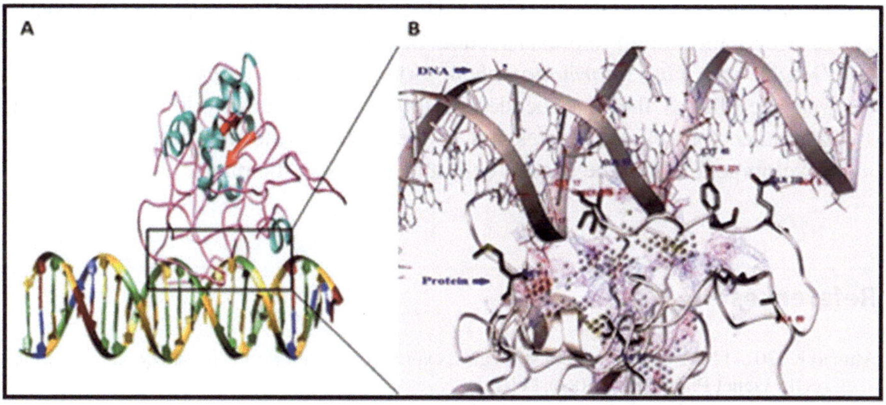

Fig. 5 Protein-DNA docking model for AP2-EREBP TF (LOC_Os03g22170) with DNA containing TCC box promoter motif of the DR-DEG and RG (LOC_Os12g 05900) obtained from HADDOCK server. Amino acid residues in pink and cyan color are interacting amino acid, while adenine in red, guanine in green, cytosine in golden, and finally thymine in blue color at chain B generated using Chimera software (B): molecular view of docked sample in Schrödinger suite (Source: Prajapati and Pandey 2015)

on molecular level which may lead to enhancement of crop productivity and better phenotypic characters. Underlying are few ways:

1. *Genomic Databases* such as NCBI, DDBJ, and EMBL have all the relevant stored information regarding any gene of interest which can be retrieved whenever needed and are free of cost.
2. *Genomic sequence* study of plants suggests functions for the genes which can be directly targeted for better crop productivity by altering it.
3. *Homology search and modeling* often indicates to a tentative identification of the sequence to a class of genes, which may help in predicting various sets of genes responsible for a particular plant phenotype.
4. *Transcription factors database*: Determination of transcription factors of any plant gene can help us to overexpress and underexpress the gene or its suppression of gene activity.
5. *In silico approaches* are important steps to analyze the genetic origin of traits and classify allelic variants using various tools which are beneficial.
6. *Genetic maps* can be used to determine which sections of the chromosome specifying particular traits can be allocated and hence help in location of the gene.
7. *Biochemical pathway databases such as KEGG* are used in studies of plant metabolism which have identified proteins critical in the functioning of maximum pathways.
8. *Phylogenetic analysis* can help in finding closely related species to one particular plant, and thus this data can be exploited in crossbreeding for better results.
9. *Protein database* provides all the data regarding shape, 3D structure, sequence of protein, and its stability criteria which can be further exploited to produce recombinant protein in plants such as Bt cotton.
10. *Databases having information about promoters and regulators* of any gene are very helpful as they are needed at the time of transcription. They help to guarantee the reliable expression of the genes at the preferred times during development and in the chosen cells and tissues of the plant.

References

Ahmad K (2014) Molecular farming: strategies, expression systems and bio-safety considerations. Czech J Genet Plant Breed 50(1):1–10

Andersen CB, Holst-Jensen A, Berdal KG, Thorstensen T, Tengs T (2006) Equal performance of TaqMan, MGB, molecular beacon, and SYBR green-based detection assays in detection and quantification of roundup ready soybean. J Agric Food Chem 54:9658–9663

Arber W, Linn S (1969) DNA modification and restriction. Annu Rev Biochem 38:467–500

Attree SM, Fowke LC (1993) Embryogeny of gymnosperms: advances in synthetic seed technology of conifers. Plant Cell Tissue Organ Cult 35(1):1–35

Bhojwani SS, Dantu PK (2013) Micropropagation. In: Plant tissue culture: an introductory text. Springer, New Delhi, pp 245–274

Brownlee C (2005) Danna and Nathans: restriction enzymes and the boon to molecular biology. Proc Natl Acad Sci 102:5909

Burke CM, Darling AE (2016) A method for high precision sequencing of near full-length 16S rRNA genes on an Illumina MiSeq. PeerJ 4:e2492

Chen X, Zehnbauer B, Gnirke A et al (1997) Fluorescence energy transfer detection as a homogeneous DNA diagnostic method. Proc Natl Acad Sci 94:10756–10761

Chopra VL, Sharma RP (1991) Plant biotechnology and agriculture. Curr Sci 60(9&10):543–547

Dummit B, Chang YH (2006) Molecular beacons for DNA binding proteins: an emerging technology for detection of DNA binding proteins and their ligands. Assay Drug Dev Technol 4:343–349

Edwards D, Batley J (2010) Plant genome sequencing: applications for crop. Plant Biotechnol J 8:2–9

Ellington AD, Szostak JW (1990) In vitro selection of RNA molecules that bind specific ligands. Nature 346:818–822

Elsayed S, Plewes K, Church D, Chow B, Zhang K (2006) Use of molecular beacon probes for real-time PCR detection of plasmodium falciparum and other plasmodium species in peripheral blood specimens. J Clin Microbiol 44:622–624

Fang X, Liu X, Schuster S, Wan W (1999) Designing a novel molecular beacon for surface-immobilized DNA hybridization studies. J Am Chem Soc 121:2921–2922

Fang X, Li JJ, Tan W (2000) Using molecular beacons to probe molecular interactions between lactate dehydrogenase and single-stranded DNA. Anal Chem 72:3280–3285

Fang J, Zhu X, Wang C, Shangguan L (2016) Applications of DNA technologies in agriculture. Curr Genomics 17(4):379–386

Ferrari EAP, Colombo RC, Faria RTD, Takane RJ (2016) Cryopreservation of seeds of *Encholirium spectabile* Martius ex Schultes f. by the vitrification method. Rev Ciênc Agron 47(1):172–177

Göhring J, Jacak J, Barta A (2014) Imaging of endogenous messenger RNA splice variants in living cells reveals nuclear retention of transcripts inaccessible to nonsense-mediated decay in Arabidopsis. Plant Cell 26(2):754–764

Gupta V, Sengupta M, Prakash J, Tripathy BC (2017) Plant biotechnology and agriculture. In: Basic and applied aspects of biotechnology. Springer, Singapore, pp 415–452

Hayward AC, Tollenaere R, Dalton-Morgan J, Batley J (2015) Molecular marker applications in plants. Plant Genotyping: Methods Protoc:13–27

Heinrichs A (2007) Making the cut: discovery of restriction enzymes. Nature Milestones

Hiyoshi M, Hosoi S (1994) Assay of DNA denaturation by polymerase chain reaction-driven fluorescent label incorporation and fluorescence resonance energy transfer. Anal Biochem 221:306–311

Kamboj A, Pateriya AK, Mishra A, Ranaware P, Kulkarni DD, Raut AA (2014) Novel molecular beacon probe-based real-time RT-PCR assay for diagnosis of Crimean-Congo hemorrhagic fever encountered in India. Biomed Res Int 2014(5):1–4. https://doi.org/10.1155/2014/496219

Kao KN, Constabel F, Michayluk MR, Gamborg OL (1974) Plant protoplast fusion and growth of intergeneric hybrid cells. Planta 120(3):215–227

Konforti B (2000) The servant with the scissors. Nat Struct Biol 7:99–100

Kostrikis LG, Tyagi S, Mhlanga MM, Ho DD, Kramer FR (1998) Spectral genotyping of human alleles. Science 279:1228–1229

Krusiński T, Wietrzych M, Grad I, Ożyhar A, Dobryszycki P (2008) Equilibrium analysis of the DNA binding domain of the ultraspiracle protein interaction with the response element from the hsp27 gene promoter—the application of molecular beacon technology. J Fluoresc 18:1–10

Lata P, Ram S, Agrawal M, Shanker R (2009) Real time PCR for the rapid detection of vanA gene in surface waters and aquatic macrophyte by molecular beacon probe. Environ Sci Technol 43:3343–3348

Lee DH, Mathew J, Pfahler W, Ma D, Valinsky J, Prince AM, Andrus L (2005) Individual donor nucleic acid amplification testing for detection of West Nile virus. J Clin Microbiol 43:5111–5116

Leister D, Ballvora A, Salamini F, Gebhardt C (1996) A PCR-based approach for isolating pathogen resistance genes from potato with potential for wide application in plants. Nat Genet 4(4):421–429

Li JJ, Fang X, Schuster SM, Tan W (2000) Molecular beacons: a novel approach to detect protein-DNA interactions. Angew Chem Int Ed Engl 39:1049–1052

Lohman TM, Ferrari ME (1994) Escherichia Coli single-stranded DNA-binding protein: multiple DNA-binding modes and cooperativities. Annu Rev Biochem 63:527–570

Marras SAE, Kramer FR, Tyagi S (1999) Multiplex detection single-nucleotide variations using molecular beacons. Genet Anal Biomol Eng 14:151–156

Marras SAE, Kramer FR, Tyagi S (2002) Efficiencies of fluorescence resonance energy transfer and contact-mediated quenching in oligonucleotide probes. Nucleic Acids Res 30:e122

Ortiz E, Estrada G, Lizardi PM (1998) PNA molecular beacons for rapid detection of PCR amplicons. Mol Cell Probes 12:219–226

Parashar D, Chauhan DS, Sharma VD, Katoch VM (2006) Applications of real-time PCR technology to mycobacterial research. Indian J Med Res 124(4):385

Pas SD, Noppornpanth S, van der Eijk AA, de Man RA, Niesters HG (2005) Quantification of the newly detected *lamivudine* resistant YSDD variants of Hepatitis B virus using molecular beacons. J Clin Virol 32:166–172

Poddar SK (1999) Detection of adenovirus using PCR and molecular beacon. J Virol Methods 82:19–26

Pohl G, Shih IM (2004) Principle and applications of digital PCR. Expert Rev Mol Diagn 4(1):41–47

Prajapati GK, Pandey DM (2015) Molecular Beacon probe based promoter motifs validation in anoxia responsive DEGs and their *in silico* interaction studies with AP2/EREBP TF in rice (*Oryza sativa* L). Int J Pharm Pharm Sci 7(3):123–130

Prajapati GK, Kashayp N, Kumar A, Pandey DM (2013) Identification of GCC box in the promoter region of ubiquinol cytochrome C chaperone gene using molecular beacon probe and its *in silico* protein-DNA interaction study in rice (*Oryza sativa* L.) Int J Comput Bioinforma In Silico Mod 2(5):213–222

Prosen D, Hatziloukas E, Panopoulos NJ, Schaad NW (1991) Direct detection of the halo blight pathogen *Pseudomonas syringae pv. phaseolicola* in bean seed by DNA amplification. Phytopathology 81:1159. (Abstr.)

Rasmussen OF, Wulff BS (1991) Detection of *Pseudomonas syringae pv. pisi* using PCR. In: Proceedings 4th international Working Group on Pseudomonas syringae Pathovars. Kluwer Academic Publishers, Dordrecht, pp 369–376

Raza K, Ahmad S (2016) Principle, analysis, application and challenges of next-generation sequencing: a review. arXiv preprint arXiv:1606.05254

Robertson DL, Joyce GF (1990) Selection in vitro of an RNA enzyme that specifically cleaves single-stranded DNA. Nature 344:467–468

Sanger F, Coulson AR (1975) A rapid method for determining sequences in DNA by primed synthesis with DNA polymerase. J Mol Biol 94(3):441–448

Smith KF, de Salas M, Adamson J, Rhodes LL (2014) Rapid and accurate identification by real-time PCR of biotoxin-producing dinoflagellates from the family gymnodiniacea. Mar Drugs 12:1361–1376

Tan WH, Wang K, Drake TJ (2004) Molecular beacons. Curr Opin Chem Biol 8:547–553

Tomlinson JA, Barker I, Boonham N (2007) Improved molecular detection of faster, simpler, more-specific methods for *Phytophthora ramorum* in the field. Appl Environ Microbiol 73:4040–4047

Tyagi S, Kramer FR (1996) Molecular beacons: probes that fluoresce upon hybridization. Nat Biotechnol 14:303–308

Tyagi S, Bratu DP, Kramer FR (1998) Multicolor molecular beacons for allele discrimination. Nat Biotechnol 16:49–53

Wang Q, Chen L, Long Y, Tian H, Wu J (2013) Molecular beacons of xeno-nucleic acid for detecting nucleic acid. Theranostics 3(6):395–408

Wu CS, Peng L, You M, Han D, Chen T, Williams KR, Yang CJ, Tan W (2012) Engineering molecular beacons for intracellular imaging. Int J Mol Imag. https://doi.org/10.1155/2012/501579

Xiang D, Zha K, Xiang W, Wang L (2014) Highly sensitive fluorescence quantitative detection of specific DNA sequences with molecular beacons and nucleic acid dye SYBR Green I. Talanta 129:249–253

Ye Q, Zhuang H, Zhou C (2009) Detection of naphthalene by real-time immuno-PCR using molecular beacon. Mol Cell Probes 23:29

You M, Yang CJ, Tan W (2013) Molecular Beacons for Detection of Single-Nucleotide Polymorphisms. In: Molecular Beacons. Springer, Berlin, Heidelberg, pp 61–74

Zargar SM, Raatz B, Sonah H, Bhat JA, Dar ZA, Agrawal GK, Rakwal R (2015) Recent advances in molecular marker techniques: insight into QTL mapping, GWAS and genomic selection in plants. J Crop Sci Biotechnol 18(5):293–308

Zimmermann U, Scheurich P (1981) High frequency fusion of plant protoplasts by electric fields. Planta 151(1):26–32

Bioformulation and Fluid Bed Drying: A New Approach Towards an Improved Biofertilizer Formulation

Pramod Kumar Sahu, Amrita Gupta, Megha Singh, Priyam Mehrotra, and G. P. Brahmaprakash

Abstract

Biofertilizers are now a popular term among agriculturists. Many improvements have been done in existing techniques, and many new techniques had evolved as well to enhance its effects on plant system. It is being used extensively for biocontrol and plant growth promotion. Bioformulation is one of the areas of prime research importance as it affects the effectiveness of inoculants. An effort has been made for further improvement in bioformulation with a vision of reducing contamination and enhancing shelf life. In this technique, fluid bed dryer (FBD) is widely used in food and drug industry. Fluidized drying by FBD also gives stress to the inoculants while constructing the formulation. This technique may have an important role in pre-exposing and hardening of inoculants to stress so that it can perform well upon application on harsh soil environment. This formulation has shown higher survival rates, reduced contamination and good plant response. FBD inoculant holds good with consortium of agriculturally beneficial microorganisms. It can be applied as seed inoculation and other popular delivery techniques. This technique can pave the way for a new kind of formulation and needs to improvise.

Keywords

Biofertilizers · Bioformulation · Fluid bed dryer · Plant growth promotion · Biocontrol

P. K. Sahu (✉) · A. Gupta · M. Singh · P. Mehrotra
ICAR-National Bureau of Agriculturally Important Microorganisms,
Mau, Uttar Pradesh, India

G. P. Brahmaprakash
University of Agricultural Sciences, GKVK, Bangalore, Karnataka, India

R. S. Sengar, A. Singh (eds.), *Eco-friendly Agro-biological Techniques for Enhancing Crop Productivity*, https://doi.org/10.1007/978-981-10-6934-5_3

1 Biofertilizers in Present Scenario

Food security for the entire population is a challenge, especially when population upturns and resources slump. The green revolution with high-input agriculture has given a boon in production, but at the same time, it had a drastic impact on soil health. Today high-input agriculture perhaps finds it insufficient to satisfy needs of billions. Many more factors come into picture when enhancing crop productivity is concerned:

- Shrinking agricultural lands
- Migration to cities
- Changing climate
- Irregular water availability
- Negative impacts due to the excessive chemicals used in the past
- Very poor availability of organic matter to soil
- Insufficient government policies

The major problem is related to plant growth and development. Upsurge in productivity requires good and sustainable nutrition. Although soil is the largest buffer of nutrients, it needs to be replenished. The chemical fertilizers alone can't beget sustainable nutrition. People are getting aware against excessive chemical use, and organic farming is much ahead of practicality. The need of an hour is to find a way in between these two. A substantial amount of agrochemicals can be replaced by potential microbes and their products. This will give high yield with sustainability. Microbes in the form of biofertilizer do have tremendous ability to support plant growth and development.

2 Effectiveness of Biofertilizers for Enhancing Crop Productivity

Microbes have magical capacity to accumulate these nutrients in its vicinity, and as we know, they colonize in rhizosphere. Active root of plants secretes exudates which attract these organisms, and the association, thus formed, enhances the nutrient availability in the active feeding area of plant roots (Ahemad and Khan 2012; Haichar et al. 2014). Apart from nutrients, these microbes secrete growth-promoting substances like IAA and GA3 for general growth promotion (Baca and Elmerich 2007; Spaepen and Vanderleyden 2011), enhancing resistance to abiotic stresses (Yao et al. 2010), protection from plant pathogens (Qi et al. 2016) and many more. These all beneficial microbes are divided into two classes – plant growth promoters and plant health promoters.

Plant Growth Promoters It provides nutrients and growth enhancers to the plant and gives good growth in absence of a pathogen, for example, N_2 fixers, P solubilizers, P mobilizers, plant growth promoters, etc. (Nehra and Choudhary 2015; Sahu and Brahmaprakash 2016; Takanashi et al. 2016).

Plant Health Promoters It sustains the plant growth in presence of a pathogen or any abiotic stress. It either blocks the pathogen directly or minimizes deleterious effects of pathogen. Entomopathogenic fungus (*Beauveria bassiana* and *Metarhizium anisopliae*), *Trichoderma*, nematode-trapping fungi and other biocontrol agents come under this class (Gorzelak et al. 2015; Labroussaa et al. 2016; Qi et al. 2016; Yan et al. 2016). Microbes also alleviate abiotic stresses by inducing/secreting different metabolites in the plants (Tiwari et al. 2011; Pandey et al. 2016).

These microbes are applied in the form of a formulation of biofertilizers. Biofertilizers are preparations containing live or latent microorganisms which enhances the plant growth (Brahmaprakash and Sahu 2012; Kumar 2014). There are many kinds of biofertilizers moving in the market. Whether it is nitrogen fixer, phosphorus solubilizer, plant growth-promoting rhizobacteria (PGPR) or any biocontrol agent, the prime requirement for its effectiveness is it should be delivered to the target in sufficient numbers and in right physiological forms. Thus, its downstream handling and the form it is being delivered (i.e. formulation) become critical issues for inoculant industry.

2.1 Bioformulation in PGPR

PGPR are a heterogeneous group of bacteria that can be found in rhizosphere, at root surfaces and in association with roots that can improve and promote the quantity and quality of plant growth characteristics (Arora et al. 2010; Bhattacharyya and Jha 2012).

Use of PGPR can play important roles in developing sustainable systems for crop production. Over the past few decades, producers have been dependent on agrochemicals and pesticides to kill pathogens, weeds, etc. which cause harmful effect on ecosystem, causing pollution, and increase pathogenicity. To overcome these harmful effects, biological control of plant pathogens has been increased due to public concern about the hazardous chemical pesticides.

To date, many bacterial genera, e.g. *Pseudomonas*, *Azospirillum*, *Azotobacter*, *Klebsiella*, *Enterobacter*, *Alcaligenes*, *Bacillus*, etc., are being reported to improve plant growth and health. These PGPR may have direct or indirect effect on plant growth by affecting symbiotic N_2 fixation, nodulation or nodule occupancy (Fuhrmann and Wollum 1989; Nehra and Choudhary 2015; Sahu and Brahmaprakash 2016).

The exact mechanism and mode of action of PGPR are not fully understood, but are thought to include the following.

Growth of plant is induced by production or induction of plant growth regulators like IAA, gibberellic acid, cytokinin and ethylene (Baca and Elmerich 2007; Spaepen and Vanderleyden 2011; Glick 2014; Gamalero and Glick 2015), symbiotic N_2 fixation (Takanashi et al. 2016), P solubilization (Walpola and Yoon 2013), nutrient mobilization (Zhang et al. 2014), moisture availability (Marasco et al. 2012), production of stress alleviators (Tiwari et al. 2011), etc. So, the formulation of PGPR bacteria may improve the plant growth characteristics which can replace the use of chemical fertilizers. The best definition of bioformulation is biologically active products containing one or more beneficial strains and economical carrier materials.

For bioformulation in PGPR, a good carrier must be taken to deliver the right number of viable cells in good physiological conditions at the right time (Arora et al. 2008). The carrier used can be soils (peat, coal, clays and inorganic soil), plant waste materials, inert materials like vermiculite, ground rock phosphate, polyacrylamide gels, liquid osmoprotectant, alginate and other polymers, etc. (Guijarro et al. 2007; Swapna et al. 2016).

The carrier may be organic or inorganic. When PGPR are formulated using inorganic or organic carrier, their stability and durability are increased. In addition, their application particularly as seed treatment becomes easier and more practical. In spite of these benefits, there are certain problems associated with bioformulation development.

2.2 Bioformulation in Biocontrol

Formulations consist of the active material which is preserved or maintained in viable condition to show its biological effect. Previously many products were tested for their capacity to serve as substrate for production of biocontrol agents such as wheat bran, paddy husk, neem cake and soybean oil. The active material is mixed with carrier materials such as water and talc so as the formulation must be safe to handle and easy to apply.

Antagonism against phytopathogenic microorganism is mediated through the production of siderophore, antibiotics, regulation of metabolism, gene responses, induced systemic resistance, etc. (Feng et al. 2013; Achari and Ramesh 2014; Yan et al. 2016; Sahu et al. 2016b). Plant to plant signal for adaptive behaviour is found to link through mycorrhizal network (Gorzelak et al. 2015).

The biocontrol activity of *Trichoderma* is important for agriculture as well as for the environment as it does not get accumulated in the food chain and does no harm to plants, animals and humans (Monte and Llobell 2003). For biomass production at different environmental conditions, the isolation, characterization and morphological descriptions of *Trichoderma* spp. are important factors to consider. Soares de Melo et al. (2014) showed parasitism of *Bionectria* sp. on plant pathogens *Rhizoctonia solani* and *Pythium aphanidermatum*. It was found that antibiosis is also one of the biocontrol mechanisms apart from mycoparasitism.

It is important to grow *Trichoderma* at suitable conditions because it is an eco-friendly biocontrol agent against other soilborne pathogen as it produces certain specific cell wall enzymes such as chitinase, xylanase, etc. Talc-based bioformulation of *Trichoderma* is used against fungal wilt pathogens affecting leguminous crops like lentils and chickpea. The type of carriers influences the viability of bacterial cells; the bacterial populations for both cellulose- and talc-based formulations were higher. *Trichoderma* spp. are widely used in agriculture as biopesticides, biostimulants and biofertilizers for a variety of plants (Gorzelak et al. 2015). Sodium alginate is cheap and has no effect on the environment so it can be used for formulation and delivery of a biocontrol fungus which would be useful for obtaining better yield of crops. *Trichoderma* species can act as biocontrol agents through different synergistic mechanisms, but it is not an easy task to predict the degree of synergism in natural environment. The *Trichoderma*-based products are very useful to the environment as they promote plant growth and induce resistance in plants. Wide use of fungicides leads to toxic residues, environmental contaminations and carcinogenic effects in humans, animals and plants (Rajavel 2000).

Biocontrol methods help to develop an eco-friendly control strategy for managing plant disease (Bharathi et al. 2004), and biocontrol agents can be used as non-phytotoxic and easily biodegradable alternative fungicide because it is necessary to increase production without harming resources and the environment. Bioformulations with *Pseudomonas fluorescens* may increase antagonistic microorganisms in soil and can produce antibiotics and other extracellular metabolites. Applications of *P. fluorescens* formulation with PGPR reduced the incidence of blister blight and also increase the yield (Saravanakumar et al. 2007). Mixed strains of *P. fluorescens* and *B. subtilis* are very effective for reducing fruit rot incidence.

The development of bioformulations to replace fungicides for the management of plant disease is of great importance. Strains of *Bacillus* are very stable because of its ability to form spores that are resistant to heat (Emmert and Handelsman 1999). The spore-forming bacteria B. *megaterium* phosphate-dissolving bacteria solubilize inorganic phosphatic compounds into soluble forms which are taken up by plant. Phosphorus-solubilizing bacteria improve plant growth by synthesizing plant growth substances such as vitamins and amino acid (Ponmurugan and Gopi 2006).

Using a single organism as biocontrol agent against plant pathogen does not result in disease control and increase yield so the bioformulations possessing mixture of bioagents are developed. Bioformulations combining *P. fluorescens* Migula strains Pf1 and AH1 and *Beauveria bassiana* Vuill. isolate B2 are developed and tested for their efficacy against leaf folder pest and sheath blight disease on rice, and those rice leaves showed a greater accumulation of enzyme lipoxygenase activity against leaf folder insect and also greater accumulation of defence enzyme like peroxidase activity against sheath blight pathogen (Karthiba et al. 2010).

Many bacterial species like *Ralstonia solanacearum* that causes bacterial wilt can survive long term in water and soil (Yabuuchi et al. 1996). Therefore, there is a need to control bacterial wilt through effective and safe methods (Lwin and Ranamukhaarachchi 2006). Many strains of *P. fluorescens* have suppressed soil-borne diseases and enhance plant yield (Hoffland et al. 1996). *P. fluorescens* can be used as a potential biocontrol agent after formulations with organic substrates like vermicompost and farmyard manure with carboxymethyl cellulose as an adhesive. It can provide better nutrient source and can increase shelf life of formulated product due to increase rates of CO_2 evolution and high dehydrogenase activity (Islam and Toyota 2004). There is interaction between soil, plant and microorganisms and also high inoculum level which result in population dynamics of biocontrol agent (Costa et al. 2006).

The colonization ability of rhizosphere by *P. fluorescens* and specific pathogen antagonist interactions are important factors in disease control (Hass and Defago 2005). Jinnah et al. (2002) found that the biocontrol agent *Pseudomonas fluorescens* was found to show positive effect on the plant growth. *T. viride* along with *Pseudomonas fluorescens* increased the biocontrol activity against stem rot of groundnut caused by *Sclerotium rolfsii* (Manjula et al. 2004). The use of fungicides cannot be avoided in controlling the foliar diseases in the absence of suitable resistant cultivars. Seed treatment with *Pseudomonas fluorescens* at 10 g per kg + foliar spray at 0.2% was useful in controlling foliar disease of cotton and also improved yield significantly in comparison to chemical fungicides (Chattannavar et al. 2010).

The effect of different substrate-based powder formulations on the population of *P. fluorescens* indicated that vermicompost and farmyard manure used as substrate carriers with carboxymethyl cellulose as an adhesive prove to be better nutrient sources and have subsequent longer shelf life in the formulated product. The talc-based strain mixture formulation of fluorescent pseudomonas was developed by Nandakumar et al. (2001). It was prepared by mixing equal volume of individual strains and blended with talc as per Vidhyasekaran and Muthamilan (1995). Talc-based strain mixtures were effective against rice sheath blight and increased plant yield. Talc- and peat-based formulations of *P. chlororaphis* and *B. subtilis* were used for the management of turmeric rhizome rot (Nakkeeran et al. 2004).

In a recent study, it was found that *P. fluorescens* and *Bacillus coagulans* strains are more effective than the common seed fungicides (carboxin) for controlling sugar beet damping off (mortality) disease (Collins and Jacobsen 2003). *Trichoderma* formulations mixed with chitin reduced vanilla disease under field conditions and increase biocontrol efficacy of the antagonist which is a result of chitinolytic activity (Radjacommare et al. 2010). The use of such bioformulations in the fields can result in the reduction of application of harmful chemicals, can protect the environment and thus can help to achieve a sustainable agricultural system.

3 Bioformulation: Vehicle for Delivery

As discussed, the form of inoculant is critical issue for effectiveness of biofertilizers. A lot has been done in the area of bioformulation. Every tick of the clock brings out an improvement over the existing. But if we see from a critic's view, a serious question comes into our mind "are these efforts able to satisfy the need of farmers?" Few of the workers have done exceptionally good work in this area, but the zeal required to deal with the issue as a whole is somewhere missing. The cause for this is the soil, the dynamic system, beyond our predictions. Making a successful knock in this dynamic system requires a lot much efforts of basic and applied research.

Many successes have been made in the past, but these improvements were mostly broken links and individual scattered efforts. It requires very firm organized efforts from a big team of good researchers. Few of the major milestone improvements in formulation are explained here.

Centuries before in history, people were aware of effects of inoculation and addition of soil from legume-grown field to the new field put under legume cultivation (Brahmaprakash and Sahu 2012). The biofertilizers initially were formulated in solid carriers. Soil was used initially as a carrier for rhizobia (Madhok 1934) since it provides hostile environment to inoculant cells.

Next to the soil, peat became the formulation of choice worldwide for decades. It was popular due to successful field results obtained under commercial cultivation (Bashan 1998). Lignite was then tried as substitute of peat as carrier material (Kandaswamy and Prasad 1971). *Rhizobium* multiplies well in lignite at 28 °C. The viability of cells could be increased when lignite was mixed with soybean powder.

Coir dust was tested by Iswaran (1972) as carrier material for growth of *Rhizobium*. It was found appropriate when mixed with an equal proportion of soil and was superior to soybean meal. Dube et al. (1980) used coal (inferior grade) as carrier material and found it superior to lignite for nodulation of soybean. The shelf life was better as compared to lignite. Later, many other modifications were tested for its suitability for the formulation. Bajpai et al. (1978) reported that FYM and tank silt with 40% moisture were better protective carrier material for *Rhizobium* than peaty soil at 20 °C when stored for a period of 10 weeks, and fly ash is also used for preparing formulation (Khan et al. 2007). Granule formulations are also reported to exhibit higher shelf life (O'Callaghan and Gerard 2005; Guijarro et al. 2007). Wettable powders are one of the good-quality bioformulations having 50–80% powder along with filler, dispersant and surfactant (Brar et al. 2006).

Apart from solid carrier-based inoculants, many other formulations have been employed like liquid- and polymer-entrapped inoculant formulations. Liquid formulations are aqueous suspension having microbial cells in emulsion with a suitable osmoprotectant (Schisler et al. 2004; Brahmaprakash et al. 2007; Velineni and Brahmaprakash 2011). Traditional peat- and lignite-based inoculants had its own drawbacks, so research has been done to harvest the possibility liquid formulations of biofertilizers. It is an aqueous, oil-based product with special cell protectant and additives that promote cell survival in storage and after application to seed or soil (Sridhar et al. 2004; Dayamani 2010; Dayamani and Brahmaprakash 2014; Lee et al. 2016). The most beneficial feature of liquid inoculants is its suitability for advanced seeding equipment and farming system. It can be produced by simple fermentation, packed directly from the fermentor aseptically, added with suitable osmoprotectant and stored for a long time without losing viability (Sharma et al. 2010; Surendra Gopal and Baby 2016).

The further progress in bioformulations brought out polymer entrapment as a method of inoculant formulation. Polymer entrapment of cells can be done after mixing it with polymer and subjected to chemical solidification. It gives uniform beads entrapping live cells inside. These beads are then fermented for further growth in polymer matrix and dried. These microbes entrapped in polymer are released in soil after degradation of polymer by the soil microbes. These polymers have demonstrated potential bacterial carriers (Deaker et al. 2004). These formulations encapsulate the living cells that protect the microorganisms against many environmental stresses. Different inert materials were evaluated as carriers like polyacrylamide gel (Singleton et al. 2002).

Newer technologies are being employed for enhancing quality of bioformulation in quest of replacing a part of agrochemicals used in farming. Nanotechnology-based carrier in microbial inoculant is being tested which contains nanoparticles made up of inorganic or organic substances (Auffan et al. 2009) and has huge potential in agriculture (Bailey et al. 2010). Swapna et al. (2016) had prepared microbial consortium in different granular inoculants and checked its efficiency in vitro. Hydrogel-based bioformulation was reported by Suman et al. (2016). It was a consortium of *Azotobacter chroococcum*, *Pseudomonas fluorescence* and *Trichoderma viride*. This is having potential to support microbial growth, good shelf life and bioefficacy.

Many other forms and carriers have been tried with varied levels of success. Every drawback of existing technique gives the clue for the development of new technique. Fluid bed dried formulation also have been designed to eliminate the drawbacks of other formulations.

4 Fluid Bed Drying (FBD)

This is a new approach for making formulation and has some obvious benefits over earlier formulations. However, the technique is yet to be standardized but has multiple benefits for a stable formulation. The possibility of FBD may be harvested for a new step towards a sustainable agriculture. In this regard, use of FBD inoculants has resulted in low contamination rate, increased survival and enhanced plant growth in vitro (Sahu 2012; Brahmaprakash and Sahu 2012; Sahu et al. 2013; Lavanya et al. 2015; Sahu and Brahmaprakash 2016; Sahu et al. 2016a).

The instability of performance is a major drawback in biofertilizer industry, and an obvious breakthrough has been achieved by using consortium of agriculturally important microorganisms. The form in which it has been applied may differ but it has certain benefit.

4.1 Working Principle

Fluid bed dryer (FBD) is a dryer that maintains substrate suspended against gravity in an upward flowing air stream. Hot air blown at terminal velocity makes the particle suspend in the air, and thus entire surface area of substrate comes into contact with hot air. This results into a high rate of moisture transfer. The blowing air is heated by electrical heaters. This hot air expands the bed of material at its critical velocity and creates turbulence in the product. This phenomenon is called fluidization and offers more surface area for drying. Since fluidization forms full agitation of particles, it causes high rate of heat transfer and uniform drying. FBD is a common machine used in food and pharmaceutical industries for making instant coffee powder and other drying operations and has potential to be used in inoculant industry (Srivastava and Mishra 2010; Brahmaprakash and Sahu 2012; Sahu et al. 2013).

4.2 Preparation of Sample

The procedure discussed here is as mentioned elsewhere (Sahu et al. 2013). Growth curves of concerned microbial isolates for consortium are drawn to determine optimum age of culture in broth for inoculation. Formulation has to be prepared with all aseptic precautions. Glassware used during inoculation, fluid bed dryer jar, centrifuge tubes, water blanks and Ringer's solution are sterilized. On the basis of growth curve drawn, the sterile broths of respective microorganisms are inoculated in such a way that it should reach to its maximum growth phase at the time of constituting consortium (Sahu 2012; Sahu et al. 2013).

Broth cultures of the concerned isolates are centrifuged at 10,000 rpm at maximum growth phase for 10 min at 4 °C. Pellets are taken out carefully after decanting the supernatant. The amount of broth centrifuged depends on the number of viable cells in broth culture, i.e. the broth containing higher number of cells requires less quantity to achieve the desired population level in pellet.

These pellets of desired microbes are dissolved in Ringer's solution and mixed to final volume of 25 ml. To this solution, CMC (carboxymethyl cellulose) is added (as fixative or sticking agent) at 0.5% of total carrier material used, and 50 g of sterilized carrier is mixed along with it. This mixture of cells of microorganisms and carrier is transferred to sterilized fluid bed dryer (FBD) jar and subjected to fluidization. The jar of fluid bed dryer is removed quickly as the drying is completed and kept in laminar flow chamber to avoid contamination. These inoculant formulations were then packed in sterilized polythene pouches and labelled (Sahu 2012; Sahu et al. 2013; Lavanya et al. 2015).

4.3 Physical Conditions of Drying Biofertilizers in FBD

FBD is a semiautomated machine and needs to set at a time-temperature regime. This depends on the culture used in particular, type of carrier, moisture content of carrier-inoculant mixture, size of the clumps formed on the FBD jar, etc. For *Pseudomonas fluorescens* in talc carrier, samples are fluidized at 40 °C for two cycles of 1 h each (Sahu 2012; Sahu et al. 2013).

4.4 Factors Affecting Drying of Biofertilizers in FBD (Sahu 2012)

4.4.1 Temperature

The type of culture used in the process determines the temperature of drying. Some organisms which can tolerate high temperature can be dried easily, but the situation is difficult in case of sensitive organisms. A temperature of ~37 °C is suitable for fluidization drying of almost all mesophiles.

4.4.2 Time

Keeping for long time in contact with hot air drastically reduces the viable cells. So, one or two cycles of 1 h are enough for the drying of sample.

4.4.3 Moisture

Ringer's solution is added to the centrifuged pellet for facilitating its proper mixing and to prevent it from osmolysis. Adding more of solution increases the moisture content of the carrier and thus enhances the drying period.

4.4.4 Type of Carrier

Friable particulate carriers can be easy to dry than the sticky carriers. Faster drying can be achieved by uniform mixing of inoculant with carrier. This also helps in preventing clumps in the final product.

4.4.5 Clogging of Screen of FBD Jar

The jar of FBD is provided with very fine perforations at its bottom for air inflow. The fine carrier particle sometimes adheres to these fine meshes, and it affects the drying. So, it should be cleaned properly to avoid ant clogging.

5 Delivery Technique of FBD Formulations

5.1 Seed Inoculation

FBD formulation can be easily applied as the seed coating (Sahu 2012). *Azospirillum* sp., *Pseudomonas* sp., *Azotobacter* sp. and phosphate microbial biofertilizer including *Bacillus coagulans* have been applied in seed inoculation by Ebrahimpour et al. (2011) in maize. One packet of the inoculant is mixed with 200 ml of rice kanji to make slurry. The seeds required for an acre are mixed in the slurry so as to have a uniform coating of the inoculant over the seeds and then shade dried for 30 min. The shade-dried seeds should be sown within 24 h. One packet of the inoculant (200 g) is sufficient to treat 10 kg of seeds (Anonymous 2010).

5.2 Soil Inoculation

As the FBD formulation is dried, it can be mixed with FYM and applied to the field. The rate of application depends on the cell load on the formulation. Other formulation when applied to the soil, four packets of the inoculant, is mixed with 20 kg of dried and powdered farmyard manure and then broadcasted in 1 acre of main field just before transplanting.

5.3 Seedling Dip

FBD formulation may also be tried with transplanted crops by dipping the roots in solution made in water. Dose can be standardized by comparing its counterpart to other formulations where two packets (400 g) of the inoculant are mixed in 40 l of water. The root portion of the seedlings required for an acre is dipped in the mixture for 5–10 min and then transplanted (Anonymous 2010).

5.4 Fertigation

Azospirillum sp., *Pseudomonas* sp., *Azotobacter* sp. and phosphate microbial biofertilizer including *Bacillus coagulans* have been applied in fertigation by Ebrahimpour et al. (2011) in maize. This same technique may also be used for FBD if it is mixed in water and applied with irrigation.

5.5 Foliar Spray

This can be a better delivery technique of FBD formulation of biocontrol agents to spread on outer plant surface. *Beauveria bassiana* may be tried in FBD formulation for foliar spray.

6 Biofertilizers: Difficulties in Adaptation

Despite brilliant efforts made in inoculant industry, the adaptation is lagging behind its actual potential. There are various factors hindering the wide adaptability in all types of farming community. Instability of field performance is the major reason behind it (Sahu 2012). Since bioformulation is a living product, utmost care is needed at all steps from the start till the end used to maintain microbial load and vigour (Sahu and Brahmaprakash 2016). Stringent quality check and mechanized processing are needed to ensure availability of quality products in market and its sustainable performance.

Registration procedure in development of new products is often complicated and expensive and results in unauthenticated products in the market. There have been guidelines for quality of different inoculants (Yadav 2009). Inadequate demand and problem of storage are main barriers faced by the producers for the production of bioformulation. Problem of storage requires special facilities and skills which most producers, shopkeepers and farmers do not possess.

Availability of cost-effective carrier materials is also a vital factor in maintaining quality of inoculant (Mishra and Dahich 2010; Bazilah et al. 2011). Instability of performance is also because of contamination which causes drift of performance from "true to the type" culture (Sahu et al. 2013). In lyophilization, water content is lowered and sample is stored for long term. Due to this, bacteria in formulation remain inactive, resistant to environmental stress and insensitive to contamination (Bashan 1998). Poor availability of nutrients to microbes in soil and various biotic and abiotic factors are various constraints that affect effective bioformulation development.

Rhizocompetance also adds to final performance and thus to adaptability. The inherent ability of inoculated culture to compete the zymogenous flora decides its success in the rhizosphere (Sahu and Brahmaprakash 2016).

7 Future Perspectives

Studies on FBD inoculant survival in the field are required. Different crops' response for FBD inoculants has to be done for checking their compatibility with crops. Formulating in different time and temperature cycles can be done to achieve better suited protocols. Studies on gene expression and protein profiling assess the effects of FBD inoculants in triggering stress-related signalling and its impact on rhizocompetance of inoculants. Test of different carriers and additives can enhance the effectiveness of the organisms. Although it is a new technique in biofertilizers, it may have promising effects on mitigating the problem of inconsistency of performance as it prevents contamination and reduces viability loss during storage. Most importantly, it gives a stress to culture so that it can be pre-exposed to harsh soil conditions.

Acknowledgements The Indian Council of Agricultural Research-National Bureau of Agriculturally Important Microorganisms (ICAR-NBAIM), Mau, India, and Department of Agricultural Microbiology, University of Agricultural Sciences, Bengaluru, India, are gratefully acknowledged for continuous support and guidance.

References

Achari GA, Ramesh R (2014) Diversity, biocontrol, and plant growth promoting abilities of xylem residing bacteria from Solanaceous crops. Int J Microbiol 2014(2014):296521. p 14

Ahemad M, Khan MS (2012) Effect of fungicides on plant growth promoting activities of phosphate solubilizing *Pseudomonas putida* isolated from mustard (Brassica compestris) rhizosphere. Chemosphere 86(9):945–950

Anonymous (2010) Biofertilizers: types, benefits and applications. www.biotecharticles.com/Agriculture-Article/Biofertilzers-Types-Benefits-and-Aplications-172.html

Arora NK, Khare E, Naraian R, Maheshwari DK (2008) Sawdust as a superior carrier for production of multipurpose bioinoculant using plant growth promoting rhizobial and pseudomonad strains and their impact on productivity of *Trifolium repens*. Curr Sci 95(1):90–94

Arora NK, Khare E, Maheshwari DK (2010) Plant growth promoting rhizobacteria: constraints in bioformulation, commercialization, and future strategies. In: Plant growth and health promoting bacteria. Springer, Berlin/Heidelberg, pp 97–116

Auffan M, Rose BJY, Lowry GV, Jolivet JP, Wiesner MR et al (2009) Towards a definition of inorganic nanoparticles from an environmental, health and safety perspective. Nat Nanotechnol 4(10):634–641

Baca BE, Elmerich C (2007) Microbial production of plant hormones. In: Associative and endophytic nitrogen-fixing bacteria and cyanobacterial associations. Springer, Dordrecht, pp 113–143
Bailey KL, Boyetchko SM, Angle TL et al (2010) Social and economic drivers shaping the future of biological control: a Canadian perspective on the factors affecting the development and use of microbial biopesticides. Biol Control 52(3):221–229
Bajpai PD, Gupta BF, Bambal IM (1978) Studies on survival of *Rhizobium leguminosarum* as affected by moisture and temperature conditions. Indian J Agric Res 12:39–43
Bashan Y (1998) Inoculants of plant growth promoting bacteria use in agriculture. Biotech Adv 6:729–770
Bazilah ABI, Sariah M, Abidin MAZ et al (2011) Influence of carrier materials and storage temperature on survivability of Rhizobial inoculants. Asian J Plant Sci 10:331–337
Bharathi R, Vivekananthan R, Harish S (2004) Rhizobacteria-based bioformulations for the management of fruit rot infection in chillies. Crop Prot 23:835–843
Bhattacharyya PN, Jha DK (2012) Plant growth promoting rhizobacteria (PGPR): emergence in agriculture. World J Microbiol Biotechnol 28(4):1327–1350
Brahmaprakash GP, Sahu PK (2012) Biofertilizers for sustainability. J Ind Inst Sci 92(1):37–62
Brahmaprakash GP, Girisha HC, Navi V, Laxmipathy R, Hegde SV (2007) Liquid *Rhizobium* inoculant formulations to enhance biological nitrogen fixation in food legumes. J Food Legumes 20:75–79
Brar SK, Verma M, Tyagi RD, Valero JR et al (2006) Recent advances in downstream processing and formulations of *Bacillus thuringiensis* based biopesticides. Process Biochem 41:323–342
Chattannavar SN, Hosagoudar GN, Ashtaputre SA (2010) Chemical and biological management of major foliar diseases of cotton. Kar J Agric Sci 23(4):599–601
Collins DP, Jacobsen B (2003) Optimizing a *Bacillus subtilis* isolate for biological control of sugar beet Cercospora leaf spot. J Biol Cont 26(2):153–161
Costa R, Cjotz M, Smalla K (2006) Effects of site and plant species on rhizosphere community structure as revealed by molecular analysis of microbial guilds. FEMS Microbiol Ecol 56:236–249
Dayamani KJ (2010) Formulation and determination of effectiveness of liquid inoculants of plant growth promoting rhizobacteria. Ph.D thesis, University of Agricultural Sciences, Bangalore, India
Dayamani KJ, Brahmaprakash GP (2014) Influence of form and concentration of the osmolyte in liquid inoculants of plant growth promoting bacteria. Int J Sci Res Publ 4(7):1–6
Deaker R, Roughley RJ, Kennedy IR (2004) Legume seed inoculation technology – a review. Soil Biol Biochem 36:75–88
Dube JN, Mahere DP, Bawat AF (1980) Development of coal as a carrier for rhizobial inoculants. Sci Cult 46:304
Ebrahimpour F, Eidizadeh KH, Damghani AM (2011) Sustainable nutrient management in maize with integrated application of biological and chemical fertilizers. Int J Agric Sci 1(7):423–426
Emmert EA, Handelsman J (1999) Biocontrol of plant disease a (Gram) positive perspective. FEMS Microbiol Lett 171:1–9
Feng H, Li Y, LiU Q (2013) Endophytic bacterial communities in tomato plants with differential resistance to *Ralstonia solanacearum*. Afr J Microbiol Res 7(15):1311–1318
Fuhrmann JJ, Wollum AG (1989) Nodulation competition among *Bradyrhizobium japonicum* strains as influenced by rhizosphere bacteria and iron availability. Biol Fertil Soil 7:108–112
Gamalero E, Glick BR (2015) Bacterial modulation of plant ethylene levels. Plant Physiol 169(1):13–22
Glick BR (2014) Bacteria with ACC deaminase can promote plant growth and help to feed the world. Microbiol Res 169(1):30–39
Gorzelak MA, Amanda KA, Pickles BJ, Simard SW (2015) Inter-plant communication through mycorrhizal networks mediates complex adaptive behaviour in plant communities. AoB PLANTS. https://doi.org/10.1093/aobpla/plv050

Guijarro B, Melgarejo P, Decal A (2007) Effect of stabilizers on the shelf life of Penicillium frequentans conidia and their efficacy as a bioagent against peach brown rot. J Food Microbiol 113:117–124

Haas D, Defago G (2005) Biological control of soil borne pathogens by fluorescent pseudomonads. Nat Rev Microbiol 3:307–319

Haichar FEZ, Santaella C, Heulin T, Achouak W (2014) Root exudates mediated interactions belowground. Soil Biol Biochem 77:69–80

Hoffland E, Halilinen J, Van Pelt JA (1996) Comparison of systemic resistance induced by avirulent and non pathogenic *Pseudomonas* sp. Phytopathology 86:757–762

Islam TMD, Toyota K (2004) Suppression of bacterial wilt of tomato by *Ralstonia solanacearum* by incorporation of composts in soil and possible mechanisms. Microbes Environ 19(1):53–60

Iswaran V (1972) Growth and survival of *Rhizobium trifoli* in coir dust and soybean meal compost. Madras J Agric Sci 59:52–53

Jinnah MA, Khalequzzaman KM, Islam MS (2002) Control of bacterial wilt of tomato by *Pseudomonas fluorescens* in the field. Pak J Biol Sci 5(11):1167–1169

Kandaswamy R, Prasad N (1971) Lignite as a carrier of rhizobia. Curr Sci 40:496

Karthiba L, Saveetha K, Suresh S, Raguchander T, Saravanakumar D, Samiyappan R (2010) PGPR and entomopathogenic fungus bioformulation for the synchronous management of leaffolder pest and sheath blight disease of rice. Pest Manag Sci 66(5):555–564

Khan MS, Zaidi A, Wani PA (2007) Role of phosphate solubilizing microorganisms in sustainable agriculture – a review. Agron Sustain Dev 27:29–43

Kumar V (2014) Characterization, bio-formulation development and shelf-life Studies of locally isolated bio-fertilizer strains. Octa J Environ Res 2(1):32–37

Labroussaa F, Zeilinger AR, Almeida RPP (2016) Blocking the transmission of a noncirculative vector-borne plant pathogenic bacterium. MPMI 29(7):535–544

Lavanya G, Sahu PK, Manikanta DS, Brahmaprakash GP (2015) Effect of fluid bed dried formulation in comparison with lignite formulation of microbial consortium on finger millet (*Eleusine coracana* Gaertn.) J Pure Appl Microbiol 9(2):193–199

Lee SK, Lur HS, Lo KJ, Cheng KC, Chuang CC, Tang SJ, Yang ZW, Liu CT (2016) Evaluation of the effects of different liquid inoculant formulations on the survival and plant-growth-promoting efficiency of *Rhodopseudomonas palustris* strain PS3. Appl Microbial Biotechnol 6:1–1

Lwin M, Ranamukhaarachchi SL (2006) Development of biological control of *Ralstonia solanacearum* through antagonistic microbial populations. Int J Agri Biol 8(5):657–660

Madhok MR (1934) The use of soil as a medium for distributing legume organism culture to cultivators. Agric Livestock India 4:670–682

Manjula K, Kishore GK, Girish AG, Singh SD (2004) Combined application of *Pseudomonas fluorescens* and *Trichoderma viride* has an improved biocontrol activity against stem rot in groundnut. Plant Pathol J 20(1):75–80

Marasco R, Rolli E, Ettoumi B, Vigani G, Mapelli F, Borin S, Abou-Hadid AF, El-Behairy UA, Sorlini C, Cherif A, Zocchi G, Daffonchio D (2012) A drought resistance promoting microbiome is selected by root system under desert farming. PLoS One 7:e48479

Mishra BK, Dahich SK (2010) Methodology of nitrogen biofertilizer production. J Adv Dev Res 1:3–6

Monte E, Llobell A (2003) Trichoderma: In: Organic agriculture. V Congreso Mundial del Aguacate, pp 725–733

Nakkeeran S, Kavitha K, Renukadevi P (2004) Induced systemic resistance and plant growth promotion by *Pseudomonas chlororaphis* strain PA–23 and *Bacillus subtilis* strain CBE4 against rhizome rot of turmeric (*Curcuma longa* L.) Can J Plant Pathol 26:417–418

Nandakumar R, Babu S, Samiyappan R (2001) A new bio–formulation containing plant growth promoting rhizobacterial mixture for the management of sheath blight and enhanced grain yield in rice. Biol Control 46:493–510

Nehra V, Choudhary M (2015) A review on plant growth promoting rhizobacteria acting as bioinoculants and their biological approach towards the production of sustainable agriculture. J Appl Nat Sci 7(1):540–556

O'Callaghan M, Gerard EM (2005) Establishment of *Serratia entomophila* in soil from a granular formulation. In: New Zealand plant protection, vol 58, 2005. Proceedings of a conference, Wellington, New Zealand Plant Protection Society, pp 122–125

Pandey V, Ansari MW, Tula S, Yadav S, Sahoo RK, Shukla N, Bains G, Badal S, Chandra S, Gaur AK, Kumar A, Shukla A, Kumar J, Tuteja N (2016) Dose-dependent response of *Trichoderma harzianum* in improving drought tolerance in rice genotypes. Planta 243:1251–1264

Ponmurugan P, Gopi C (2006) In vitro production of growth regulators and phosphatase activity by phosphate solubilizing bacteria. Afr J Biotechnol 5(4):348–350

Qi X, Xiaofeng S, Guo H, Qi J, Cheng H (2016) VdThit, a thiamine transport protein, is required for pathogenicity of the vascular pathogen *Verticillium dahlia*. MPMI 29(7):545–559

Radjacommare R, Venkatesan S, Samiyappan R (2010) Biological control of phytopathogenic fungi of vanilla through lytic action of *Trichoderma* spp. and *Pseudomonas fluorescens*. Arch Phytopathol Plant Prot 43:1–17

Rajavel R (2000) Seed borne *Colletotrichum capsici* (Syd). Butter and Bisby and its management. M.Sc. thesis, Tamil Nadu Agricultural University, Coimbatore

Sahu PK (2012) Development of Fluid Bed Dried (FBD) inoculant formulation of consortium of agriculturally important microorganisms (AIM). M.Sc. thesis, University of Agricultural Sciences, Bangalore, India

Sahu PK, Brahmaprakash GP (2016) Formulations of biofertilizers–approaches and advances. In: Microbial inoculants in sustainable agricultural productivity. Springer, New Delhi, pp 179–198

Sahu PK, Lavanya G, Brahmaprakash GP (2013) Fluid bed dried microbial inoculants formulation with improved survival and reduced contamination level. J Soil Biol Ecol 33(1&2):81–94

Sahu PK, Lavanya G, Gupta A, Brahmaprakash GP (2016a) Fluid bed dried microbial consortium for enhanced plant growth-a step towards next generation bioformulation. Vegetos, Ms. No. SciTech161098. https://doi.org/10.4172/2229-4473.1000165 (in press)

Sahu PK, Sharma L, Gupta A, Renu (2016b) Rhizospheric and endophytic beneficial microorganisms: treasure for biological control of plant pathogens. In: Santra S, Mallick A (eds) Recent biotechnological applications in India. ENVIS Centre on Environmental Biotechnology, University of Kalyani, West Bengal, pp 50–63

Saravanakumar D, Vijayakumar C, Samiyappan R (2007) PGPR-induced defense responses in the tea plant against blister blight disease. Crop Prot 26:556–565

Schisler DA, Slininger PJ, Behle RW, Jackson MA (2004) Formulation of *Bacillus* spp. for biological control of plant diseases. Phytopathology 94:1267–1271

Sharma MP, Sharma SK, Dwivedi A (2010) Liquid biofertilizer application in soybean and regulatory mechanisms. Agric Today 44–45

Singleton P, Keyser H, Sande E (2002) Development and evaluation of liquid inoculants. In: Herridge D (ed) Inoculants and nitrogen fixation of legumes in Vietnam. ACIAR proceeding 109e. Australian Centre for International Agricultural Research, Canberra, pp 52–66

Soares de Melo I, Montes Peral Valente AM, Nessner Kavamura V, Vilela D, Simoni E, Faull JL (2014) Mycoparasitic nature of Bionectria sp. strain 6.21. J Plant Prot Res 54(4):327–333

Spaepen S, Vanderleyden J (2011) Auxin and plant-microbe interactions. Cold Spring Harb Perspect Biol 3(4):a001438

Sridhar V, Brahmaprakash GP, Hegde SV (2004) Development of a liquid inoculant using osmoprotectants for Phosphate solubilizing bacteria. Kar J Agric Sci 17:251–257

Srivastava S, Mishra G (2010) Fluid bed technology: overview and parameters for process selection. Int J Pharm Sci Drug Res 2(4):236–246

Suman A, Verma P, Yadav AN, Srinivasamurthy R, Singh A, Prasanna R (2016) Development of hydrogel based bio-inoculant formulations and their impact on plant biometric parameters of wheat (*Triticum aestivum* L.) Int J Curr Microbiol App Sci 5(3):890–901

Surendra Gopal K, Baby A (2016) Enhanced shelf life of *Azospirillum* and PSB through addition of chemical additives in liquid formulations. Int J Sci Environ Technol 5(4):2023–2029

Swapna G, Divya M, Brahmaprakash GP (2016) Survival of microbial consortium in granular formulations, degradation and release of microorganisms in soil. Ann Plant Sci 5(5):1348–1352

Takanashi K, Sasaki T, Kan T, Saida Y, Sugiyama A, Yamamoto Y, Yazaki K (2016) A dicarboxylate transporter, LjALMT4, mainly expressed in nodules of *Lotus Japonicas*. MPMI 29(7):584–592

Tiwari S, Singh P, Tiwari R, Meena KK, Yandigeri M, Singh DP, Arora DK (2011) Salt-tolerant rhizobacteria-mediated induced tolerance in wheat (*Triticum aestivum*) and chemical diversity in rhizosphere enhance plant growth. Biol Fertil Soils 47(8):907–916

Velineni S, Brahmaprakash GP (2011) Survival and phosphate solubilizing ability of *Bacillus megaterium* in liquid inoculants under high temperature and desiccation stress. J Agric Sci Tech 13:795–802

Vidhyasekaran P, Muthamilan M (1995) Development of formulations of *Pseudomonas fluorescens* for control of chickpea wilt. Plant Dis 79:782–786

Walpola B, Yoon MH (2013) Phosphate solubilizing bacteria: assessment of their effect on growth promotion and phosphorous uptake of mung bean (*Vigna radiata* [L.] R. Wilczek). Chil J Agric Res 73(3):275–281

Yabuuchi E, Kosako Y, Yano L, Hotta H, Nishiuchi Y (1996) Validation of the publication of new names and new combinations previously effectively published outside the IJSB. Int J Syst Bacteriol 46:625–626

Yadav AK (2009) Glimpses of fertilizer (Control) order, 1985 for biofertilizers (amendment, November 2009), National center for organic farming, Department of Agriculture and Cooperation, Government of India. Biofertilizer News 17(2):11–14

Yan Q, Philmus B, Hesse C, Max K, Jeff HC, Joyce EL (2016) The rare codon AGA is involved in regulation of pyoluteorin biosynthesis in *Pseudomonas protegens* Pf-5. Front Microbiol. https://doi.org/10.3389/fmicb.2016.00497

Yao L, Wu Z, Zheng Y, Kaleem I, Li C (2010) Growth promotion and protection against salt stress by *Pseudomonas putida* Rs-198 on cotton. Eur J Soil Biol 46:49–54

Zhang L, Fan J, Ding X, He X, Zhang F, Feng G (2014) Hyphosphere interactions between an arbuscular mycorrhizal fungus and a phosphate solubilizing bacterium promote phytate mineralization in soil. Soil Biol Biochem 74:177–183

Analysis of Chlorophyll Fluorescence: A Reliable Technique in Determination of Stress on Plants

M. K. Adak

Abstract

Photosynthetic research has been adopted the most informative and still remained unanswered in many aspects for plant growth and development. The basic and applied research for photosynthesis under ambient as well as natural or induced stress condition is well characterized in different plant species by the use of fluorescence technique. The description of photosynthetic process in terms of photochemistry and its quenching in heat and other forms has added much information to characterize the responses of plant genotypes. The principle, experimental setup, standardization with materials to materials, application and exercise, and finally derivation of data have been under revision for improved instrumentations. The imaging analysis of leaf under photosynthetic condition and its any deviation under stress condition has warranted the fluorescence technique as most reliable for in vivo, however, noninvasive and reliable in understanding plant's behavior. Moreover, the amalgamation of fluorescence study through pulse amplitude modulation technique with CO_2 gas exchange device has added advantages to the simultaneous recording of light-generated products in carbon reduction cycle. In reference to appropriate application of PAM measured florescence and imaging of chlorophyll fluorescence in determination, quantification and steady monitoring are customized according to instruments variation. Therefore, in this mini review, a specific focus has been set for an overview of fluorescence methodology with useful parameters, its proper use in plants' responses, allied external factors affecting the photochemistry, the relationship with carbon reduction with acquired light energy, and further scope for up-gradation.

M. K. Adak (✉)
Plant Physiology and Plant Molecular Biology Research Laboratory, Department of Botany, University of Kalyani, Kalyani, West Bengal, India

R. S. Sengar, A. Singh (eds.), *Eco-friendly Agro-biological Techniques for Enhancing Crop Productivity*, https://doi.org/10.1007/978-981-10-6934-5_4

Keywords
Chlorophyll excitation · Maximum fluorescence · Pulse amplitude modulated fluorescence · Light harvestibg complex · Photosynthetic efficiency · Photon flux density · Plants irradiation stress

1 Introduction

Plants' responses to abiotic stress have been documented in many dimensions for the assessment of their survival, sustainability, and productivity, particularly, in relation to agricultural scenario. Responses of the plants through various interfaces of physiological, cellular, and molecular proportions have been evaluated depending on types, magnitude, and duration of stress. Out of a number of screening indices like physiological parameters, chlorophyll fluorescence has been attributed as the most authentic, realistic, reliable, and popular technique for plant physiological researches (Adams and Demmig-Adams 2004). Though it has been the practices as a traditional method to characterize the plant genotypes, still, the modification as well as sophistication with newer inputs of model had added its modern state of the art. The plant physiologists and breeders, thus, are more interested to probe it for genotypic identification in photosynthetic research as noninvasive and nondestructive technique. Plant's performances facing the various interfaces of abiotic stresses are to get more insights in detail information regarding photosynthetic light reactions coupled with carbon assimilation. Photosystem II (PSII) and accessory light harvest complexes (LHCs) through imaging fluorescence have conceived many information in actively photosynthesizing tissues and its any alteration under abiotic stress (Adams and Demmig-Adams 2006). Chlorophyll fluorescence is the documentation of data for utilization, distribution, and dissipation of light energy between two photosystems either for photochemistry or quenching into heat and other forms (Fig. 1). Imaging chlorophyll fluorescence has nourished the predominant roles of photosynthesis in relation to light energy conversion and its simultaneous transformation into reduced carbon reaction mechanism when both are in irreversible way coupled with infrared gas analyzer (IRGA). The chlorophyll.

Fluorescence has an added advantage to measure the crop's water relation in its improvement under drought for screening in linking genomic information with physiological responses (Allen and Mullineaux 2004). Chlorophyll fluorescence mainly detected with pulse amplitude modulation (PAM) imaging being noninvasive, in vivo, and rapid screening of the photosynthetic rate per se gives high resolution of information when the data undergo digitization. This is followed by simulation in computer-based modeling (Baker et al. 2007). A relationship including the photosynthetic performances with plant yield/biomass, however, is not always straightforward or even linear. Still, repeated observations through PSII imaging analysis have been customized for photosynthetic research even in agronomic purposes in plant selection where optimization of photosynthetic rate becomes the most crucial (Barbagallo et al. 2003). This is more saturated with

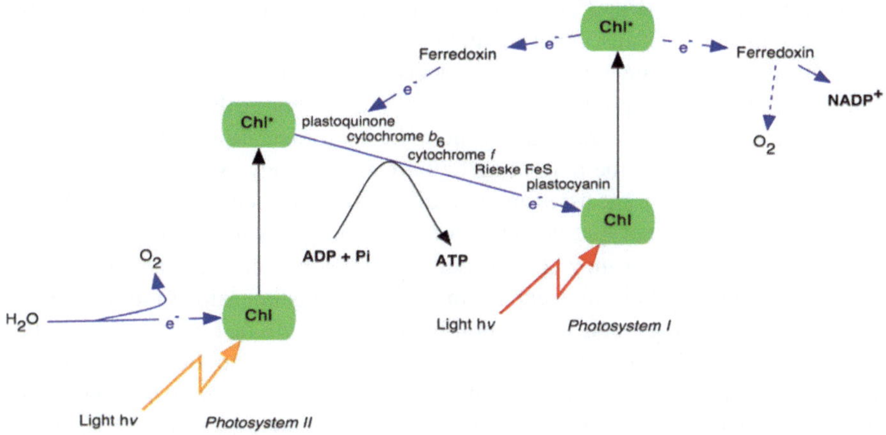

Fig. 1 Different components of photosystems (I and II) and light-induced transport of electrons. The quenching of light energy by PSII into quinones makes all the available reaction centers (RC) closed or reduced

merits of PAM instrumentation where a high number of measurements at a time are exercised in stipulated and shortened period of time with maximum rapidity. However, the choice of different parameters for characterization of chlorophyll fluorescence is another clue for the user to make easy according to plant species, types of treatments, etc. as per their needs. Besides this, the portability, user friendliness, repeatability, and reliability of this instrument have set it more suitable for chlorophyll fluorescence either in normal field or in closed-door laboratory setup (Baker et al. 2007). Chlorophyll fluorescence has been performing a pivotal role in different aspects of crops and its sustenance and productivity. Changing the climate and its fluctuations or variability plays determinant roles in crop functioning. During the last twenty-first century, it would be a major thrust for selection of plant types as resilient to environmental fluctuations (Farage et al. 2006). The ability to tolerate the environmental hazard is demanding for new types of crop products that ought to be increased for productivity by employing improved anticipation of physiological attributes. Chlorophyll fluorescence among those attributes is regarded as a tool for selection of plant types in relation to reactions of genotypes to abiotic stresses. The leaf growth at different developmental stages coupled with adequate photosynthetic performances has been reviewed with focus on chlorophyll fluorescence (Laisk et al. 2006). The fluorescence monitoring is related to the photosynthetic yield since any allocations of energy or its modified distribution may alter the carbohydrate synthesis. Apart from loss of energy by heat dissipation or re-emission with longer wavelengths, maximization of energy is utilized predominantly for photochemistry. This eventually becomes more as determining factor for plant's photobiological responses under any sort of abiotic interference that diverge its normal path of light utilization more toward non-photosynthetic quenching than photochemistry (Kramer et al. 2004). Therefore, plant workers have been paying attention to relate fluorescence parameter and any of plant's responses as a function of stress either evaluating in situ or ex situ experimentation.

In this paper, it is intended to give a concise revision regarding the fundamentals and basics of the chlorophyll fluorescence techniques in details. This review is also considered as a simple mechanical layout of the instrumentation with its operational view frequently used in the field of agriculture and ecology. Additionally, a flat but précised description for the assessment of different fluorescence parameters through PAM studies for phenotypic detection of crop species is incorporated. With this background, this chapter may be considered as a comprehensive account as well as a revision for chlorophyll fluorescence in vivo with the dimension of update knowledge in ecophysiology and even photobiology. It is more focused on practical aspects of protocol standardization for PAM analysis, its application in stressed leaves as well as in different developmental stages. Thus, it establishes that in vivo chlorophyll fluorescence for PSII activities and PAM images would be a modern and progressive sophistication in the research of photosynthesis and its allied domain.

2 Basics and Fundamental Attributes: PSII Fluorescence in Plants

It is the PAM that is granted in almost every advanced physiological laboratory to detect the interplays of abiotic stressors on plants in relation to light energy exploitation. Those data also would be correlated with concomitant reflection in photosynthetic performances with plant growth in a specific developmental stage. In PAM fluorescence measurement, a continuous light is not favored; rather a high intensity of light just mimicking the solar light intensity is alternatively applied (Leipner et al. 2001). A detector is used to measure the fluorescence emission and thus provides a strong as well as efficient system to measure the chlorophyll fluorescence (PSII). In theory, the chlorophyll molecule (of PSII) after absorbing the photosynthetic active radiation (PAR) has undergone three consecutive molecular changes: either utilization of the photon for a complete photochemistry or release of energy in the form of light with higher wavelength or excess heat form (Li et al. 2000). These processes share the decay of chlorophyll excitation state for photochemistry and thus lower the fluorescence yield of PSII. Collectively all the processes that are capable to minimize the chlorophyll fluorescence are responsible for the event of quenching of chlorophyll molecules (Li et al. 2004). Initially if a dark-adapted leaf is placed under strong light active to photosynthesis (i.e., actinic light), a peak in fluorescence yield is recorded. This is accompanied with the reduction of electron acceptor molecules (as quinone) of PSII called closed state. This is due to transfer of energy by LHCII to plastoquinone, Q_A^-, and itself is reduced by charge separation. Closed state of Q_A^- is explained with the idea that it cannot accept e^- any more until it passes the first to another carrier, often as Q_B (Fig. 2).

This is also referred to as Kautsky effect and determines that at closed state the photochemistry falls and thereby induces a characteristic rise in yield of

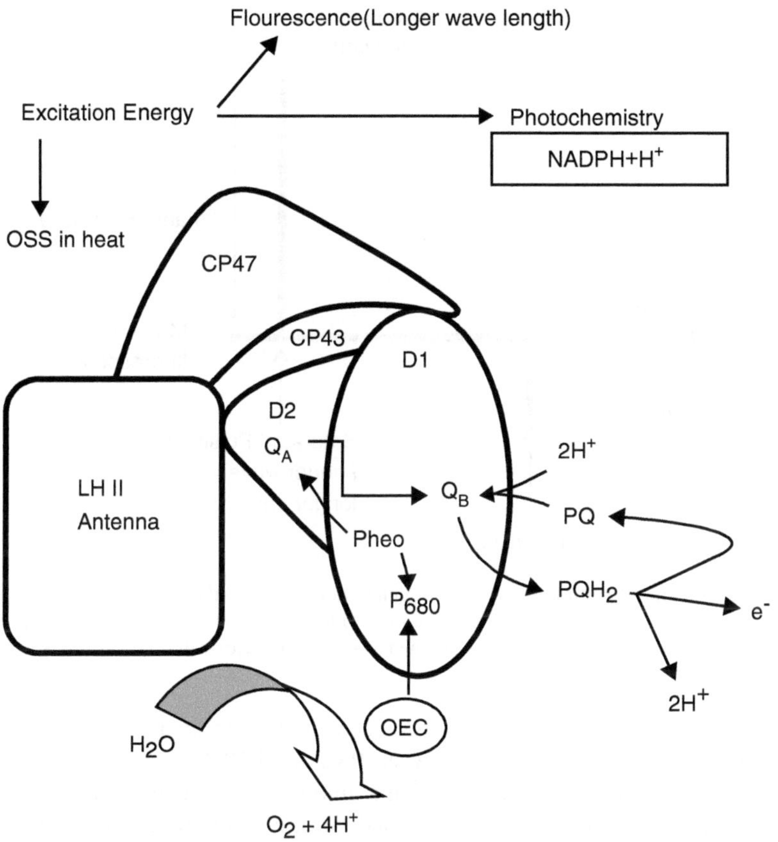

Fig. 2 The molecular phenomena for energy transfer coupled with reduction of PSII allied electron acceptors. Photons are either channelized to photochemistry (to release an e^- from core pigment, i.e., P680$^+$in RC of PSII) or fluorescence or loss of heat. These processes are in rivalry and resolve the PSII efficiency

fluorescence (Long and Bernacchi 2003). Following this time, the fluorescence again comes down with a typical shape, and this could be explained based on quenching phenomena. The transfer of e^- from PSII under illumination is due to light activation of enzyme cascade in Calvin cycle for subsequent conversion of CO_2 into reduced carbon residues. The later is also preluded by the entry of stomatal aperture to let the CO_2 enter. This induces the phenomenon of photochemical quenching (qP). In a light-induced leaf, this process is also accompanied by loss of energy into dissipated heat with the phenomenon of non-photochemical quenching (qNP). qNP otherwise predicts the efficiency of plants to minimize any photoenergy-induced damages of reaction centers, particularly, when stomata are closed to limit the entry of CO_2 under drought condition. From the technical point of view, qNP has got few reflectants with variably interacting with loss of excess energy according to its relaxation period (Maxwell and Johnson 2000). qNP is the first of

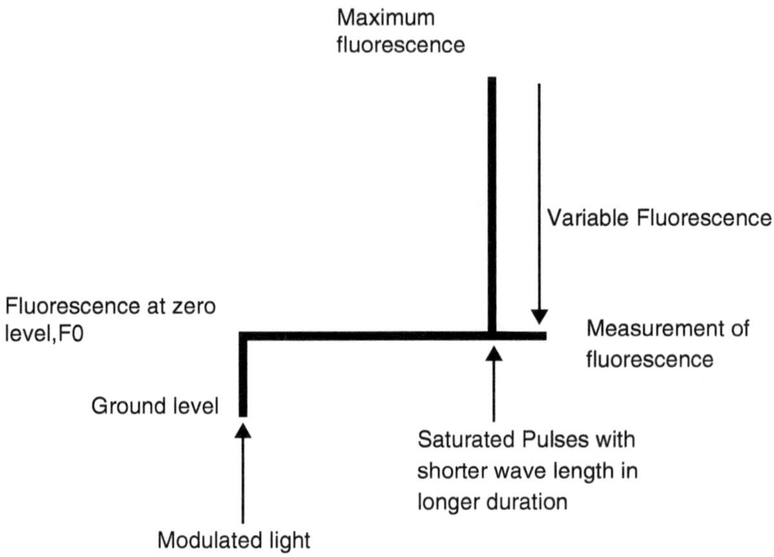

Fig. 3 The principle for fluorescence signal generation: the modulated light that cannot induce photochemistry, but enough to elicit the basal or zero level fluorescence (F_0) when all the RC are oxidized. The maximum fluorescence is attained at saturated pulses while all the RC is reduced

those being rapid which is a pH-/energy-dependent component. The second one is due to detachment or uncoupling between PSII and LHCII and takes a minute for relaxation called qT. The last one takes maximum period for relaxation and thus is more relaxed; however, it is less determined. Still, it is interrelated to photo-inhibition of PSII and often referred as qI. Therefore, the maximum and minimum fluorescence yields symbolized as F_{max} and F_0 on a dark-adapted leaf, respectively, are the two primaries to evaluate the quenching coefficient of PSII (Melis 1999). A typical saturated light in regular, however, in short pulse intervals is used employing F_{max} and F_0 in assessment of photochemical and non-photochemical quenching (NPQ). The method used in this field is popularly known as saturated pulse amplitude modulation method. This simply is clarified as detection of fluorescence yield of PSII by consecutive use of a measuring/modulated light (ML) by switching on and off (i.e., in modulated pulse); however, with high frequency and thus on simulation, the detector reads the correct fluorescence yield (Meng et al. 2001) (Fig. 3).

In practice, a healthy/stressed leaf is allowed to be dark adapted for maximum period to enable all the RC closed (i.e., reduced state). Then by using a low/weak intensity light (ML), the basal fluorescence (F_0) is measured. This is immediately followed by measuring the peak or maximum values of fluorescence with a saturated pulse (SP) with strong intensity but short intervals to lift the fluorescence value in maximum (F_{max}). Similarly, another maximum fluorescence is recorded during the application of SP called F_m' (Fig. 4). Similarly, the difference between F_0' and F_m' is documented as F_v', however, under illuminated condition. This time PSII experiences a complete repression of photochemical quenching. However, the

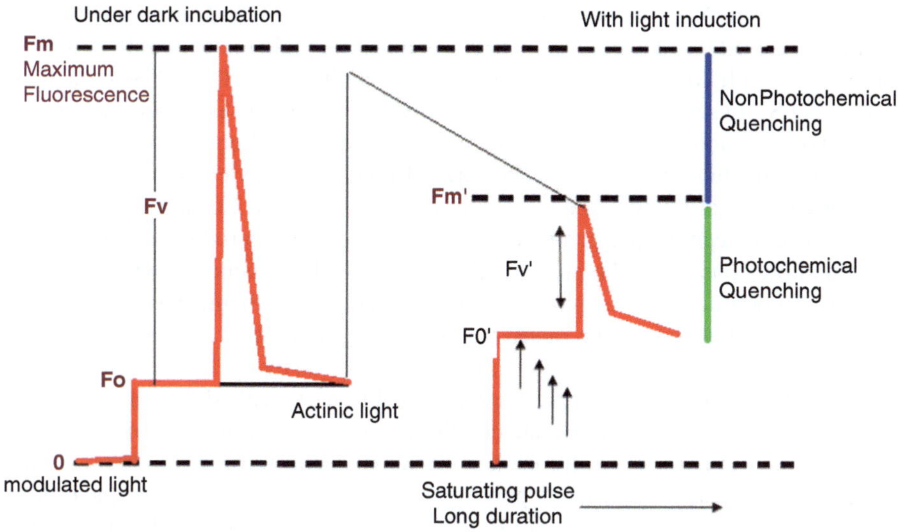

Fig. 4 Succession of fluorescence event in reading of photochemical and non-photochemical quenching: a dark-adapted plant raised the maximum fluorescence (F_{max}) when transferred to actinic light; the saturation pulse with 20–30 min (long duration) raised the fluorescence to F_m', however, under light condition. The fluorescence achieved immediately prior to saturating pulse is termed as transient fluorescence F_t. In PAM study, turning off the actinic illumination or even with far-red light directs to score the zero level fluorescence

difference between F_m and F_m' is laid with the events of dark- and light-adapted condition, respectively, on PSII. Thus, the values F_m' is much lower than for F_m. The point to note is that with short SP there undergoes no change in NPQ and thus the decrease in *F*m directly correlates the NPQ, particularly, under stress condition. Therefore, plants under any sort of stresses with damages/inhibition of PSII structure/function, respectively, record a decrease in F_m values and ensure an alternative release of energy through increased NPQ (Berger et al. 2007).

3 Techniques for PAM Studies: Derivations of Data

As already mentioned, initially a leaf is dark adapted for a minimum period (of around 30 min) and called dark adaptation. A weak beam of light, however, not sufficient to induce photochemistry but could induce fluorescence called modulated light (ML) is used to detect the ground fluorescence (F_0). However, definitely it would be in darkness. Photochemistry is referred to as release of e^- from reaction center of PSII and its subsequent reduction to Q_A (Fryer et al. 1998). This is immediately followed by an application of saturating pulse with 8000–10000 μM/m^2/s during 1–2 s in succession or series.

The fluorescence induced by SP reaching the maximum value is called maximum fluorescence (F_m'). This measurement is used also for the PSII photochemistry as F_v/F_m, where F_v is the variable fluorescence ($F_v = F_0 - F_m$). This parameter is of immense importance to achieve the intrinsic quantum yield under actinic light (AL) which is the photosynthetic light immediately applied after the initial light pulse (ML) and SP in series as turned on repeatedly. This again raises the fluorescence to the maximum, and since it is the maximum fluorescence at light-adapted leaf called "effective quantum yield" $\left(F_m^{'}\right)$. The dark-adapted leaf that yielded maximum fluorescence (F_m) is also understood as "optimum quantum yield." Under this condition, all the reaction centers are completely closed, i.e., remain in reduced state. The AL would allow continuing the photosynthesis. This is then quenched with the initiation of competition between photochemical and non-photochemical events (Kramer et al. 2004). The peak of the raised fluorescence gradually is attained to the local fluorescence maximum (i.e., becomes plateau) followed by a decline. A steady-state competition is attained under actinic light condition and called "steady-state fluorescence" (F_s) or F'. In steady-state condition, the electron transport and photosynthesis remain unchanged within a short duration (Hendrikson et al. 2004). It also reveals that at steady state though photosynthesis is continued, however, the proportion of oxidized to reduced Q_A $\left(QA^+ / Q_A^-\right)$ is constant. To attend the steady state with AL, it takes a longer duration (<20 min). However, it may vary according to condition and types of the standardization for plant genotypes and instrument, respectively. Under AL condition, the short pulses of SP are repeatedly turned on with short intervals to let the fluorescence to be maximum. Since it is from light adopted leaves, thus, maximum fluorescence is read as $F_m^{'}$. Definitely, the value of the $F_m^{'}$ must be lesser than F_m, and this is both from the conjugation of non-photochemical quenching (NPQ) and PSII partial reduction. Once the actinic light is switched off, the fluorescence drops down the ground state level, and that is equivalent to the dark-adapted F_0. Therefore, under illumination, the F_0 may be redesignated as $F_0^{'}$.The application of far red (F_R) with weak intensities is also used measuring F_0 and $F_0^{'}$. In fact, F_R is used to induce PSI, thereby making Q_A^- fully oxidized (Johnson et al. 2004). The difference between $\left(F_m^{'} - F_0^{'}\right)$ is the variable fluorescence under light $\left(F_v^{'}\right)$.

$$F_v^{'} = F_m^{'} - F_0^{'}$$

$$F_v^{'} / F_m^{'} = \begin{pmatrix} \text{Maximum efficiency} \\ \text{of PS} \end{pmatrix} - \begin{pmatrix} \text{II under light adapted leaf} \\ [\text{when the } A_L \text{ is withdrawn}] \end{pmatrix}.$$

The operating efficiency of PSII photochemistry is considered as $F_q^{'}$ or ΔF.

$$F_q^{'} = \Delta F = F_m^{'} - F_0^{'} / F_m^{'}.$$

This is also established as maximum quantum efficiency of PSII (ΦPSII).

$$\Phi\text{PSII} = F_m^{'} - F_0^{'} / F_m^{'}.$$

Another important parameter studying the photochemistry of PSII is the rate of electron flow. This parameter is of immense importance as dealing with the rate of electron flow through PSII. Under stress condition, the value of $F_q^{'} / F_m^{'}$ is most important under light-adapted condition (Krause and Jahns 2004). $F_q^{'} / F_m^{'}$ can be dissected into two important components:

$$\text{Photo chemical quenching of PSII} = \text{PSII} = F_q^{'} / F_v^{'}.$$

$$\text{PSII maximum efficiency under light adapted state} = F_v^{'} / F_m^{'}.$$

This parameter describes maximum operating efficiency of PSII under light-adapted state, and decrease of this parameter essentially reflects an increase in NPQ. So, when the ratio of $F_v^{'}$ to $F_m^{'}$ is lowered, that indicates the plants are losing maximum light for photochemistry and leading a dissipation of excess energy under intense illumination. Therefore, this could be no less to say that any stress condition, deleterious to PSII structure and function, can supersede the NPQ in leaf (Lawson et al. 2002). In practice, the maximum quantum yield is the most commonly and widely used parameter to study the imposition of stress factor. Thus, maximum quantum yields of PSII photochemistry:

$$\text{O}_{\text{Po}} = F_v / F_m, \text{ otherwise could be } 1 - F_v / F_m.$$

The variable fluorescence (F_v) in simplest form can be expressed as

$$F_v = F_m - F_0.$$

The precise ratio as F_v/F_m is also a determinant of maximum efficiency of PSII, if all the RC remains oxidized or open in dark-adapted leaf sample. Another important parameter deals with the rate of e$^-$ transport with values ØPSII, thus indicative of the whole photosynthesis.

$$\text{OPSII} = F_m^{'} - F_0 / F_m$$

With the above equation, the photochemical quenching (qP) could also be derived with slight modification as

$$\text{qP} = F_m^{'} - F / F_m^{'} - F_0$$

This also holds some information of amount of RC oxidized or open (however, in approximation).

Ø $_{\text{Po}}$ is related to the quantification of maximum photochemical efficiency of oxidized or open RC of PSII. The values of Ø $_{\text{Po}}$ are variable to different plant species, and under unstress condition it may be stable around 0.832. Still, under PSII influencing stress condition, it comes down significantly, however, variable with other facets of plant's genotypic potential (Miyake et al. 2004). Whatever the cases might be, the decline in values of Ø $_{\text{Po}}$ is an outcome of fluorescence from PSI causative to the F_0 level. The origin of fluorescence is based on after de-excitation of the first singlet state stationed chlorophyll a molecule. On the contrary, chlorophyll b is not much efficient to release fluorescence since energy does not withhold and rapidly transferred to chlorophyll a. Thus, an estimation of PSII fluorescence is revealed about 2–3% of the absorbed energy, and also it accounts 90% by the PSII coupled with LHCII (Morison et al. 2005). Under normal physiological condition, fluorescence is very meager from PSI, and in addition variable fluorescence is only obtained from PSII. For the quantification of fluorescence yield in terms of photon absorption in PSII, it suggests the ratio of total number of photons emitted (n_F) to the total number of photons absorbed (n_A). In another form of quantum yield, the rate of all the concerned events along with their rate constants is employed with those attributes for turnover of excited ground state of energy (Nedbal and Whitmarsh 2004).

$$O_F = n_F / n_K = k_F / \sum k_i$$

The rate constant (k_F) is denoted by the number of constants in the competing processes, e.g., for photochemistry events in reaction center II. The rate constant k_P, k_D for heat dissipation of energy from LHCII through NPQ events, kq for the events of quenching by different accessory pigments (like carotenoid and others), and k_T for nonfluorescent pigments in energy transfer (Oxborough 2004). Still, k_D is predicted as the summation of all other non-radiative de-excitation events, and thus for convenience three basic quantities are adopted. These are quantum yield of photochemistry ($Ø_P$), the maximum (Øm) and minimum ($Ø_0$) yield in photochemical e^- ejection from PSII, and finally quantum yield of non-radiative energy dissipation through heat or simply NPQ.
With this more simplified way, it could be expressed as

$$O_P = k_P / \left(k_p + k_F + kd_D\right)$$

The excitation energy absorbed by PSII as per conservation law could not be missing and thereby is converted into photochemical and NPQ. With these findings it could also be hypothesized that the total $Ø_P$ for all the processes like photochemistry, NPQ, and fluorescence should be one in summation [i.e., $Ø_P + Ø_F + Ø_D = 1$]. However, in simplest way and routine analysis of chlorophyll fluorescence, the parameters mostly dealt with plant's performances in photochemical reaction are tabulated (Table 1).

Table 1 The derivations and their definitions in PSII analysis with PAM studies

F_v/F_m	$(F_m{-}F_0)/F_m$	Maximum quantum efficiency of PSII for dark-adapted plants for photochemistry
F_v'/F_m'	$F_m'{-}F_0'/F_m'$	Maximum efficiency of PSII photochemistry under light-adapted leaf
Photochemical quenching qP	$F_q'/F_m'{=}F_m'{-}F'/F_m'$	Quantum efficiency of PSII e^- transport in light
Non-photochemical quenching (NPQ)	$F_m{-}F_m'/F_m'$	Non-photochemical quenching or heat loss by PSII
qL = fraction of open reaction center of PSII	$(F_q'/F_v')/(F_0'/F')$	Fraction of open PSII center

4 Oxidative Stress: Implication in Fluorescence Studies

Plants are under influence of stress factors in relation to oxidative stress in light. A decrease in F_v/F_m is registered as a common phenomenon. This happens as a symbolic or instance but simple and rapid way addressing the stress. However, the fall in F_v/F_m and its correlation is not a simple rather complex phenomenon in stressed photosynthetic cells. Under the condition of oxidative stress, the stressed tissues undergo quenching of light more in non-photochemical processes, and that lowers the value of F_m even under dark adaptation (for limited or prolonged time). Under this condition, it cannot retrieve the decrease in F_v/F_m. The increase in NPQ on the other hand leads to photoinactivation of PSII reaction centers (Baker and Rosenqvist 2004). This is accompanied by more release of heat for dissipation of excitation energy since photochemistry is curtailed. Photo-inhibition is a process which is furnished by over-excitation of photosystems under illumination in excess. This in turn is manifested as oxidative damages and decline in PSII reaction centers and its functionalities. This is also associated with a high value of F_0 (dark adapted). Therefore, the decrease in F_m or increase in F_0 as an outcome of stress imposition must be taken care, while possible explanation is given. In addition, the fluorescence and its possible changes are required to be studied both by biophysical properties of leaves and optical properties of photosystems, respectively. It is quite a common experience that changes of the leaf water potential under abiotic stress also influence the assimilation rate of leaf mostly through changing of fraction of PSII in its closed status (Maxwell and Johnson 2000). This predominantly happens in case of ROS-mediated damages the thylakoid membrane and through structure-function relationship. In most of the cases, a fall in F_v/F_m indirectly is understood the decline of photosynthetic performances, however, under ambient light condition. Still, this is not always true since the quantum yield of PSII photochemistry, in most of the cases, remains well below of F_v/F_m value. The latter is regarded as the maximum quantum yield of PSII photochemistry. It is emphasized that maximum quantum yield of PSII not necessarily circumvents the yield which signifies under ambient light (Melis 1999).

Under oxidative stress the detection of F_v/F_m or F_q' / F_m' which employs the possible changes in PSII operating efficiency (in maximum) may be erroneous with attainment of PSI fluorescence. In most of the cases, the PSII photochemical yield is measured to be derived from PSII solely. This is true only when fluorescence is measured beyond 700 nm which is contributed by PSI, though in negligible amount. For C_3 and C_4 plants, the estimated value of F_0 at around or above the wavelength 700 nm can modify the same value from PSII (Nejad et al. 2006). Therefore, a decrease in F_q' / F_m may not necessarily be given the PSII efficiency in real under wavelength above 700 nm. So, below 700 nm, the activity of fluorescence study is rationalized to reduce the PSI activity as fluorescence signal. Another possibility of error in F_q' / F_m' may occur when saturating light pulses is used. This is because a multiple turnover of PSII reaction centers under saturating pulse. This is accompanied along with reduction of quinone (Q_A) could convert the plastoquinone (PQ) to plastoquinol (PQOH). The former behaved as a quencher of chlorophyll fluorescence. Therefore, due to this interconversion, a shortfall of PQ under saturated light pulse can bring the lowering in quenching and thus overestimation of $F_m^{'}$, the result (Oxborough 2004). Leaves with high ratio of plastoquinone to plastoquinol are more prone with such error before the application of saturated light pulse. So, under oxidative stress it is the plant's genetic plasticity to modulate the photochemistry such a way that minimizes the reduction of Q_A. Under oxidative stress as the plants are over-oxidized with PSII due to some impairment of e^- flow from Q_A to Q_B, leaves are more tuned to dissipate energy by NPQ. Thus, NPQ parameter is marked as an indicator for excess-radiant energy dissipation to heat. The PSII antenna complexes coupled with LHCII are also similarly cascaded with this release of energy. More accurately, the xanthophyll cycle through its de- peroxidation can quench the excess light energy. This also includes the attenuation of LHCII antenna, PSII deactivation, and others (Pascal et al. 2005). It is already stated that at the same time ØPSII implicates an approximation of photochemical quenching (qP) and effective quantum yield (F_v/F_m); those processes alter the efficiency. During over-excitation with high irradiance (a condition for oxidative stress), the reduction in RC as a consequence of a high intensity light may alter the value of qP. Thus, plants are trailed with maximally for photochemistry and subsequent generation of reducing equivalents (NADPH + H^+). In application the chlorophyll fluorescence with reference to photochemical quenching under treatment with 1-methylcyclopropane (1-MCP) in banana peel de-greening could be sited (Li et al. 2002). A delay in loss of fluorescence yield was scored with the treatment of 1-MCP, but ethylene decreased the greenness and $Ø_{Po}$. This clearly indicates the changes in maximum fluorescence (F_m). Ongoing incubation with 1-MCP had regain F_m values quite a few times more, thus suggesting maintenance of fluorescence. The later might have disconcerted with oxidative exposure under ethylene treatment.

5 Some Limitation with Dark Adaptation in Plants for NPQ Measurement

It is well understood that dark adaptation and its duration of plant are not always the same. In fact, NPQ, in general, is granted in analysis of PSII fluorescence. In practical, the basal level of fluorescence ($F_0{'}$) is measured when AL is switched off. And thereby the fluorescence comes down to the minimum. Under the same condition, a light-adapted leaf is required of its PSII fully opened either via dark adaptation (switching off the AL) or using a pulse of F_R light. The imposition of SP in regular intervals causes the relaxation of $F_m{'}$ in dark until it reached its dark-adapted value. The NPQ can be dissected in different components that can relax the kinetics as recorded in PAM studies. In most of the cases, NPQ in high-energy-state quenching is more available. The quenching by state transitions (qT) is the shifting of peripheral LHCII from PSII to PSI. The photo-inhibitory quenching (qI), however, remains small at moderate and low light, but at high saturating light levels, it induces a significant fall in photosynthesis (Niyogi et al. 1998). In fact, qI can't be eliminated by dark adaptation and thus also described as sustained quenching. Photo-inhibition is an outcome of a number of processes for inactivation of proper photochemistry. Still, spillage of e^- from PSII being the most predominant of that. In addition, it is also admissible for plants that any sort of photoinactivation is also repairable. Mostly, the replacement and de novo synthesis of D1 protein in the core region of RC are the most important. The latter even undergoes reassembly in PSII-RC (Oxborough et al. 2000). Under dark adaptation, the plants are also accelerated with accumulation and sustenance of zeaxanthin for prolonged quenching. To attain the peak of F_0, the plants are also offered with some chances of photo-inhibitory damages. But, it is not necessarily with the sustenance of zeaxanthin. Since NPQ and effective quantum yield (F_q'/F_m'), two parameters, are to coincide each other in nonlinear fashion, the dark-adapted opening of RC can be easily related to photoinactivation (Quilliam et al. 2006).

6 More Insights in Dark-Adapted State for PSII Measurement

There opens another problem whether dark adaptation is obligatory for measuring the PSII activity. This really contradicts, in fact, with some common observations: (1) Is the photo-inhibition or some continued downregulation of PSII activities really required? (2) When dark inactivation is allowed, the values of effective quantum yield decrease and whether photosynthetic activities is needed at all, the question (Scharte et al. 2005). Any kind of stress makes it more acute and becomes an initial for NPQ analysis. To sustain drought condition, the value of F_v/F_m is also utilized for the PSII maximum activity. Before the dawn comes, the proportion of F_v to F_m may be indicative of photo-inhibition through downregulation of PSII. The stomatal regulation is set as a key for the decline in F_v/F_m at least, at the onset of drought stress, however, compounded with other factors. Still, to recover the F_v/F_m value, the diurnal photo-inhibition may take a positive role. But, when the stress is relieved, the dark-adapted plants start actual functioning followed by changes in PSII fluorescence (Schreiber 2004). Theoretically, a leaf under dark condition will be detained with the rate of the electrons transport at immediate. The relaxation of NPQ becomes more delayed because the loss of energy through other processes rather than photochemistry starts in action. As already mentioned earlier, xanthophyll cycle becomes the predominant compound for this protective functioning. To open all the RC in PSII and thereby bring the value of NPQ as nil, a sufficient period for dark adaptation is required. To start the photosynthesis under AL or SP, gradual changes in F_{max} are recorded, more specifically when it reaches to the plateau (Swarbrick et al. 2006).

7 Relationship Between PSII Fluorescence Events and Downstream Process in Carbon Reduction

The inactivation of PSII reaction centers is manifested as oxidative injuries and substantial amount of heat loss with the increase in F_0. The cellular functioning, particularly, the water relation under drought condition, the F_m and F_0 will be distorted along with absorption of light by leaf. A drop in F_v/F_m thus directly is correlated with the photosynthetic compromization in carbon reduction reactions. Still, it does not necessarily mean that F_v/F_m would correlate the intrinsic quantum yield carboxylation as well as oxygen evolution. If a light response curve is made, the maximum quantum yield for CO_2 is obtained from its linear phase. Under saturated light condition or with depleted values of F_v/F_m, it is also possible that a leaf might not conciliate the overall photosynthetic performances. A pitfall in photosynthetic rate per se is a feature under either light-limited or light-oversaturated illumination. This essentially includes the fall in CO_2 reduction by NADPH + H^+ along with deviation of F_v/F_m values. Under stress condition this would deal with the maximum quantum yield of PSII, but the application of stress in distal parts of plants away from the leaves may score hardly any changes in F_v/F_m. The rate of PSII efficiency is otherwise related to bottom-up regulation in the ratio of NADPH + H^+ to ATP generated in photosynthetic phosphorylation. In fact a number of factors are contributed to the linear rate of e^- flow through PSII that could be related to Calvin cycle (Baker 2008). The efficiency of Rubisco, regeneration of RUBP 1,5-bisphosphate, the flow of air or atmospheric CO_2 to the active site of enzyme, rate of gas exchange through stomata, photosynthetic carbon oxidation vis-à-vis photorespiration, and finally rate of reduced carbon from chloroplast are the determinants for NADPH +H^+ and ATP accumulation (Buchanan and Balmer 2005). These all are consequently manifested into PSII efficiency. It is well admitted that imposition of stress factors is also the variable to make use of any significant contribution to any of those on F_v/F_m. Any significant impacts of stress regardless of its nature can initially establish some sort of water-deficit stress (Baker et al. 2001). The dehydration can lead to limit the water use efficiency (WUE), stomatal conductance (g_S), and control of intercellular CO_2 concentration (Ci) and translocation of solutes from source to sink. These all collectively address the regulation and impairment of proper PSII functioning. Still, the rate-limiting step in regulation of any metabolite accumulation in carbon reduction paths coupled with precise behavior of PSII activity is yet to be deciphered (Baker and Rosenqvist 2004). The higher and lower in NADPH and ATP, respectively, would also downregulate the e^- flow from PSII, and that communicates in decrease in QA reduction. In simple PAM analysis, this could also be detected by F_v/F_m data as in subdued order (Ashraf et al. 2007). It is fairly in conformity that gathering of H^+ in thylakoid lumen with associated increase in ATP synthesis is linked to the NPQ. So, a steady decrease in F_v/F_m under any sort of abiotic stress may also predict the NPQ by photosynthetic phosphorylation under higher illumination.

The efficiency of PSII (F_q'/F_v') is governed by two essential attributes. These are F_v'/F_m' and F_q/F_v. The variable fluorescence (F_v) denotes the $F_m' - F_0$ under irradiated condition of leaf where F_0 is the basal fluorescence under light. It is also noted that F_v/F_m could estimate the maximum quantum yield of PSII photochemistry to be obtained from fully oxidized QA under illumination. Under similar condition this also derives the participation of NPQ from PSII in leaves. Under different environmental conditions, the efficiency of PSII is highly variable and could also predict the proportion of photochemical and non-photochemical quenching and thus is also called PSII efficiency factor. The latter is essentially connected, however, nonlinearly with the fraction of RC that remains still open or oxidized (Calatayud 2006). The photochemical quenching (qP) is the actual proportion of light energy that is transuded to generate NADP(H) + H^+ out of photolysis of water. On the other hand, F_q'/F_v' is reflected with the potential of chloroplast to maintain QA in open or oxidized state so that it could be accepted with more electrons from the quinine pool via PSII. Therefore, the QA with its possibility of interconversion in oxidized and reduced stats is most likely to be offered as a determinant for PSII photosynthetic efficiency or maximum PSII efficiency. The latter is amenable through the derivation of F_v'/F_m' and F_q'/F_v'. Still, their direct determination is rather non-simplified and almost the cases it is done by irradiation of the leaf to a pulse of far-red (FR) or immediately switching off the AL. This will let reach the QA in maximum oxidized condition (Chaerle et al. 2006). Sometimes F_0' could be overestimated during far-red light pulse, and NPQ is also relaxed partially, and the PSII operating efficiency may be curtailed if the photosynthetic photon flux density (PPFD) is increased. This otherwise will decrease those F_v/F_m and F_q'/F_v'. On the other hand, the latter two can change significantly with increasing PPFD. In a straightforward relation under excess light, irradiation over the saturated level initiates the NPQ, and those could be detected by a decreasing F_v/F_m. This otherwise also indicates a decrease to turn over oxidized QA from its reduced state (Ashraf et al. 2007). Under the situation of increasing NPQ, PSII operating efficiency falls when a dark-adapted leaf is exposed to photosynthetic/actinic light that is also associated with higher value in F_q'/F_v' but hardly any significant alteration in F_v/F_m.

8 Important Keys in Determination of PSII Operating Efficiency

It is well admitted that PSII operating efficiency, (F_q/F_m) in maximum, has got the two most important attributes: F_v/F_m and F_q/F_m. It was already stated that F_v represents the variable fluorescence ($F_m - F_0$) in light-adapted leaf. F_0 is the minimal or basal fluorescence in the light when QA is opened or maximally oxidized. On the other hand, F_v/F_m estimate quantum yield of PSII in maximum for photochemistry. It otherwise also gives the ratio of absorbed light to be utilized in photochemical quenching and thus may alternatively predict the rate of e^- through PSII. This is based with the knowledge that that leaf with its maximum absorption of light and thus photosynthesizes with maximum. As a result, this would be accessed for the aid of NPQ to changes in the PSII operating efficiency in the light. F_q'/F_v' behaved as a powerful tool to estimate the photochemical quenching (qP). This is derived from the basic data of ($F_m' - F'/F_m' - F_0'$) and denotes the ratio of PSII maximum efficiency in operating efficiency and related nonlinearly to the sizes of RC; those are open. The latter behaved as a meaning of the relative rates of QA reduction to oxidation. Under stress condition, however, F_v/F_m is set as a reliable factor to assay the redox of PSII and allied LHCII; still, F_q'/F_v' and F_v'/F_m' are also equally or more importantly used. Those describe that any decrease of the later would reflect the increase in NPQ (Baker 2008)). It is quite obvious that calculation of F_v/F_m and F_q/F_v essentially demands the F_0 value. F_0, in general, is measured either in two ways: a complete dark incubation of leaf sample or by a pulse of weak F_R, when the AL is removed to get the maximally oxidized QA (Miyake et al. 2004). In practice, switching off the AL permits to remeasure the basal fluorescence F_0' In fact, to get the actual value, users take it as advantages with F_R application. By application of saturated pulse (SP) in regular interval, it is easy to get the relaxation of F_m' value in the dark, so long it obtains its dark-adapted data. NPQ being the paramount attribute to characterize the plant's fluorescence under stress condition has other components also. This means to get the NPQ; however, in majority of cases, the high-energy-state quenching (qE) is adopted. Still, other important components do also occur. One of those is the migration of peripheral LHCII from PSII to PSI. One advantage of this is to get under low light condition, and under high light it is considered as small. Another is photo-inhibitory quenching (qI) which is small under low and even moderate illumination, still, becomes predominant under saturated condition. One important feature of qI is that it could not be removed even under dark adaptation, i.e., it essentially refers to the sustained quenching. Photo-inhibitory quenching (qI) is an outcome of reversible photooxidation of PSII-RC. This is followed by repair through partial dissembling and denovo synthesis of D1 protein (Leipner et al. 2001). Quenching in prolonged period essentially involves accumulation and sustenance of zeaxanthin as photo-protective pigments dispersed in PSII-RC. Notwithstanding, a rise in F_0 may indicate photo-inhibitory damages but must not with preservation of zeaxanthin. It is noteworthy that any changes in NPQ may be read by F_v'/F_m' since these coincide nonlinearly; still, it may be inexact or imprecise at higher values of NPQ. On the contrary, the F_v'/F_m' would possibly be more accurate to contribute NPQ making to F_q'/F_m', i.e., the quantum efficiency of PSII exclusively in light. The overestimation of F_0 due to FR pulses and partial relax of NPQ can be prevailed over otherwise. This is from calculation of F_0 from F_m during the measurement and values of F_0 and F_m from dark-adapted leaves with the derivation: $F_0 = F_0/\{(F_v/F_m) + (F_0/F_m)\}$. It is also commented that this equation holds more validity under normal condition (non-stressed) of leaves, when photo-inhibition remains minimized (Long and Bernacchi 2003).

9 Imaging of Chlorophyll Fluorescence: An In Vivo Tool for Stress Analysis

Out of a number of advantages in chlorophyll fluorescence, few are most important. However, those include predominantly two effective attributes: for screening of a huge number of plant samples at a time and to assess the heterogeneity of the leaves, those might very often go changes in fluorescence also (Aldea et al. 2006). It not only reflects the physiological heterogeneity of a leaf, not even under stress, but also the patchy activities of stomata. A large number of variable plant genotypes are to come as example to be indexed reliably with chlorophyll imaging under such stress condition. This is also encompassed with the genotypic variations in higher productivity. A number of examples could be cited for early detection of plant's responses with chlorophyll fluorescence imaging. The most important parameters for studying the fluorescence like F_v/F_m, photochemical quenching, and electron transport rates all are accessed to evaluate even under immature leaf condition exposed to various stressors. So, quantification of different fluorescence parameters is set as a basic in PAM studies for damaged leaves under stress to reflect the photochemical quenching or its any anomaly (Baker and Rosenqvist 2004). For a ready reference, PSII could be damaged with UV radiation with turnover of antenna pigments as well as LHCII. This could be detected with a rapid fall or decline in maximum quantum efficiency as well as more light energy to non-photochemical quenching (NPQ). It still means that quenching of fluorescence is the fall in fluorescence yield either detected by F_v/F_m or ØPSII in collection or in separate through photochemistry and loss of energy by releasing long wavelength. A steady-state competition between these processes may reflect the decay of the chlorophyll in excited state. An increase of any one can downregulate the chlorophyll fluorescence and its yield in a number of attributes so indexed in the PAM analysis (Calatayud et al. 2008). This was very closely adjoined with the states of electron transport chain between PSI and PSII. This has well been documented that a plant or even a leaf would raise the fluorescence when transferred from dark adaptation to light. This is accompanied and characterized with acceptance of electrons by the acceptors and its concomitant reduction. The RC is now attained as reduced or closed and must not be reduced again until the e^- passes to next carrier. Quinone (Q_A) is the most initial to accept the electrons and its reduced state denotes the RC closed. PAM instrumentation initiates imaging the fluorescence in the closed state of RC to detect the chlorophyll fluorescence and its yield in increased manner (Chaerle and Straeten 2000). In technicality, the definition of the chlorophyll imaging imprints the detection of fluorescence yield with different attributes through various color codes normalized to arbitrary color bar provided. There set a false color scale within the limit with 0.0 to maximum 1.0. Another important terminology is the non-photochemical quenching (NPQ). In fact after getting the all RC to be closed the level of fluorescence though attained maximum, it immediately comes down. Through the induction of light-mediated enzymes in Calvin cycle and stomatal regulation, the ready supplement of CO_2 for carboxylation of RUBP is accomplished. These cause the electrons to be transferred away from PSII (photochemical quenching, qP). With this dissipation of heat energy rather than photochemistry (non-photochemical quenching, qNP) becomes the measuring process for PAM imaging. Generally, three subsets are adhered for the imaging of fluorescence in accordance with chronological events and status of the plants. The most important is photoinhibition (qI) with others like state transition (qT) as well as energy-dependent quenching (qE) (Chen et al. 2009).

Along with different component of abiotic stress imposed to plants and its relevance to impact using image analysis, oxidative stress is probably the more sensitive to monitor. In fact, irrespective of abiotic stresses oxidative damages is the common to bring significant and irreversible changes of cellular physiology in plants (Cornic and Fresneau 2002). In the chloroplast, excited chlorophyll molecules could bring some sets of redox reaction accompanying with photolysis of water. Along with generation of reactive oxygen intermediates (ROI) or reactive oxygen species (ROS), chloroplast may also develop the oxidant/oxidants as PSII drawing the electrons from H_2O to O_2. There is ample probability with this oxidant to oxidative degradation of different biomolecules (Dai et al. 2007). The lipid and proteins integrated with chloroplast membrane is equally prone to be damaged and thereby may affect the PSII along with LHCII. Therefore, chloroplast and other e^- harboring cell fractions are housed to be the site of stress perception as well as to monitor the impact of stress. Thus, imaging of chlorophyll fluorescence has been set as a consistent trait for ROS/ROI and PSII maximum efficiency. An arbitrary relationship exists between damaged leaf generating ROS and changes in maximum quantum yield (ØPSII) in photochemistry (Biswal and Biswal 1999). In addition with abiotic stress factors, the wound caused by biotic stress has also got implication with imaging techniques. With the invasion of fungal pathogen as well as bacterial infection, the mechanical injuries or damages of the tissues are accompanied by a collection of cellular events: loss of water, electrolyte, and solutes (DeEll and Toivonen 2000). This causes more entry of the pathogens into vascular system of the plants. To combat the invading pathogens, the plants secure the metabolic flux more toward the wound portion as one of the measures, and thereby modified sink strength is the result. With the imaging technique in plants, a steady-state decline in ØPSII immediate after infection in the peripheral tissues surrounding the wound was featured (Ehlert and Hincha 2008). As the duration of invasion proceeds, the leaf tissues recorded differential expression in ØPSII activity and a differential expression with PAM, thereby suggesting sensitivity of imaging for finding the tissue lysis. A consistent rise in qP to dissipate maximum heat-generating processes as well as qL (fraction of open PSII) may be correlative since stomata might restrict the gas exchange in carboxylation. Thus, image analysis has been established as modified sink strength in pathogenic stress as a reliable instrumentation aid. In more sophisticated way, a further development in fluorometer with pulse for short duration facilitates the visualization in photosynthetic activity. This in true sense depicts the heterogeneity in space of photosynthetic functionalities with reference to modules of fluorescence. The sources in this heterogeneity could be of natural phenomena like leaf growth and environmental stresses and even under pathogenic invasion also (Fambrini et al. 2010). This way photosynthetic activity could properly be justified with the information of the heterogeneity. In more advancement machine, fluorometer systems with high-performance imaging extend the facilities for analysis of a lone chloroplast and whole leaves. In the late 1980s, the first report came with CCD technologies, adopted to monitor the fluorescence in sunflower leaves under sulfur dioxide (SO_2) exposure. A differential tissue-specific change in fluorescence images in leaves was detected to show the variability in SO_2 sensitivity as a result of differential stomatal conductance. In a similar way, the penetration of some selective herbicides in leaves as a function of tissue heterogeneity could also be visualized by PAM image analysis. The pattern of inhibition by 2,6-dichloromethyl monourea (DCMU) has markedly separated the tissue sensitivity for photosynthetically efficient, less efficient, and even non-efficient also (Feng and Cao 2005). More relevant images under water deficit were obtained in fluorescence imaging with heterogeneous pattern of quenching events. Thus, the changes in $(F_m' - F_s)/F_m'$ in water-stressed leaves were accredited as heterogeneous behavior of stomatal regulation. Even in multi-stress factor imposed to plants, an alteration of steady-state blue, green, red, and far-red fluorescence is recorded. Such visualization of imaging fluorescence with sophisticated instrumentation has been a reliable indicator under plant's responses to any elicitor under abiotic stress also (Fig. 5). In maize

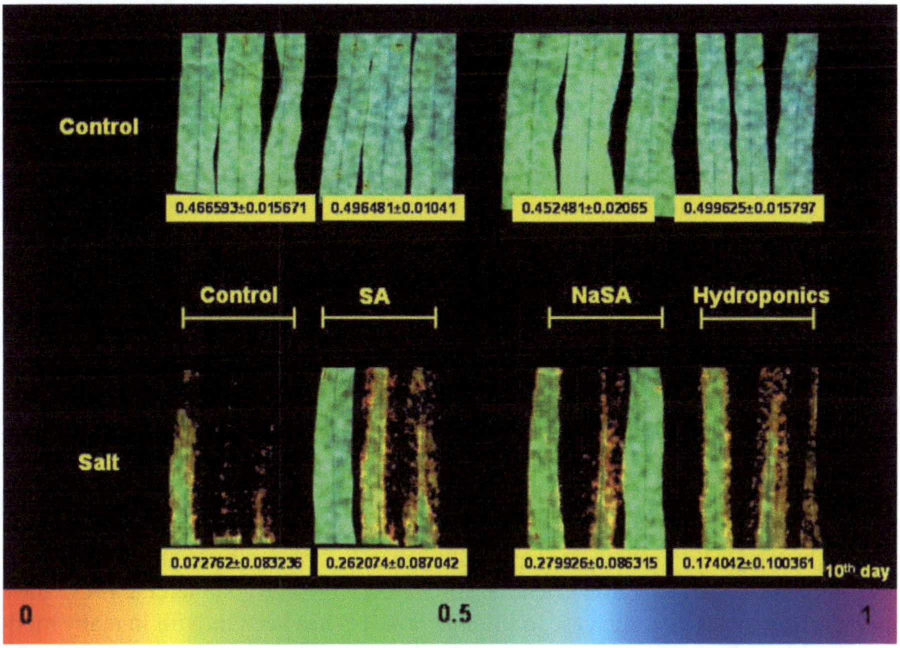

Fig. 5 A PAM images for chlorophyll fluorescence in terms of optimum and effective quantum yields to characterize plant's responses to different treatment with salicylic acid (SA) and sodium salicylate (NaSA) under salt stress. Under illumination of photosynthetic active radiation (~1000 μmol m^{-2} s^{-1}), it images a two-dimensional distribution in quantum yield $\left(\Phi PS-\text{II} = F_m' - F_0' / F_m'\right)$ (unpublished data)

multicolor fluorescence imaging remained more efficient in probing for detection of nitrogen and iron deficiency but not so much significant for other metals (Fambrini et al. 2010).

10 PAM Image Analyzer: Instrumentation Brief

PAM light with different intensities and duration is primarily used in detection of chlorophyll imaging. Initially, a number of instruments were employed for detection of fluorescence since the days of 1980 and onward (Li et al. 2002). In true sense PAM employs an encoded message in the form of signal amplitudes for a series of pulses. The kinetics in fluorescence rise and its decay within the LHCI and LHCII of chloroplast-bound photosystems and accessory pigments are typically monitored by PAM fluorometry. This fluorometer is principally employed with three major light sources; however, it varies in different dimensions of models. Initially, it starts with a short flush of modulated or measuring light (ML) with intensity of 15 μmol photon $m-^{2}$ $s-^{1}$ under wavelength of 650 nm for 1000 millisecond range. This fluorescence-inducing light (preferably with red) is applied in repeat (600 Hz–20 KHz) with low, still, non-variable intensities. This chlorophyll fluorescence is detected nonlinearly with a photodiode detector. Thus, minimum fluorescence as earlier mentioned (F_0) is recorded. This is immediately followed by the use of actinic light (AL) or photosynthetic light at the same wavelength with intensity of around 100 μmol photon m^{-2} s^{-1}. Halogen lamp is used in frequent as light source (Fambrini et al. 2010). This initiates the primary photochemical processes by the chlorophyllous tissues and changes the value of F_0. Finally, a series of saturated pulses of short duration is applied with very high photon (up to 10,000 μmol photon m^{-2} s^{-1}) with a white halogen lamp. With these pulses, all the RC and electron carriers (plastoquinone) are attained completely closed or in reduced state. Therefore, application of saturation pulse with stipulated intervals on a light adapted (both by measuring and actinic light) to measure the fluorescence was also called as light-doubling method (Flexas et al. 2002). Both quenching of PQ as photochemical and non-photochemical paths are estimated with this doubling method. With the saturated pulses, the photochemical quenching is reduced with complete reduction of Q_A. Thus, photochemical quenching attains the value of zero while the non-photochemical quenching to be the maximum. Consequently, it lets the photochemistry almost nil, and all the energy would be channelized to fluorescence and some sorts of heat. This also brings the maximum fluorescence on its values both at dark- and light-adapted condition as depicted F_m and F_m', respectively. The automation and utilization of most user-friendly software in mostly commercial PAM have popularized its use. In each case, a suitable protocol has been in-built in many modern machines. In those cases some modifications are still there to get the meaningful data in robust. One of those is that the leaf is held horizontally to the AL and ML to avoid the diffused or heterogeneous apparent of light over the areas of the leaf (Flexas et al. 2004). In this regard two important rudiments are more added with standardization of PAM. First, there should be an even distribution of light from source over the leaf to be imaged and, second, the capability of light source to provide saturated pulse (SP) in closing of all RC in PSII. This will provide the measurement of F_m'. However, it might be sufficient for recording Fm in dark-adapted plants, however, lower in case of real F_m' under AL. In addition the intensity of SP also varies with variation of plant species to image. For maize and sorghum like C4 plants, higher intensities of luminance are recommended to reach the SP within the threshold of illumination. Therefore, a recommended trait for installing a PAM instrument is that the capability of wider ranges in SP and to the extent of evenly distribution of light over the leaf would be in uniform. In modern state of the art, the fiber-optic systems are mostly chosen against conventional imaging system. The key variation of these two is laid on how the measurement is accomplished in F_0' on which the following rest of the quenching attributes depend. Use of FR pulse is the distinct feature of much fiber-optic system incorporating light-emitting diode (LED) system. The variation in commercialization is based on fitting of a panel of LED measuring the AL. These are consisted with one wave band and by default are incapable to hold out FR. This is very much particularized where plants receive a significant amount of photo-inhibition under stressful condition to record any changes of F_0' (Maxwell and Johnson 2000). Still, it opens question for its soundness. Because the requirement to calculate the F_0' is exclusively based on few rationales, PSII-RC are to open at the time of recording F_0', the irreversibility of downregulation between the measurement of F_0 and F_m' as well as no turnaround of photo-inhibition between the measurement of F_m' and F_m (Oxborough 2004).

11 Principle of Adoption of PAM for Analysis of Stress in Plants

In relation to PAM studies, excellent insights are obtained for plants responding to abiotic stress as well as its acquired tolerance. It is also possible to relate the qualitative and quantitative assay for damages of chloroplast and associated photosynthetic accessory sets with stress imposition as well as its recovery (Gilmore and Govindjee 1999). Regardless of the cases, stress may either block or alter the passage of electrons between PSI and PSII, and very certainly photochemistry is diverted. This otherwise dissipates more energy into NPQ with heat and fluorescence. The most classical example for employing the chlorophyll fluorescence in plants' response to biotic stress is set with water-deficit condition and its manifestation into photosynthetic characterization (Grassi and Magnani 2005). In photosynthesis the role of H_2O is the determined roles under light excitation to generate a strong reductant (NADPH) in a pH-dependent electron movement over the chloroplast membrane. Water stress induces both loss of membrane functionality in terms of depletion of energy turnover for photochemistry from PSII and generating highly oxidizing intermediate moieties as ROS (Guidi and Degl'Innocenti 2008). However, it is well established that water stress is more earlier to impact on PSII-based changes of F_m' and/or ØPSII rather than to deplete the carboxylation and subsequent reactions in Calvin cycle (Hogewoning and Harbinson 2007). This is willingly accepted in recording with a delayed fall in values of effective quantum yield (F_0/F_m) or increase in qNP. More so, the impact of stress is also variable according to C_4 as well as C_3 species with more stomatal closure in later reducing the water use efficiency and finally lowers the intercellular CO_2 concentration. Thereby, in C_3 plants it ensures a ready thrust for plants to dissipate more energy in qNP trailing the photochemistry with rapid electron flow to oxygen (Kangasjarvi et al. 2005). Thereby, chances of generating the ROS are more prevailed in excess along with documentation of changed fluorescence properties.

12 Integration of Chlorophyll Fluorescence with Other Photosynthetic Analyzing Techniques

The combination or even amalgamation of PAM instrumentation with gas exchange-aided machine conceives a lot of information in photosynthetic research. The most readily used technique is the combination of infrared gas analyzer (IRGA) coupled with fluorescence imaging (i.e., PAM) system. The principle behind this instrument is dealt with a simultaneous as well as in vivo correlation of CO_2 assimilation rate through gas exchange in open/closed system with PSII efficiency (for F_v/F_m). This is essentially accomplished with complete elimination or a significant reduction in photorespiratory rate under an LED-based illumination with controlled PAR (Lawson et al. 2002). Moreover, systems are devised such a way that a reduction in O_2 or an increase of CO_2 concentration is possible by the user to construct a calibration for either light or CO_2 response curve for sample. This is so sophistically designed to make an easy and anticipatory assumption in determination of gas fluxes in leaves of different species, imaging of CO_2 diffusion pattern in leaves, variation in venation for gas exchange, and finally the fluxes on carbon acquisition in tissues (Morison et al. 2005). This system has an added advantage with controlled gas flow in leaf chamber with CO_2-absorbing materials like soda lime (making the CO_2 as zero level) that provide the information on photorespiratory mutant (Barbagallo et al. 2003). Other factors like leaf temperature, relative humidity, etc. are also required to verify whether light-induced changes of fluorescence could bring any alteration for net CO_2 assimilation rate (Furbank et al. 2009). In this way it was practically derived that carbon reduction in Calvin cycle reactions is the predominant sink sustenance of photosynthetic e^- transport as evident from ØPSII and other modules. In details it is more elucidated that the relative humidity is the major drive for stomatal regulation and that also indirectly linked to photosynthetic e^- transport. The closing of stomatal aperture reduces the intercellular CO_2 concentration, and thereby photosynthetic rate is minimized vis-à-vis the photosynthetic energy transfer between PSII and PSI (Chaerle and Straeten 2001). The plant phenotyping with other phenological traits has been customized with chlorophyll fluorescence imaging and other imaging technologies. In matchup to chlorophyll quenching of photonic energy in LHCII before charge separation, thermoluminescence (TL) gives unique information on the energetic of the charge stabilization in photochemistry (Maxwell and Johnson 2000). TL, in principle, employs a technique with rapid cooling followed by the progressive warming of a pre-illuminated leaf. In recording it reveals the different types of charge pairs in the form of successive emission bands. The latter are resolved and compared with corresponding decay phases at constant temperatures. This is more advantageous when chlorophyll fluorescence is merged with thermoluminescence in a single device to monitor possible changes in energy dissipation under stress condition of a leaf against a control. With more stringency, if this experiment is conducted, the most supporting data characterized plants photosynthetically (like intrinsic imaging water use efficiency, iWUE) (Meroni et al. 2009). For biotic stress, where plant pathogen makes its score by degradation of chlorophyll, the fluorescence imaging combined with hyper-spectral imaging enables for early detection of the disease, rather than visible syndromes. The head blight disease in wheat may be cited as a good index with this imaging phenotyping (Bauriegel et al. 2011).

13 Inferring Comments

With the presentation of this review, the principle and methodologies of chlorophyll fluorescence have been attempted to document in most precise manner. With the advent in plants or crop sciences, the selection of potential species by phenotyping and monitoring the photobiological processes, chlorophyll fluorescence, and its imaging has been a good and successful technique. The present review is expected to cover an update for photosynthetic research through working principles and methodologies in sophisticated manner. Undoubtedly, PAM imaging technique lends itself outstandingly sound to the modern state of the arts for improvement of crop varieties mainly for speedy, noninvasive, and in vivo and with easy measurement producing lot of significant data. In combination with gas exchange and even with thermoluminescence, the usefulness becomes prudent in some sort of variations, in real for plants under stress. However, the illustrated scope with remote sensing of passive fluorescence by satellite aided technologies happens to be another effective input for automation in crop selection.

References

Adams WW III, Demmig-Adams B (2004) Chlorophyll fluorescence as a tool to monitor plant response to the environment. In: Papageorgiou GC, Govindjee (eds) Chlorophyll a fluorescence: a signature of photosynthesis. Springer, Dordrecht, pp 583–604

Adams WWIII, Demmig-Adams B (2006) Energy dissipation and photoinhibition: a continuum of photoprotection. In: Demmig-Adams B, III WWA, Mattoo AK (eds) Photoprotection, photoinhibition, gene regulation and environment. Springer, Dordrecht, pp 49–64

Aldea M, Frank TD, DeLucia EH (2006) A method for quantitative analysis of spatially variable physiological processes across leaf surfaces. Photosynth Res 90:161–172. ISSN 0166-8595

Allen JF, Mullineaux CW (2004) Probing the mechanism of state transitions in oxygenic photosynthesis by chlorophyll fluorescence spectroscopy, kinetics and imaging. In: Papageorgiou GC, Govindjee (eds) Chlorophyll a fluorescence: a signature of photosynthesis. Springer, Dordrecht, pp 447–461

Ashraf M, Nawazish S, Athar HUR (2007) Are chlorophyll fluorescence and photosynthetic capacity potential physiological determinants of drought tolerance in maize (Zea mays L). Pak J Bot 39(4):1123–1132
Baker NR (2008) Chlorophyll fluorescence: a probe of photosynthesis in vivo. Annu Rev Plant Biol 59:89–113. ISSN 1543-5008
Baker NR, Rosenqvist E (2004) Applications of chlorophyll fluorescence can improve crop production strategies: an examination of future possibilities. J Exp Bot 55:1607–1621
Baker NR, Oxborough K, Lawson T, Morison JIL (2001) High resolution imaging of photosynthetic activities of tissues, cells and chloroplasts in leaves. J Exp Bot 52:615–621
Baker NR, Harbinson J, Kramer DM (2007) Determining the limitations and regulation of photosynthetic energy transduction in leaves. Plant Cell Environ 30:1107–1125
Barbagallo RP, Oxborough K, Pallett KE, Baker NR (2003) Rapid noninvasive screening for perturbations of metabolism and plant growth using chlorophyll fluorescence imaging. Plant Physiol 132:485–493
Bauriegel E, Giebel A, Geyer M, Schmidt U, Herppich WB (2011) Early detection of Fusarium infection in wheat using hyper-spectral imaging. Comput Electron Agric 75:304–312
Berger S, Benediktyovâa Z, Matoués K, Bonfig K, Mueller MJ (2007) Visualization of dynamics of plant-pathogen interaction by novel combination of chlorophyll fluorescence imaging and statistical analysis: differential effects of virulent and avirulent strains of P. syringae and of oxylipins on A. thaliana. J Exp Bot 58:797–806
Biswal B, Biswal UC (1999) Photosynthesis under stress: stress signals and adaptive response of chloroplasts. In: Pessarakli M (ed) Handbook of plant and crop stress. Marcel Dekker Inc., ISBN0-8247-1948-4, New York, pp 315–336
Buchanan BB, Balmer Y (2005) Redox regulation: a broadening horizon. Annu Rev Plant Biol 56:187–220
Calatayud A, Gorbe E, Roca D, Martinez PF (2008) Effect of two nutrient solution temperatures on nitrate uptake, nitrate reductase activity, NH4+ concentration and chlorophyll a fluorescence in rose plants. Environ Exp Bot 64(1):65–74
Chaerle L, Van Der Straeten D (2000) Imaging techniques and the early detection of plant stress. Trends Plant Sci 5:495–501. ISSN 1360-1385
Chaerle L, Van Der Straeten D (2001) Seeing is believing: imaging techniques to monitor plant health. Biochim Biophys Acta 1519:153–166. ISSN 0304-4165
Chaerle L, Leinonen I, Jones HG, Van Der Straeten D (2006) Monitoring and screening plant population with combined thermal and chlorophyll fluorescence imaging. J Exp Bot 58:773–784. ISSN 0022-0957
Chen CP, Frank TD, Long SP (2009) Is a short, sharp shock equivalent to long-term punishment? Contrasting the spatial pattern of acute and chronic ozone damage to soybean leaves via chlorophyll fluorescence imaging. Plant Cell Environ 32:327–335. ISSN 0140-7791
Cornic G, Fresneau C (2002) Photosynthetic carbon reduction and oxidation cycles are the main electron sinks for photosystem II activity during a mild drought. Ann Bot 89:887–894. ISSN 0305-7364
Dai F, Zhou M-X, Zhang G-P (2007) The change of chlorophyll fluorescence parameters in winter barley during recovery after freezing shock and as affected by cold acclimation and irradiance. Plant Physiol Biochem 45:915–921. ISSN 0981-9428
DeEll JR, Toivonen PMA (2000) Chlorophyll fluorescence as a non-destructive indicator of broccoli quality during storage in modified atmosphere packaging. Hortscience 35:256–259. ISSN 0018-5345
Ehlert B, Hincha DK (2008) Chlorophyll fluorescence imaging accurately quantifies freezing damage and cold acclimation responses in Arabidopsis leaves. Plant Methods 4:12. ISSN 1746-4811
Fambrini M, Degl'innocenti E, Cionini G, Pugliesi C, Guidi L (2010) Mesophyll cell defective1, a mutation that disrupts leaf mesophyll differentiation in sunflower. Photosynthetica 48:135–142. ISSN 0300-3604

Farage PK, Blowers D, Long SP, Baker NR (2006) Low growth temperatures modify the efficiency of light use by photosystem II for CO2 assimilation in leaves of two chilling tolerant C4 species, Cyperus longus L. and Miscanthus × giganteus. Plant Cell Environ 29:720–728

Feng Y-L, Cao K-F (2005) Photosynthesis and photoinhibition after night chilling in seedlings of two tropical tree species grown under three irradiances. Photosynthetica 43:567–574. ISSN 0300-3604

Flexas J, Bota J, Escalona JM, Sampol B, Medrano H (2002) Effects of drought on photosynthesis in grapevines under field conditions: an evaluation of stomatal and mesophyll limitations. Funct Plant Biol 29:461–471. ISSN 1445-4408

Flexas J, Bota J, Loreto F, Cornic G, Sharkey TD (2004) Diffusive and metabolic limitations to photosynthesis under drought and salinity in C3 plants. Plant Biol 6:269–279. ISSN 1435-8603

Fryer MJ, Andrews JR, Oxborough K, Blowers DA, Baker NR (1998) Relationship between CO2 assimilation, photosynthetic electron transport and active O2 metabolism in leaves of maize in the field during periods of low temperature. Plant Physiol 116:571–580

Furbank RT, Von Caemmerer S, Sheehy J, Edwards G (2009) C4 rice: a challenge for plant phenomics. Funct Plant Biol 36:845–856

Gilmore AM, Govindjee (1999) How higher plants respond to excess light: energy dissipation in photosystem II. In: Singhal GS, Renger G, Irrgang KD, Govindjee SS (eds) Concepts in photobiology: photosynthesis and photomorphogenesis. Narosa Publishing House, New Delhi, pp 513–548. ISBN 0-7923-5519-9

Grassi G, Magnani F (2005) Stomatal, mesophyll conductance and biochemical limitations to photosynthesis as affected by drought and leaf ontogeny in ash and oak trees. Plant Cell Environ 28:834–849. ISSN 0140-7791

Guidi L, Degl'Innocenti E (2008) Ozone effects on high light-induced photoinhibition in Phaseolus vulgaris. Plant Sci 174:590–596. ISSN 0306-4484

Hendrikson L, Furbank RT, Chow WS (2004) A simple alternative approach to assessing the fate of absorbed light energy using chlorophyll fluorescence. Photosynth Res 82:73–81

Hogewoning SW, Harbinson J (2007) Insights on the development, kinetics, and variation of photoinhibition using chlorophyll fluorescence imaging of a chilled, variegated leaf. J Exp Bot 58:453–463. ISSN 0022-0957

Johnson G, Kiirats O, Edwards GE (2004) New fluorescence parameters for determination of QA redox state and excitation energy fluxes. Photosynth Res 79:209–218

Kangasjarvi J, Jaspers P, Kollist H (2005) Signalling and cell death in ozone-exposed plants. Plant Cell Environ 28:1021–1036. ISSN 0140-7791

Kramer DM, Avenson TJ, Kanazawa A, Cruz JA, Ivanov B, Edwards GE (2004) The relationship between photosynthetic electron transfer and its regulation. In: Papageorgiou GC, Govindjee (eds) Chlorophyll a fluorescence: a signature of photosynthesis. Springer, Dordrecht, pp 251–278

Krause GH, Jahns P (2004) Non-photochemical energy dissipation determined by chlorophyll fluorescence quenching: characterization and function. In: Papageorgiou GC, Govindjee (eds) Chlorophyll a fluorescence: a signature of photosynthesis. Springer, Dordrecht, pp 463–495

Laisk A, Eichelmann H, Oja V, Rasulov B, Ramma H (2006) Photosystem II cycle and alternative electron flow in leaves. Plant Cell Physiol 47:972–983

Lawson T, Oxborough K, Morison JIL, Baker NR (2002) Responses of photosynthetic electron transport in stomatal guard cells and mesophyll cells in intact leaves to light, CO2 and humidity. Plant Physiol 128:1–11

Leipner J, Oxborough K, Baker NR (2001) Primary sites of ozone-induced perturbations of photosynthesis in leaves: identification and characterization in Phaseolus vulgaris using high resolution chlorophyll fluorescence imaging. J Exp Bot 52:1689–1696

Li XP, Bjorkman O, Shih C, Grossman AR, Rosenqvist M (2000) A pigment-binding protein essential for regulation of photosynthetic light harvesting. Nature 403:391–395

Li XP, Muller-Moule P, Gilmore AM, Niyogi KK (2002) PsbS dependent enhancement of feedback de-excitation protects photosystem II from photoinhibition. Proc Natl Acad Sci U S A 99:15222–15227

Li XP, Gilmore AM, Caffarri S, Bassi R, Golan T (2004) Regulation of photosynthetic light harvesting involves intrathylakoid lumen pH sensing by the PsbS protein. J Biol Chem 279:22866–22874
Long SP, Bernacchi CJ (2003) Gas exchange measurements, what can they tell us about underlying limitations to photosynthesis? Procedures and sources of error. J Exp Bot 54:2392–2401
Maxwell K, Johnson GN (2000) Chlorophyll fluorescence: a practical guide. J Exp Bot 51:659–668. ISSN 0022-0957
Melis A (1999) Photosystem II damage and repair cycle in chloroplasts: what modulates the rate of photodamage in vivo? Trends Plant Sci 4:130–135
Meng Q, Siebke K, Lippert P, Baur B, Mukherjee U, Weis E (2001) Sink-source transition in tobacco leaves visualized using chlorophyll fluorescence imaging. New Phytol 151:585–595
Meroni M, Rossini M, Guanter L, Alonso L, Rascher U, Colombo R, Moreno J (2009) Remote sensing of solar-induced chlorophyll fluorescence: review of methods and applications. Remote Sens Environ 113:2037–2051
Miyake C, Shinzaki Y, Miyata M, Tomizawa K (2004) Enhancement of cyclic electron flow around PSI at high light and its contribution to the induction of nonphotochemical quenching of chl fluorescence in intact leaves of tobacco plants. Plant Cell Physiol 45:1426–1433
Morison JIL, Gallouet LT, Cornic G, Herbin R, Baker NR (2005) Lateral diffusion of CO2 in leaves is not sufficient to support photosynthesis. Plant Physiol 139:254–266
Nedbal L, Whitmarsh J (2004) Chlorophyll fluorescence imaging of leaves and fruits. In: Papageorgiou GC, Govindjee (eds) Chlorophyll a fluorescence: a signature of photosynthesis. Springer, Dordrecht, pp 389–407
Nejad AR, Harbinson J, van Meeteren U (2006) Dynamics of spatial heterogeneity of stomatal closure in Tradescantia virginiana altered by growth at high relative humidity. J Exp Bot 57:3669–3678
Niyogi KK, Grossman AR, Bjorkman O (1998) Arabidopsis mutants define a central role for the xanthophyll cycle in the regulation of photosynthetic energy conversion. Plant Cell 10:1121–1134
Oxborough K (2004) Using chlorophyll a fluorescence imaging to monitor photosynthetic performance. In: Papageorgiou GC, Govindjee (eds) Chlorophyll a fluorescence: a signature of photosynthesis. Springer, Dordrecht, pp 409–428
Oxborough K, Hanlon ARM, Underwood GJC, Baker NR (2000) In vivo estimation of photosystem II photochemical efficiency of individual microphytobenthic cells using high-resolution imaging of chlorophyll a fluorescence. Limnol Oceanogr 45:1420–1425
Pascal AA, Liu Z, Broess K, van Oort B, van Amerongen H (2005) Molecular basis of photoprotection and control of photosynthetic light-harvesting. Nature 436:134–137
Quilliam RS, Swarbrick PJ, Scholes JD, Rolfe SA (2006) Imaging photosynthesis in wounded leaves of Arabidopsis thaliana. J Exp Bot 57:55–69. ISSN 0022-0957
Scharte J, Schon H, Weis E (2005) Photosynthesis and carbohydrate metabolism in tobacco leaves during an incompatible interaction with Phytophthora nicotianae. Plan Cell Environ 28:1421–1435. ISSN 1365-3040
Schreiber U (2004) Pulse-Amplitude-Modulation (PAM) fluorometry and saturation pulse method: an overview. In: Papageorgiou GC, Govindjee (eds) Chlorophyll a fluorescence: a signature of photosynthesis. Springer, Dordrecht, pp 279–319
Swarbrick PJ, Schulze-Lefert P, Scholes JD (2006) The metabolic consequence of susceptibility and the activation of race-specific or broad-spectrum resistance pathways in barley leaves challenged with the powdery mildew fungus. Plant Cell Environ 29(6):1061–1076

The LISA and SOCRATEES© Approach for Sustainable Crop and Soil Management

Athanasios C. Gertsis and Christos Vasilikiotis

Abstract

This chapter outlines in the first part a holistic approach for sustainable agro-ecosystems management, using miscellaneous "tools" and the concepts of precision agriculture as underlying principles. Education and applied research are integrated to provide an efficient approach, through the concept of "LISA (Low-Input Sustainable Agriculture) and SOCRATEES (Soil Crop Atmosphere and Technology Educational Evaluation Systems)." Organic management approaches are also outlined in the second part.

Keywords

LISA · SOCRATEES© · Low inputs · Integrated management · Organic agriculture · Cover crops · No-till

1 Part A. LISA and SOCRATEES[1]©: An Introduction

An educational and research scheme developed for achieving sustainable crop production and copyrighted by the first author (A. C. Gertsis) will be outlined in this chapter, with emphasis on "bridging" **educational and research approaches at tertiary levels of educational institutions**. This proposed scheme incorporates a number of components within the entire "living" and "nonliving" (biotic and abiotic) continuum, where biological processes take place, but also human

[1] SOCRATEES (Soil Crop Atmosphere and Technology Educational Evaluation Systems) is an International copyrighted acronym by Dr. Athanasios C. Gertsis.

A. C. Gertsis (✉) · C. Vasilikiotis
Department of Agro-Environmental Systems Management, Perrotis College/American Farm School, 54 Marinou Antypa St., 60097, Thermi 57001, Greece
e-mail: agerts@afs.edu.gr; cvasil@afs.edu.gr

R. S. Sengar, A. Singh (eds.), *Eco-friendly Agro-biological Techniques for Enhancing Crop Productivity*, https://doi.org/10.1007/978-981-10-6934-5_5

interventions are simultaneously affecting most of the system's functions. The approach is considered herein as an expansion and upgrade on the old SPAC (soil-plant-atmosphere continuum) concept and incorporates new tools and techniques for enhancing crop productivity which became widely available in the last few years through the use of cutting-edge technology and the advent of precision agriculture applications. The proposal considers the following aspects in its development, to "enhance" crop productivity and develop a holistic or "systems approach" (Peart and Shoup 2004). The "enhancement of crop productivity" meant, in the entire text by the authors, as efforts to facilitate crop production management, strongly consider along the pathway any effects on the agro-ecosystem (environment of crop productivity), and increase the input use efficiency (IUE), which will lead to more profit to farmer level and hopefully to lower price at the consumer's level along with food safety:

- A lot of attention in education and applications of cutting-edge technologies is given during the last few years in developing "environmentally sustainable management systems"; thus, any proposed methodology should be very much focused on providing solutions at multiple levels of agricultural production. A "systems approach" or holistic scheme is critical and absolutely necessary to be integrated in any new management technique.
- Higher education programs (curricula) should be adjusted in the most current situations and market demands, and more emphasis should be given to integrated and multidisciplinary approaches and curriculum building.
- Environment-related legislation is becoming more demanding in terms of short-term accomplishments and more qualitative in goals; systems which are in line with future legislation should be developed.
- Cutting-edge technology application in environmental management advances in a high rate, but, at the same time, the cost of acquisition and use by the broader public and end users is decreasing at a similar rate.
- The modern era work in education and research requires more multidisciplinary teams and similar approaches to be developed, in order to handle the complex systems and problems for which solutions with an increasingly higher efficiency are required. An old approach in environmental situations where pollution problems are requesting "temporary" solutions was humorously stated as the "dilution is the solution to the pollution!" However, such an approach is not acceptable any more, since environmental and agricultural production-related problems have been excessively researched and better than "temporary" solutions are required and demanded for sustainability reasons.
- Agricultural management systems to be developed should focus on "the end user" but also should strongly consider the "education and technology carrier" who, in most cases, is the student interested to "transfer technology applications at the farmer's level." There have been a lot of efforts devoted on this "transferability pattern" without much success, however.

Low-Input Sustainable Agriculture (**LISA**) systems of agricultural production are being considered as management systems which optimize production input efficiency. They maximize the internal resources of a farm, eliminate or minimize environmental impacts, and increase farmer's profit by reducing purchased inputs (Lockeretz 1988). Sometimes these systems are considered as synonymous to **organic or biological agriculture**, but they do differ in many and significant aspects. In LISA systems, synthetic fertilizers and other production inputs can be used in conventional agriculture but in reduced amounts and in timely applications. Timing of input application as well as optimum amount is the key to successful application of LISA systems (Gertsis et al. 1988). The leaf water potential is a key indicator for water use efficiency (Gertsis et al. 1989). To do so, understanding the interrelationships of processes taking place within the entire continuum of plant's life (**soil-plant-atmosphere continuum**, **SPAC**) and the critical growth and developmental stages of each plant species is essential to provide optimum inputs. Unfortunately, each plant species possesses its own physiology and reacts differently than other plants at a given set of environmental parameters. This makes decision choices more difficult, since different sets of directions should be developed for each plant species.

The **SOCRATEES©** scheme (**So**il **Cr**op **A**tmosphere and **T**echnology **E**ducational/**E**valuation **S**ystems) helps make LISA systems feasible and more efficient. Educating students to become future researchers, agricultural systems managers, and/or successful farmers and environmentally sensitive managers in general involves the development of a sound and integrated knowledge background. Strong consideration to the currently available and cutting-edge technology should be given, for a curriculum to be successful. Outlining the major factors affecting crop growth and development is the first step to be followed by more insight information and understanding critical processes within each factor and the magnitude of their effect. An example scheme is outlined below, in order to demonstrate the complexity and the dynamic nature of the system where plants grow and develop.

Example
Soil factors affecting crop growth and development:

1. *Soil moisture*
2. Soil reaction (pH)
3. Soil's cation exchange capacity (CEC)
4. Soil aeration
5. Other factors:
    1.1. *Soil moisture effect on crop growth*
        1.1.1. Soil water availability (total water amount vs. available to crop water amount)
            1.1.1.1. Soil water potential critical levels for each species
            1.1.1.2. Soil water moisture characteristic curve
            1.1.1.3. Critical potential for photosynthesis (species dependent)

1.1.2. Seed germination requirements
  1.1.2.1. Optimum water for germination vs. excessive or too little water
  1.1.2.2. Management of water depth and preplanting conditions
  1.1.2.3. Soil conditions in the germination soil volume (microorganisms)
  1.1.2.4. Soil fertility levels in the germination soil volume
1.1.3. Initial seedling establishment requirements
1.1.4. Critical four-leaf stage (species depended)
1.1.5. Initial root system establishment
1.1.6. Continuous supply to meet the crops' demands (actual ET) during the entire biological cycle
1.1.7. Adequate supply at critical stages (vegetative vs. reproductive growth)

This list can be drawn to higher levels of complexity, and in reality this is almost the methodology used for building a crop-soil simulation model!

The SOCRATEES© approach aims to extend the **three-pillar system of SPAC** into a **four-pillar system** by adding the **technology** component, which became very significant in developing current management strategies. The technology is the key element of the new approach to environmental management and agricultural production, as expressed by precision agriculture techniques and technology use. Also the use of innovative technology is a key component to modern academic curricula (Gertsis et al. 1998, 2001, 2002; Graves et al. 2002). The **education** component is considered as the vital link to provide sustainability and efficiency on the entire approach. Educated and well-trained farmers will pursue and promote the best sustainable systems! Ultimately the entire approach can serve as **an evaluation tool** for identification of possible gaps within the system. The concept provides a multidimensional as well as multidisciplinary methodology for assessing all major aspects of the system under consideration, i.e., a cyclic process and a holistic view.

As mentioned in the preface, the "LISA and SOCRATEES©" scheme proposed, herein, is in fact a personal "philosophical" approach; therefore, the name of the great ancient Greek philosopher, **Socrates**, was used, in a symbolic way! This scheme is proposed to be used in updating and/or upgrading the educational background coursework currently required and methodologies used in most academic curricula and education in the tertiary education level, to satisfy professional skills and knowledge needed in life science subject areas. It also aims in encouraging researchers to use a holistic view approach by enhancing participation of many disciplines in the development of efficient management systems and enhance decision-making by a more global evaluation of the entire system.

2 Crop and Soil Simulation Models

Crop simulation models (CSMs) are important tools for agricultural systems management, and they are expected to become an indispensable part of the future decision-making process for sustainable management. As mentioned in the introduction of this publication, the author feels that the expression CSM is quite misleading for the wider audience or users, since it provides an incomplete picture of what actually CSMs are. Most of the CSMs are integrating processes and functions of the entire soil-plant-atmosphere continuum, and furthermore the programming code is mainly simulating soil processes. They could be more realistically called "soil-crop simulation models" (**SCSMs**), which better reflect their potential use and capacities.

They were developed in computer programs, using extended databases on crop growth and development; soil and atmosphere processes affecting plants, obtained in field; and growth chamber and greenhouse research, by multidisciplinary teams of scientists. Their potential use is also multidimensional, since they can be used for research (Whisler et al. 1986), educational (Gertsis et al. 1998), management (Gertsis and Symeonakis 1998), monitoring, evaluation and consultancy purposes (Gertsis and Whisler 1998).

Major reasons for building crop simulation models in biological systems are outlined by Whisler et al. (1986), in one of the first and most thorough reviews of this subject, as shown below:

- To aid in interpreting results from experiments (under field or control growth conditions)
- To provide additional agronomic research tools
- To provide farmers with an agronomic tool

Boote et al. (1996) add to the above reasons the following:

- For crop system decision management
- For policy analysis

Other possible uses and objectives of simulation models in physical and biological systems include, in the authors' view and experience:

- To enhance education providing new tools to understanding systems
- To identify knowledge gaps
- To integrate these tools with additional hardware and software technology and implement precision farming (as discussed more in the next chapter)
- To interface GIS with agronomic modeling (Hartkamp et al. 1999; Singh 1993)

A thorough review of the use and abuse and other aspects of crop simulation models was presented recently in a special issue of *Agronomy Journal* (September to October 1996, Vol. 88, No.5, pp. 689–716) by pioneers and leading scientists in the area of crop simulation (Baker, J., Passioura, J. Monteith, J., Sinclair, T.,

Seligman, N., Boote, K., Jones, J., Pickering, N.). A second set of papers appeared also in *Agron. J.* (93:650–670 (2001)) as a follow-up of the previous set, to include papers from Eastern Europe and the former Soviet Union.

The building process of a model consists of the following major steps in general:

1. Creating a conceptual model of the system under evaluation (Peart and Shoup 2004).
2. Identifying major variables involved in system's functions (research review)
3. Developing relationships between and among variables (literature review databases) and best-fit **mathematical equations (models**) to describe the above relationships as well as interactions
4. Writing the program code (development of the **source code**) to integrate the above relationships in a physical and sound sequence
5. Verifying the programming (debugging-**verification**)
6. Adjusting certain model parameters to make the model represent more realistically a process and the end output (**calibration**)
7. Validating the model with independent datasets under different conditions (an ongoing process model **validation**)

Some of the advantages of simulation models are that they:

- Enhance multidisciplinary action and collaboration.
- Enhance teaching of complex systems management.
- Predict processes and reduce cost of application.
- Reduce field experimentation to the most required level.
- Identify knowledge gaps in systems evaluation.
- Facilitate decision-making and optimize management.
- Enhance policy design and implementation in various scales.

Disadvantages of simulation models can be that they:

- Require extensive validation (Gertsis and Hodges 1985; Gertsis et al. 1996, 1997, 1998; Gertsis 1997) to become from site-specific to more global models).
- Have an initial high cost of establishment (require multidisciplinary teams).
- Present a risk of abuse by some users (when used by not adequately trained users).
- Require extensive and reliable data recording and analytical (correct statistical modeling) work.
- Used to require powerful computer and software, but lately (last decade) this is not any more a disadvantage.

Most of the above stated disadvantages are mainly related to the "abuse by humans" as is the case of most tools when not used appropriately. And let's not forget that with computers, it always happens that "garbage given as inputs will produce garbage outputs!"

There are currently a large number of models developed to evaluate the performance of crops grown under various environments. Also models exist to evaluate major processes in entire ecosystems or large land sectors. Most of the models still remain a tool useful mostly to researchers and scientists and have not been widely available to the end users such as the farmers and the students. A model widely used in the past by cotton farmers is the GOSSYM/COMAX set of a simulation model and an expert system (Gertsis et al. 1996, 1997; Baker et al. 1983; Lemmon 1986; McKinion et al. 1989; Reddy et al. 2002), which also has been one of the most universally validated crop growth simulation models. This model actually "trained" the farmers for a number of years to become efficient producers of cotton, and they are now capable to work by themselves and develop optimization strategies in cotton production.

A practical and easy model used by the farmers for soil organic management is the Soil Carbon Manager a decision support tool to assist in developing management strategies for sustainable agroecosystems. Version 1.00 (CRCSLM 1998) is the latter name of SOCRATES (Grace et al. 2006). The model predicts SOC changes under winter rainfall farming systems, for a range of soil types, climatic conditions, crops, and management practices. **SOCRATES, or Soil Organic Carbon Reserves And Transformations in EcoSystems**, is a process-based simulation model designed to estimate changes in topsoil SOC with a minimum dataset of soil, climate, and biological inputs (Grace et al. 2006). Main considerations in the development of SOCRATES were that it should be based on generic concepts of carbon cycling and biogeochemistry and easy to use and widely applicable. It would also not require detailed fractionations of carbon pools as inputs.

In general, the crop-soil simulation models have to be viewed as potential tools for enhancement of education, research, management, and policy decisions and result in enhancement of crop productivity.

3 Precision Agriculture

3.1 Introduction and Definitions

Precision agriculture (PA), precision farming (PF), site-specific management (SSM), site-specific farming (SSF), site-specific crop management (SSCM), farming by the foot, and precision management (PM) are all terms **almost synonymously used** to describe a new way of agricultural production management. In this publication, the term precision agriculture (PA) will be preferred over the others and used to describe the practices and approaches underlined in this Scheme. A *Searching Nomenclature for Precision Agriculture Resources on the Internet* is provided by Donna Schenck-Hamlin and Wenyan Zhan at Kansas State University, Digital Libraries Department (*http://www.oznet.ksu.edu/pr_prcag/links.shtml*) for those interested to extensively search related items in the web.

One of the first formal definitions of PA comes from the US House of Representatives (1997) and stated that "PA is an integrated information-and production –based farming system that is designed to increase long term, site-specific and whole farm production efficiency, productivity and profitability while minimizing unintended impacts on wildlife and the environment." In the European Union (EU), it is receiving significant attention lately, since Europe is lacking behind in farm applications of PA, and a study was submitted to EU officials to promote and enhance the use of PA.

In a recent report prepared for the EU agricultural sector and strongly considered for the next Common Agricultural Policy (CAP 2021–2018), PA is defined as "Precision agriculture (PA) or precision farming, is a modern farming management concept using digital techniques to monitor and optimize agricultural production processes" [EPRS (European Parliamentary Research Service) Scientific Foresight study, PE 581.892, December 2016]. This definition uses a statement, mostly focusing on optimization processes, while missing the environmental component of the earliest definition.

Precision agriculture encompasses a set of technology elements or tools (hardware and software) used to firstly **define and map** any **variability** existing within the field system (soil and crop), then quantify the variability into **management zones**, and finally **apply precisely** the required production inputs within each zone to obtain optimum results and **achieve system's sustainability**. Many similar definitions could be phrased in here, but all include the important keywords stated above and goals such as "system's optimization and sustainability." The meaning of optimization and sustainability is of paramount importance in the management practices used in PA. The result of PA application is an **improvement in the efficiency and environmental impact** of crop production systems.

In **conventional** or even **organic agriculture** practices, all inputs into the soil-plant system are still provided almost equally per unit area. Here lies the major difference in the new management provided by PA approach. It is well known that the variability within soil is tremendous, even when very small areas are considered (parts of a farmer's field). The variability of a crop grown at a soil type is usually less per unit area than that of the same soil area, mainly because the plant "buffers out" part of the soil's variability in fertility and water availability through its physiological mechanisms of adaptation, tolerance, and resistance. However, in case of a disease, pest problem, or weeds, the variability is significant and must be considered in any management system used. It is the **total variability of the system** that PA techniques are designed to manage in an optimal way.

In most publications related to PA, the main emphasis is given to the soil's inherent variability. It is, however, of outmost importance to consider the variability shown at the crop level, due not only to soil-related factors and processes but also to the environmental variables affecting plant growth and development and to other biological factors. In this aspect the "LISA and SOCRATEES©" approach is used in conjunction with the principles of PA. This publication emphasizes a more integrated approach to define variability within the entire system of crop productivity.

Recent research in soil-related properties has resulted in production of **soil properties** sensoring devices which can relatively fast and with low cost identify variability within the soil medium. **Crop properties sensors** also provide information on the crop status per unit land. **Yield monitors** used to create **yield maps** are coupled with maps showing soil and crop variability Next, relationships are evaluated and assessed between soil and crop productivity variables, and multiple criteria are established to provide optimum management applications and ultimately **precise input application**, through **variable-rate technology** (**VRT**).

3.2 Major System Components of Precision Agriculture

In the last decade, PA has emerged from a conceptual model and multiple-task operations implemented into a farmer's field-scale operation level. As stated earlier, PA consists of hardware and software components. The most important of these components are briefly described below.

- Accurate positioning systems in land areas is the foreground of PA. This is achieved with the **Global Navigation Satellite Systems (GNSS)** such as the US system, Global Positioning Systems (GPS), the Russian GLONASS, the European EGNOS, etc., where the global coordinate system of longitude (x-axis) and latitude (y-axis) is used to provide a unique position (coordinates) for each land unit. The GPS is a combination of software and hardware parts. The current cost of a handheld GPS unit is in the order of 100 $–300 $.[2] The precision of these systems is continuously improved.
- Ground and aerial data collection is done with specialized **sensors,** which monitor soil and crop parameters, such as soil moisture, soil electrical conductivity, soil pH, leaf chlorophyll, crop chlorophyll and reflectance levels (such as the NDVI – normalized difference vegetation index), crop temperature, weed and disease rate, and **yield monitors**. Other data collection which describe properties within the SPAC are the **aerial** (*variously called UAS* (*unmanned aerial systems*) *or UAVs* (*unmanned aerial vehicles*) *or drones*) **and satellite images** obtained at certain periods during the biological cycle of a crop species and present valuable information on the health status and yield potential of the crop.
- All the previous data and information collected are managed with the help of specialized software called **geographic information systems (GIS)**. These software are capable of analyzing the data in multiple layers and levels and provide optimal decisions through integration of all data and properties. **Digital maps** are created to present and visualize any variability exhibited in the crop production system, which reveals **potential risk areas** for management practices to be applied (Westetvelt and Reetz 2000) and yield maps provide essential information on the production potential. An example is presented from the first wheat yield map produced in Greece, at the Perrotis College's educational program (Gertsis et al. 2004).

[2]*An exchange rate of ca. 1,14 $ = 1 € is a current estimate by the time this publication was written.*

- With all the above materials and methods, management zones are established based on significant variability shown. The concept of **management zones** is considered as the major goal of PA techniques. Therefore, special attention should be given in the **criteria used to develop zones** if any variability exists. These criteria should include both **soil and crop parameters** affecting the productivity of the agricultural system. A free access software that facilitates investigation of optimum number of management zones was created by Fridgen et al. (2003) and is called Management Zone Analyst (MZA). Recently, EZZone (Lowrance 2012), also a similar open access software, was developed in the University of Georgia, United States. This is a free online resource for creating Agricultural Management Zones from univariate georeferenced data. The goal is to create zones that are useful to farmers by simplifying zone creation and creating zones that conform to row spacing and equipment specifications [http://ezzone.pythonanywhere.com].
- The ultimate application of PA is the **variable-rate technology** used to provide accurate (precise) inputs within each management zone as defined in earlier steps.

3.3 Management Zones

"Management zones are regions of a farm field that have been differentiated from the rest of the field for the purpose of receiving individual management attention." The simplicity of this definition and the wide breadth of interpretation are key to the power of this management tool. For those who want a "cookbook" approach to precision agriculture, it is also a source of some confusion. **There are no set rules for creating a management zone**. The specifics of a particular management zone can only be determined on a site-specific basis and must include all the factors that affect management on a farm. This means that the zones are not solely functions of field's physical properties. Chemical soil properties and crop characteristics are also considered along with the microclimate in a spatial mode. In addition management zones must strongly consider all relevant inputs from the sociopolitical-economic-agroecological system that is a modern farming operation (Vellidis and Gertsis 2002; Ess and Morgan 2003).

This is one of the fundamental elements of PA and the major difference from all conventional and organic agriculture production systems. Various questions need to be answered before accepting the final management zones area. According to Fridgen et al. (2003) *three questions arise when considering managing by zones*:

One, *what information should be used as a basis for creating different management zones within a field?*

Two, *how can information be processed into unique management units (i.e. procedures for classification)?*

And **three**, *how many unique zones should a field be divided into?*

To the author's experience, some additional considerations should include the **crop species** to be grown as well as the **rotation schemes** to be followed. It is very important to consider the rotation scheme, when designing the zones, since the working system we are called upon to investigate is very complex; we need to provide fast but efficient enough procedures, and at the same time, the cost should be considered. One of the objectives of this publication is to propose such simple fast low-cost procedures for management zone delineation.

A major gap in the MZ approach is the lack of considering more than one important variable to establish MZ with multiple criteria. A research project in **Perrotis College (Greece)** introduced the concept of integrated management zones (IMZ) through a software "Digital Precision Agriculture" (Gertsis 2012). The software was developed through a field research project to attempt establishment of integrated management zones (IMZ) which include simultaneously more than one fundamental soil property affecting crop growth and yield. The model requires inputs from fundamental but easily measured soil properties affecting crop growth and yield taken from sampling an appropriate number of points per farm. Currently the model accepts data on four soil properties significantly affecting crop growth and productivity: **pH, SOC, EC, and soil texture**. The software provides also for future use of climatic factors variables, and so **rainfall and temperature** data could be implemented in the next version of it, since there is provision in the source code of the model for these two variables. The current version of the model considers the variability of the soil system and provides an integrated methodology to weigh each parameter measured, in order to provide the "integrated management zones" (IMZ) and a map of each field. The critical values and ranges for each of the four soil properties were established based on soil science literature-related review but are adjustable at the user's choice. It outputs integrated digital maps of relative homogeneous management zones (MZ) of the input soil properties, which can be used for optimization of input use efficiency (IUE) and therefore provide higher income and cause less adverse environmental effects. In addition the software inputs the used crop species and determines the appropriateness of the given crop under the soil properties, based on an initial database of crop species. It suggests alternative crop species that are better adaptable at the given soil environment.

3.4 Criteria for Establishment of Management Zones

Unfortunately not much work is done in establishing fixed criteria on the magnitude of soil or crop parameters for differentiating the management zones. Some criteria used are outlined below:

- **Soil texture**: based on the 12 classes outlined in the USDA scheme (or other systems such as the IUSS (International Union of Soil Sciences), 10 classes in base system).

- **Soil water holding capacity**: each textural class provides different water holding capacity, but a generalization can reduce the SWHC classes to about 4–5 classes. The critical number however remains the range of soil moisture within each textural class. For instance, if a soil is a **sandy loam** texture, then what is the level of difference (% moisture) that can differentiate two soil moisture zones within this textural class? One answer can be derived from the FC and PWP range of this soil, which determines the available to plant soil water. But this approach still does not take full under consideration the stress to be produced to crops grown under a range of soil moisture regime within the same texture, because the AWC defined as the difference between FC and PWP is not a realistic approach; it is the critical soil water potential for each crop species which makes the difference in stress levels.
- **Soil pH**: the chemical reaction of soils is very important since it greatly affects soil nutrient availability and defines the optimum range for each plant species. Information of optimum pH ranges are available for most of the cultivated species, so this soil property can be used as a standard criterion for a management zone differentiation. The other advantage of using this property as a standard criterion is that it does not change in time, unless significant management practices (mainly fertilization) are used.
- **Soil organic matter**: some sensors are under development to measure this very important topsoil property, based on light reflection of the soil.
- **Crop canopy color**: this is a parameter which can be used to identify stress conditions occurring in the crop system. Strongly associated with this property is **crop temperature**. Farmers empirically decide on management practices very often based on these crop properties. Therefore any system used to establish management criteria should consider strongly these crop factors. Satellite imagery and aerial pictures are used to assess the crop conditions based on canopy color.
- **Leaf chlorophyll level**: this crop parameter indicates nutrient status, mainly nitrogen, in the entire crop. It was shown to be highly correlated with the final yield in many species, since N is the major plant nutrient used for management decisions.
- Sensors which measure **other important soil and crop variables** may be developed in the near future, and they can assist in more integrated evaluation efforts.
- Lastly, **atmospheric parameters** such as air temperature, relative humidity, solar radiation, wind velocity, and rainfall are of immense importance in management decisions. Fortunately, these parameters are currently very easily monitored using inexpensive equipment, as compared to the near past.

3.5 Variable Rate Technology (VRT)

The final stage of PA strategies, based on well-defined management zones, is the precise input application given at each management zone. It involves the **most expensive part of technology** of all the previous components, and this is where a lot of work is currently in progress.

There are two basic types of VRT, **sensor-based and map-based**. Sensor-based or "on-the-go" systems have the ability to sense the level of input which is required as the equipment is actually moving across a field and are recently becoming very much commercially used.

3.6 Low-Cost But High-End Technology Used

More and more low-cost technology is available to the public for small- or large-scale application of PA methodologies. The cost in general has been reduced by multiple times the last 10 years in major components of PA.

The initial notes of this text included the progress of PA in the middle 2000s. In the last few years (>2014), there has been a monumental development in commercial applications which are expected to further grow in the next 5 or more years. More of the major agricultural equipment companies already developed "custom" equipment for PA applications, such as nitrogen sensors for precision N application using "on-the-go" systems, precision herbicide sensors, auto-steer or auto-guidance systems, and agricultural robotics. The very recent advent and intense commercialization of UAVs (unmanned aerial vehicles or UAS systems or drones) expanded the remote sensing facilities from the expensive level of satellite images to very economic and flexible aerial pictures of high reliability for crop management practices.

3.7 PA Applications in Crop Species

Initially PA was applied in field crops, but recently more crop species (field crops, vegetables, trees (Gertsis et al. 2013) are added in the long list. The large area farms are the cases where PA has shown the most financial benefits, although successful stories at small farm scale are feasible. The reduced costs of PA technologies are becoming attractive for small-scale farmers.

4 Part B. Organic Agriculture Approaches for Sustainable Management

4.1 Organic Agriculture and Sustainable Management of Agriculture

Organic agriculture is based on a systems approach to agriculture and views crop production as an integrated part of the surrounding environment. Similarly with LISA systems, the goal of organic agriculture is to optimize crop production while, at the same time, preserving the function of natural ecosystems and minimizing environmental impacts. In addition, organic farming strives to enhance natural resources and improve the livelihoods of farmers, farmworkers, and rural communities.

As organic agriculture prohibits the use of synthetic fertilizers and pesticides, it relies primarily on the efficient use of internal resources. The use of naturally occurring substances for fertility and the avoidance of synthetic fertilizers, which have a high embodied energy, result in more energy-efficient farming systems.

In addition to LISA and SOCRATEES systems and the precision agriculture technologies mentioned in the previous sections, organic agriculture is employing two farming practices that contribute to the long-term sustainability of organic farming systems. The use of cover crops has been a basic practice of organic farming since its inception, while the use of no-till practices in organic agriculture has only been developed recently.

Cover Crops

Two key issues in organic agriculture are weed management to prevent competition with cultivated crops and efficient use of internal resources for maintaining soil fertility. The use of chemical herbicides is disallowed in organic agriculture, creating a great need to enlist cultural or mechanical methods to control weeds. Similarly, for maintaining soil fertility, organic farmers cannot rely on chemical fertilizers, which have become the dominant form of soil fertility in conventional agriculture. A widely used cultural method for weed control and soil fertility management based on ecological principles is the use of cover crops (Hartwig and Ammon 2002). Cover crops are crops planted with the sole purpose of protecting the soil, competing with weeds and improving soil quality and fertility (Dabney et al. 2001).

Cover crops have been used for both arable farming systems and orchards. Cover crops can be integrated into a variety of cropping systems during the periods when no main cash crop is grown. They can also be used in vegetable production systems as part of crop rotation. Perennial cover crops are commonly used in perennial cropping systems such as orchards, while annual cover crops are used in annual cropping systems (Hartwig and Ammon 2002). Annual cover crops can be either cool-season annuals, which are planted in the fall and terminated in the spring, or warm-season annuals, which are planted in the spring and grow through the summer. Cool-season cover crops commonly used are rye (*Secale cereale* L), triticale (×Triticosecale Wittm.), oats (*Avena sativa* L), ryegrass (*Lolium* spp.), barley (*Hordeum vulgare* L), hairy vetch (*Vicia villosa* Roth.), common vetch (*Vicia sativa*), Austrian peas (*Pisum sativum* spp. arvense (L.) Poir. or *Pisum arvense* L), and annual clovers (*Trifolium incarnatum* L, *Trifolium squarrosum*, *Trifolium berseem*) (Clark 2008). Warm-season cover crops are buckwheat (*Fagopyrum esculentum* Moench), phacelia (*Phacelia tanacetifolia* Benth.), and Sudan grass (*Sorghum bicolor* var. sudanese) (Miyazawa et al. 2014; Brust et al. 2014 a, b). Cover crops can either consist of single-species or multi-species mixtures of grasses and legumes.

The dominant cover crop systems in the Mediterranean region are single-species cover crops. Oats, ryegrass, rye, triticale, and barley are quite common graminoid crops. Common leguminous cover crops are field peas, common vetch, hairy vetch, and clovers (Dabney and Delgado 2010; Campiglia et al. 2010). Vetches are usually planted in mixtures with a small percent of a graminoid crop (oat or barley) to provide structural support and uptake excess nitrogen from the soil. The traits that are important for cover crop species are quick establishment, high biomass production, high nitrogen contribution, competitiveness with weeds, cold tolerance, and ability to be terminated with mechanical methods if they are used in organically managed systems.

Cover crops help control weeds, improve soil structure and quality, and enhance biological control by increasing biodiversity and attracting beneficial insects. Legume cover crops enrich the soil with nitrogen through nitrogen fixation and enhance biological control by increasing habitat for beneficial insects (Parr et al. 2011). Brassica cover crops can suppress soilborne diseases and plant-parasitic nematodes (Snapp et al. 2005).

Grasses and brassicas improve N-cycle and scavenge residual nitrogen from the soil preventing leaching and pollution of water sources (Dabney and Delgado 2010). Cover crops play a significant role in protecting topsoil. Soil erosion is a serious problem worldwide, particularly in the United States, where an estimated 30% of topsoil has been lost in the past 200 years due to farming practices that keep the soil bare during the off-season (Tyler et al. 1994). In the Southeastern United States, cover crops were found to reduce soil erosion by as much as 62%, compared to bare soil (Langdale et al. 1991). Cover crops can also help conserve soil moisture by reducing water evaporation and increasing soil infiltration, reducing water stress for many crops at critical stages of their development increasing yields.

Grass cover crops have low cost and provide quick weed suppression and high biomass production. In addition, they can uptake nitrogen and reduce leaching (Tonitto et al. 2006). Legumes increase soil fertility through their symbiotic relationship with rhizobacteria. Some cover crop species, such as rye, triticale, barley, and brassicas, have strong allelopathic properties and reduce germination and growth of weed species (Dhima et al. 2006). This is advantageous for weed control, and even though it can potentially interfere with seed germination of a following crop, it was not found to interfere with corn germination and growth (Dhima et al. 2006). Both hairy vetch and common vetch have also been shown to have a significant reduction of weeds when used as cover crops. In a study of organic pepper production following cover crops, hairy vetch was found to produce the highest pepper yields (Isik et al. 2009). Ryegrass and oats were also found to be competitive with weeds, while ryegrass produced the highest biomass (Isik et al. 2009). In an evaluation study of cover crop mixtures in vegetable systems in Ohio, rye, barley, crimson clover, and hairy vetch were found to be competitive with seeds and overwinter reliably and were easily terminated (Creamer et al. 1997).

Cover crop effectiveness depends on the consistent and optimal performance of the selected cover crop species. To increase consistency and resilience, an ecologically sound cover crop system should involve a more diverse mixture of both legumes and graminoids. A multi-species cover crop would have the advantage of combining a diversity of species, each with different functional traits contributing to a more efficient resource use and providing a higher degree of weed control. Multi-species cover crops can exhibit a faster canopy closure, have a higher LAI, and produce higher aboveground biomass when compared with single cover crop species, resulting in a better competitive ability against weeds (Uchino et al. 2011).

Cover crop mixtures can provide a more efficient utilization of resources. Deep-rooted cover crops grown with shallow-rooted cover crops can uptake water and nutrients more efficiently by exploring a larger section of the soil profile. Nutrient cycling can also be enhanced by the use of cover crop mixtures. Combining cereals and grasses with a high C:N ratios with legumes that have a lower C:N ration can improve mineralization rates of crop residues (Dabney and Delgado 2010). Timing cover crop kill is critical to prevent soil moisture depletion by the actively growing cover crop. This should be balanced with the need to achieve optimal N accumulation by legume cover crops.

Most research on multi-species cover cropping systems is limited to two-species cover cropping systems. A recent study on spring sown multi-species cover crop mixtures for the western corn belt found that mixtures were more productive than the individual cover crops and exhibited an enhanced ecological resilience and adaptability (Wortman et al. 2012). Another study of warm-season cover crops found that even though multi-species mixtures are over-yielding compared to the individual species grown as single cover crops, they did not result in increased weed suppression, biomass stability, or enhanced productivity of the subsequent crop (Smith et al. 2014). Further research is necessary to determine optimal mixture combinations of cover crops under different climates and management methods. Finally, the timing and method of cover crop termination is critical for their successful adoption as a weed suppression method organic agriculture (Wortman et al. 2013).

Although cover crops provide beneficial functions to arable farming systems and orchards, they have potential drawbacks that need to be carefully assessed before incorporating cover crops in a farming system. Cover crops require additional cost for establishment, termination, and management of their residues. Including cover crops in a cropping system can delay planting of the following main crop, result in nitrogen immobilization, and provide habitat for pests or diseases (Dabney et al. 2001). While cover crops increase soil water holding capacity, in arid and semiarid regions, they can compete for soil moisture and reduce available soil moisture for the following crop (Dabney et al. 2001). This can be an issue with warm-season cover crops, but not in the case of cool-season cover crops in Mediterranean-type climates, where there is adequate precipitation and soil moisture during their growth season.

4.2 Organic No-Till Agriculture

One of the most pressing problems facing modern agriculture is soil degradation and erosion. The FAO estimates that nearly 220 million hectares of land in Europe are moderately to severely degraded (FAO 1998). The destruction of forests, the excessive use of pesticides and fertilizers, the spread of monocultures, and the intensive tillage practices all contribute to this problem. The cost of soil degradation and erosion in Europe is estimated by the FAO to be at about 10% of agricultural production value each year.

Throughout the world, the problem of soil erosion has led to the development of new farming technologies, termed conservation agriculture. One of the most promising of these new technologies is the practice of no-till (no-tillage or zero-tillage). Intensive tillage for planting preparation and weed control and the lack of protective plant cover are both important factors, which contribute to soil erosion and soil degradation. In no-till farming, a cover crop (or green manure) is planted in the fall to provide continuous soil cover during winter and early spring. In the spring, the cover crop is terminated and a main cash crop is planted into the plant residue that remains on top of the soil as protective mulch. This high-residue mulch preserves soil moisture, reduces soil erosion, and enhances soil building (Sapkota et al. 2012). In addition, it prevents weed growth, thus minimizing the need for herbicides and field cultivation. During the past few decades, no-till farming practices have been successfully applied worldwide with more than 90 million hectares currently under no-till production. The leading country is the United States, where 23 million ha are under no-till production, accounting for 16% of its arable land. Europe has been much slower in adopting these new farming practices.

No-till farming practices have a number of advantages over conventional tillage, including increased soil moisture retention, reduced soil erosion and loss of top soil, higher soil organic matter, and increased soil microbial diversity. No-till farming provides the additional benefits of weed suppression and minimizing inputs, as well as reduced fuel use (no-till farmers can limit tractor traffic to only one pass, compared to the six to eight passes conventional tillage requires). However, no-till farming applications have relied so far on the use of herbicides for killing the cover crop which increase the costs of production and have a negative impact on the environment and human health. Herbicides are prohibited in organic farming, preventing organic growers from adapting these soil conservation practices. In an effort to eliminate chemical inputs in agriculture and to allow the use of no-till in organic farming, researchers in the United States and Brazil have been experimenting with mechanical ways to kill the cover crop (Creamer and Dabney 2002; Sayre 2003). In addition to the traditional ways of killing cover crops by mowing, they have developed a variety of roller-crimper attachments that allow farmers to knock down the cover crop without cutting it, forming thick mulch (Creamer and Dabney 2002; Sayre 2003).

This innovative farming practice has allowed organic farmers to adopt no-till farming and help conventional farmers to apply no-till farming with minimal use of herbicides. This roller-crimper can be combined with a no-till planter that can direct seed into this high-residue mulch, allowing for knockdown of the cover crop and planting of main crop in a single pass. The use of a roller-crimper in no-till farming has the added advantage that the resulting high-residue mulch decomposes much slower than a mowed cover crop, providing better weed control and soil building. The resulting reduced tractor traffic in the fields minimizes soil compaction and prevents the creation of a hardpan below the "plow layer," as well as decreased fuel use (and cost) for the farmer. Any technology that would allow farmers to minimize equipment traffic can have an important impact on their economic sustainability.

Ongoing research at the USDA has found promising applications of no-till. When hairy vetch was killed in no-tillage systems, the resulting high-residue mulch was shown to protect the soil from erosion, improve conservation of soil moisture during summer months (Clark et al. 1997), and suppress weeds (Teasdale and Rosecrance 2003). In organic vegetable production, cover crop use for weed suppression has shown a great promise, but success of maintaining high yields in organic no-till is mixed (Carr et al. 2013). Winter annual cover crops, such as rye and hairy vetch, terminated with a roller-crimper were found to provide efficient weed suppression and increase soil quality, but vegetable yields were reduced (Leavitt et al. 2011).

One of the biggest concerns with no-till is the incomplete knockdown of the cover crop which would then regrow and have negative impact on the main crop yields. Teasdale et al. have found that using a stalk chopper can effectively kill a hairy vetch cover crop (Teasdale and Rosecrance 2003). The use of a roller-crimper was also shown to be very effective in killing rye, wheat, and black oat cover crops (Clark and Panciera 2002; Kornecki et al. 2009). There was little cover crop regrowth, and rye mulch effectively suppressed weeds early in the corn season (Clark and Panciera 2002). The growth stage of the cover crop plays a crucial role in the ability of mechanical killing. When a hairy vetch cover crop was rolled down during its vegetative stage, only 19% was killed, whereas if the cover crop was rolled down after flowering, more than 90% was killed (Creamer and Dabney 2002).

No-till trial for organic production at the Rodale Institute has been refining their roller-crimper design, and they have been successful in growing corn and soybeans organically with a no-till system (Laura Sayre 2003). They have been using legumes and legume/grass mixes as cover crops combined with corn as the summer crop or a straight grass such as rye, wheat or barley followed by soybeans. Their research has shown that the optimal goal is not to cut the stems of the cover crop, just to crimp them and knock them flat on the ground. The most critical factor affecting the ability of the cover crop to regrow and become a weed is the growth stage of the cover when it is rolled. Research at the Rodale Institute has shown that the cover crop must reach full flowering stage; otherwise it is still in vegetative growth stage and it would regrow (Creamer and Dabney 2002). The ideal stage is when there is a 50–75% bloom, even though crops knocked down when at 20% bloom also gave relatively good results (Sayre 2003). Another concern is to make sure that seeds have not reached maturity so they cannot create a weed problem later. More recent studies have shown that using cover crop thermal-based phenological models can predict the most effective roller-crimper termination dates (Mirsky et al. 2009, 2012).

5 Conclusions

As outlined in the above parts, there are enough technological tools to enhance crop productivity in current years with a very promising trend for the next few years. An almost exponential growth of PA-related technologies exists in the last few years, with a trend to further increase for at least the next 5 years (to 2022).

Sustainable crop productivity must be enhanced, simultaneously, with environmental and food safety considerations to feed the increased human and animal population in Earth. The new management approaches are necessarily linked with the use of cutting-edge technology. PA will soon become a "one-way" approach to achieve all the mentioned goals in crop productivity. Equally valued and significant technological applications already exist for enhancing animal production systems, although this is not part of this chapter's discussion. Aiming at increasing crop, input use efficiency and sustainability through the use of holistic and low-input approaches (i.e., LISA and SOCRATEES©) must be utilized. Needless to mention that the "soil" component of the continuum system must receive greater attention, in terms of management and education. The new technologies can be applied to any system of agricultural production (integrated or organic) and at any scale, despite the fact that – currently at least – it is more feasible financially in large-scale agriculture. Technologies coupled with an environmental sensitivity attitude will end in the most optimum result for all society. Complementary weed management and soil conservation strategies, such as the use of cover crops and no-tillage, show a great promise for both integrated crop management and organic agriculture. Both forms of agricultural production systems (integrated crop management and organic agriculture) must coexist to provide a positive symbiotic action in nature and provide incentives to both types of producers. Technologies can be used in both systems and be equally effective. However, the value of "education of users in PA" is of outmost importance (Kitchen et al. 2002; Gertsis et al. 2001) for a more efficient application of the above!

References

Baker DN, Lambert JR, McKinion JM (1983) GOSSYM: a simulator of cotton crop growth and yield, Technical bulletin 1089. South Carolina Agricultural Experiment Station, Clemson

Boote KJ, Jones JW, Pickering NB (1996) Potential uses and limitations of crop models. Agron J 88:704–716

Brust J, Weber J, Gerhards R (2014a) Do cover crop mixtures have the same ability to suppress weeds as competitive monoculture cover crops? Julius Kühn Archiv (443):422–430

Brust J, Claupein W, Gerhards R (2014b) Growth and weed suppression ability of common and new cover crops in Germany. Crop Prot 63:1–8

Campiglia E, Caporali F, Radicetti E, Mancinelli R (2010) Hairy vetch (*Vicia villosa* Roth.) cover crop residue management for improving weed control and yield in no-tillage tomato (*Lycopersicon esculentum* Mill.) production. Eur J Agron 33(2):94–102

Carr P, Gramig G, Liebig M (2013) Impacts of organic zero tillage systems on crops, weeds, and soil quality. Sustainability 5(7):3172–3201

Clark A (ed) (2008) Managing cover crops profitably. Diane Publishing

Clark S, Panciera M (2002) Cover crop roll-down for weed suppression in no-till crop production. Fruit and vegetable crops research report

Clark AJ, Decker AM, Meisinger JJ, McIntosh MS (1997) Kill date of vetch, rye, and a vetch-rye mixture: I. Cover crop and corn nitrogen. Agron J 89(3):427–434

CRCSLM (1998) Soil carbon manager. A decision support tool to assist in developing management strategies for sustainable agro-ecosystems. Cooperative Research Centre for Soil and Land Management, Adelaide. http://www1.crca.asn.au/resource_materials/Issue3.pdf

Creamer NG, Bennett MA, Stinner BR (1997) Evaluation of cover crop mixtures for use in vegetable production systems. HortScience 32(5):866–870
Creamer NG, Dabney SM (2002) Killing cover crops mechanically: review of recent literature and assessment of new research results. Am J Altern Agric 17:32–40
Dabney SM, Delgado JA, Reeves DW (2001) Using winter cover crops to improve soil and water quality. Commun Soil Sci Plant Anal 32(7–8):1221–1250
Dabney SM, Delgado J (2010) Using cover crops and cropping systems for nitrogen management. In: Advances in nitrogen management. Soil and Water Conservation Society, Ankeny, pp 230–281
Dhima KV, Vasilakoglou IB, Eleftherohorinos IG, Lithourgidis AS (2006) Allelopathic potential of winter cereals and their cover crop mulch effect on grass weed suppression and corn development. Crop Sci 46(1):345
EPRS-European Parliamentary Research Service –. Scientific foresight study PE 581.892, December 2016
Ess D, Morgan M (2003) The precision farming guide for agriculturists. Deere & Company, One John Deere Place, Moline, ILL., 61625, USA
FAO (1998) Soil protection needs more attention in Europe http://www.fao.org/WAICENT/OIS/PRESS_NE/PRESSENG/1998/pren9836.htm Online Press Release. Last accessed 20 Mar 2017
Fridgen JJ, Kitchen NR, Sudduth KA, Drummond ST, Wiebold WJ, Fraise CW (2003) Management Zone Analyst (MZA): software for subfield management zone delineation. Agron J 96:100–108
Gertsis A (1997) Optimizing cotton nitrogen fertilization for sustainable production in Greece. In: Proceedings of 11th World Fertilizer Congress of CIEC, vol III, 7–13 September 1997, Gent, Belgium, pp 292–300
Gertsis (2012) "Digital precision agriculture", a software to manage soil information for establishment of integrated management zones. In: Agro environ, international conference, Wageningen, October 2012. https://library.wur.nl/ojs/index.php/AE2012/article/view/12417
Gertsis AC, Hodges H (1985) Validation of GLYCIM-a soybean crop simulator. In: Agronomy abstracts, annual meetings of ASA, CSSA, SSAA, December 1–6, 1985 Chicago, p 11
Gertsis A, Symeonakis A (1998) Efficient cotton production in Greece with the use of GOSSYM-a cotton crop simulation model. International symposium MODEL IT, Wageningen, The Netherlands, November 29–December 2, 1998. ActaHorticulturae 476, pp 307–312. ISBN90 6605 940 0
Gertsis A, Whisler FD (1998) GOSSYM: a cotton crop simulation model as a tool for the farmer. In: International symposium MODEL IT, Wageningen, The Netherlands, November 29–December 2, 1998. ActaHorticulturae 476, pp 213–217. ISBN90 6605 940 0
Gertsis A, Krieg DR, Hatfield JL (1988) Management strategies for maximum cotton production on the southern high plains of Texas. III. Water use and water use efficiency considerations. In: Proceedings – Beltwide cotton production research conference, New Orleans, pp 73–75
Gertsis AC, Krieg DR, Hatfield JL (1989) Leaf water status of cotton as affected by water supply and evaporative demand. In: Agronomy abstracts, annual meeting of ASA, CSSA, SSAA, Las Vegas, October 15–20, 1989, NV, p 112
Gertsis AC, Galanopoulou-Sendouca S, Papathanasiou G (1996) Use of a cotton growth simulator – GOSSYM – to evaluate leaf area development as affected by seasonal accumulation of heat units. In: Proceedings of "International symposium on Applied Agrometeorology and Agroclimatology" Volos, Greece 24–26 April 1996. ISBN92-828-4137-5, EIR 18328 EN (Office of Official Publications of the European Communities)
Gertsis AC, Galanopoulou-Sendouca S, Papathanasiou G (1997) Use of GOSSYM – a cotton growth simulation model – to manage a low input cotton production system in Greece. In: Proceedings of first European conference for information technology in agriculture, Copenhagen 15–17 June 1997, Denmark, pp 359–363
Gertsis A, St Galanopoulou, Whisler FD, Papathanasiou G , Symeonakis A (1998) Evaluation of GOSSYM – a cotton crop simulation model-under the soil, climatic and cultural conditions

in Greece. In: Proceedings of 2nd world cotton research conference, 16–22 September 1998, Athens, Greece

Gertsis A, Vellidis G, Vergos E, Kanemasu E (2001) The establishment of precision agriculture course in the regular curriculum of a two-year post secondary agricultural program in Greece. In: Book of abstracts, 3rd European conference on precision agriculture, 18–20 June 2001, Montpellier, France, pp 26–27

Gertsis A, Vellidis G, Vergos E (2002) Principles of precision agriculture: establishment of a new course in an innovative curriculum at the Dimitris Perrotis College of Agricultural Studies. In: Sideridis AB, Yalouris CP (eds) Proceedings of 1st conference of Hellenic Association of ICT in Agriculture, Food and Environment (HAICTA), 6–7 June, Athens Greece, pp 88–92

Gertsis A, Vellidis G, Rains G (2004) Applications of precision agriculture in Ellas (Greece): wheat yield mapping. In: Vlachopoulou M, Gertsis A, Manthou V, Illiadis L, Salampasis M (eds) Proceedings of the 2nd international conference of HAICTA, 17–19 March 2004, Thessaloniki, Greece, pp 293–298

Gertsis A, Fountas D, Arpasanu I, Michaloudis M (2013) Precision agriculture applications in a high density olive grove adapted for mechanical harvesting in Greece. Procedia Technol 8:152–156

Grace P, Ladd JN, Robertson GP, Gage SH (2006) SOCRATES – a simple model for predicting long-term changes in soil organic carbon in terrestrial ecosystems. Soil Biol Biochem 38:1172–1176

Graves AR, Hess T, Matthews RB, Stephens W, Middleton T (2002) Crop simulation models as tools in computer laboratory and classroom-based education. J Nat Resour Life Sci Educ 31:48–54

Hartkamp AD, White JW, Hoogenboom G (1999) Interfacing geographic information systems with agronomic modelling: a review. Agron J 91:761–772

Hartwig NL, Ammon HU (2002) Cover crops and living mulches. Weed Sci 50(6):688–699

Isik D, Kaya E, Ngouajio M, Mennan H (2009) Weed suppression in organic pepper (*Capsicum annuum* L.) with winter cover crops. Crop Prot 28(4):356–363

Kitchen NR, Snyder CJ, Franzen DW, Wiebold WJ (2002) Educational needs of precision agriculture. Precis Agric 3:341–351

Kornecki TS, Price AJ, Raper RL, Arriaga FJ (2009) New roller crimper concepts for mechanical termination of cover crops in conservation agriculture. Renew Agric and Food Syst 24(03):165

Langdale GW, Blevins RL, Karlen DL, McCool KK, Nearing MA, Skidmore EL, Thomas AW, Tyler DD, Williams JR (1991) Cover crop effects on soil erosion by wind and water. In: Hargrove WL (ed) Cover crops for clean water. Soil and Water Conservation Society, Ankeny, pp 15–22

Leavitt MJ, Sheaffer CC, Wyse DL, Allan DL (2011) Rolled winter rye and hairy vetch cover crops lower weed density but reduce vegetable yields in no-tillage organic production. Hortscience 46(3):387–395

Lemmon H (1986) COMAX: an expert system for cotton crop management. Science 233:29–33

Lokeretz W (1988) Open questions in sustainable agriculture. Am J Altern Agric 3(4):174

Lowrance C (2012) Open source hardware and software in agriculture: an autonomous sap flow measurement wireless network & a user friendly management zone delineation tool. MSC thesis (supervised by Dr. G. Vellidis), University of Georgia, GA, USA

McKinion JM, Baker DN, Whisler FD, Lambert JR (1989) Application of the GOSSYM/COMAX system to cotton crop management. Agric Syst 31:55–65

Miyazawa K, Takeda M, Murakami T, Murayama T (2014) Dual and triple intercropping: potential benefits for annual green manure production. Plant Prod Sci 17(2):194–201

Mirsky SB, Curran WS, Mortensen DA, Ryan MR, Shumway DL (2009) Control of cereal rye with a roller/crimper as influenced by cover crop phenology. Agron J 101(6):1589–1596

Mirsky SB, Ryan MR, Curran WS, Teasdale JR, Maul J, Spargo JT, Moyer J, Grantham AM, Weber D, Way TR, Camargo GG (2012) Conservation tillage issues: cover crop-based organic rotational no-till grain production in the mid-Atlantic region, USA. Renew Agric Food Syst 27(01):31–40

Parr M, Grossman JM, Reberg-Horton SC, Brinton C, Crozier C (2011) Nitrogen delivery from legume cover crops in no-till organic corn production. Agron J 103(6):1578

Peart RM, Shoup WD (2004) Agricultural systems management. Optimizing efficiency and performance. Marcel Dekker, Inc, New York

Reddy KR, Kakani VG, McKinion JM, Baker DN (2002) Applications of a cotton simulation model, GOSSYM, for crop management, economic and policy decisions. In: Ahuja L, MA L, Howell TA (eds) Agricultural system models in field research and technology transfer. Lewis Publishers (CRC Press In.), Hoboken, pp 33–54

Smith RG, Atwood LW, Warren ND (2014) Increased productivity of a cover crop mixture is not associated with enhanced agroecosystem services. PLoS ONE 9(5):e97351. https://doi.org/10.1371/journal.pone.0097351

Sapkota TB, Mazzoncini M, Bàrberi P, Antichi D, Silvestri N (2012) Fifteen years of no till increase soil organic matter, microbial biomass and arthropod diversity in cover crop-based arable cropping systems. Agron Sustain Dev 32(4):853–863

Sayre L (2003) Introducing a cover crop roller without all the drawbacks of a stalk chopper. Rodale Institute. http://www.newfarm.org/depts/NFfield_trials/1103/notillroller_print.shtml. Accessed 20 Dec 2014

Singh U (1993) Linking crop models with geographic information systems to assist in decision making: a prototype for the Indian semiarid tropics. Intl. Ferti. Paper series P-15. New Delhi, India

Snapp SS, Swinton SM, Labarta R, Mutch D, Black JR, Leep R, Nyiraneza J, O'Neil K (2005) Evaluating cover crops for benefits, costs and performance within cropping system niches. Agron J 97(1):322–332

Teasdale JR, Rosecrance RC (2003) Mechanical versus herbicidal strategies for killing a hairy vetch cover crop and controlling weeds in minimum-tillage corn production. Am J Altern Agric 18:95–102

Tonitto C, David M, Drinkwater L (2006) Replacing bare fallows with cover crops in fertilizer-intensive cropping systems: a meta-analysis of crop yield and N dynamics. Agric Ecosyst Environ 112(1):58–72

Tyler DD, Wagger MG, McCracken DV, Hargrove WL, Carter MR (1994) Role of conservation tillage in sustainable agriculture in the southern United States. In: Conservation tillage in temperate agroecosystems. Lewis Publishers Inc, Boca Raton, pp 209–229

Uchino H, Iwama K, Jitsuyama Y, Ichiyama K, Sugiura E, Yudate T (2011) Stable characteristics of cover crops for weed suppression in organic farming systems. Plant Prod Sci 14(1):75–85

Vellidis G, Gertsis A (2002) Principles of precision agriculture. Lecture notes for the DPACS, AFS, Thessaloniki, ELLAS (Greece)

Westetvelt JD, Reetz HF Jr (2000) GIS in site-specific agriculture. Interstate Publishers Inc, Danville

Whisler FD, Acock B, Baker DN, Fye RE, Hodges HF, Lambert JR, Lemmon HE, McKinion JM, Reddy VR (1986) Crop simulation model in agronomic systems. Adv Agron 40:141–208

Wortman SE, Francis CA, Lindquist JL (2012) Cover crop mixtures for the Western corn belt: opportunities for increased productivity and stability. Agron J 104(3):699

Wortman SE, Francis CA, Bernards MA, Blankenship EE, Lindquist JL (2013) Mechanical termination of diverse cover crop mixtures for improved weed suppression in organic cropping systems. Weed Sci 61(1):162–117

Integrated Farming Systems for Sustainable Agriculture

N. A. Dar, Bilal Ahmad Lone, B. A. Alaie, Z. A. Dar, Gulzafar, F. A. Bahar, S. A. Haque, and K. N. Singh

Abstract

In the mid-1960s, there was little interaction between technical scientists (who were mostly on experiment stations) and social scientists (who tended to be concentrated in planning units). The green revolution was beginning to have a great deal of success in Asia and Latin America, being based on good climate (i.e. plenty of water) and soils, very homogeneous and favourable production environments and the adoption of improved varieties of wheat, maize and rice that were very responsive to fertilizer. Improved inputs were also readily available, and there was an accessible market for the products. However, in most of sub-Saharan Africa, and certain parts of Latin America and Asia, there has been no green revolution. This is because climatic conditions are often not as favourable (i.e. too much or too little rainfall and limited amounts of irrigation), soils are generally poor, production environments are very heterogeneous and poor, and the input and output markets are poorly developed. Not surprisingly, there has been great difficulty in developing improved technologies that are attractive to farmers in such areas.

Keywords

IFS · Sustainability

N. A. Dar (✉) · B. A. Lone · B. A. Alaie · Z. A. Dar · Gulzafar · F. A. Bahar · S. A. Haque
K. N. Singh
Shere –e- Kashmir University of Agricultural Sciences and Technology,
Srinagar, Jammu and Kashmir, India
e-mail: pckvkbudgam@gmail.com; alonebilal127@gmail.com

R. S. Sengar, A. Singh (eds.), *Eco-friendly Agro-biological Techniques for Enhancing Crop Productivity*, https://doi.org/10.1007/978-981-10-6934-5_6

1 Background of Integrated Farming Systems

In the mid-1960s, there was little interaction between technical scientists (who were mostly on experiment stations) and social scientists (who tended to be concentrated in planning units). The green revolution was beginning to have a great deal of success in Asia and Latin America, being based on good climate (i.e. plenty of water) and soils, very homogeneous and favourable production environments and the adoption of improved varieties of wheat, maize and rice that were very responsive to fertilizer. Improved inputs were also readily available, and there was an accessible market for the products. However, in most of sub-Saharan Africa, and certain parts of Latin America and Asia, there has been no green revolution. This is because climatic conditions are often not as favourable (i.e. too much or too little rainfall and limited amounts of irrigation), soils are generally poor, production environments are very heterogeneous and poor, and the input and output markets are poorly developed. Not surprisingly, there has been great difficulty in developing improved technologies that are attractive to farmers in such areas.

In green revolution areas, farmers were able to benefit from the improved technologies even if they did not do things quite right and the inputs they used were very divisible (e.g. they could use a little or a lot of improved fertilizer or seed). However, in areas with less hospitable environments (e.g. low rainfall areas that are found in many countries in Africa and Latin America), farmers have to do things exactly right if they are to benefit (e.g. planting in good soil moisture) and also they need lumpy inputs (e.g. control over traction). In addition, yield increases are not so good – they tend to involve incremental rather than major (i.e. revolutionary) changes in yields.

Thus, in the green revolution areas, because of the spectacular nature of the technology, experiment station-based technical scientists were very successful in their work. However, the lack of success in using a similar approach in poorer agricultural areas (i.e. with resource-poor farmers) led to the evolution of the FSR approach, in which there is close cooperation between technical and social scientists (Norman 1993).

Integrated farming systems (IFS) seem to be the possible solution to the continuous increase of demand for food production, sustainability, stability of income and improvement in nutrition for the small and marginal farmers with limited resources. Integration of different agriculturally related enterprises with crop activity as base will provide ways to recycle produces and waste materials of one component as input through another linked component to bring in improvement in soil health and reduce the cost of production of the products which finally raise the total income of the farm.

Farming system is a resource management strategy to achieve economic and sustained agricultural production to meet diverse requirements of the farm household while preserving the resource base and maintaining high environmental quality. The advantages of integrated farming systems include pooling and sharing of resources/inputs, efficient use of family labour, conservation, preservation and utilization of farm biomass including nonconventional feed and fodder resources, effective use of FYM, regulation of soil fertility and health, income and employment generation for many people and increase in economic status by utilization of underutilized resources in an efficient and remunerative manner. Population pressure on land leading to division and fragmentation of land holdings necessitates identification and adoption of suitable farming systems (Korikantimath and Manjunath 2008).

Farming system models of different situation could enhance the productivity of the farm, improve the profitability of the farmer and sustain the productivity of the soil through recycling of organic source of nutrients from the enterprises involved. IFS also helps to realize money round the year by sale of milk, egg, meat, edible mushroom, bee keeping, etc. to the resource-poor farmers. Unlike crop activity, where the involvement of labourers for individual operations are bulky with limited number of operations at wider interval. The enterprise linkage provides good opportunity for day-to-day operations with limited labourers. This helps the family labourers of the rainfed farmers to work for the farm round the year. The standard of living of the farmer will also improve substantially by the linkage of biogas, quality food commodities for the family use, availability of money round the year, etc.

Farming system approach envisages the integration of agroforestry, horticulture, dairy, sheep and goat rearing, fishery, poultry, pigeon, biogas, mushroom, sericulture, apiculture and by-product utilization of crops with the main goal of increasing the income and standard of living of small and marginal farmers. The challenge is to upgrade the technological and social disciplines on a continuous basis and integrate these disciplines to suit the region and the farm families in a manner that may ensure increased production with stability, ecological sustainability and equitability.

Research studies carried out in different situations, viz. low land, irrigated upland and upland, have demonstrated the technical feasibility and economic viability of the integrated farming systems. Besides facilitating cash income, these farming system models generate additional employment for family labour and minimize the risk associated with conventional cropping system.

2 Global Perspective

Population in the most tropical countries is pressing hard for producing more food from the limited areas available for cultivation. The goal of sustainable agriculture can be achieved by seeking the optimal use of internal production inputs in a way that provide acceptable levels of sustainable crop productivity and livestock production resulting in economically profitable return. Integrated farming system approach is not only a reliable way of obtaining fairly high productivity but also a concept of ecological soundness leading to sustainable agriculture.

The Technical Advisory Committee of the CGIAR defined sustainable agriculture as successful management of resources for agriculture to satisfy changing human needs while maintaining or enhancing the quality of the environment and conserving natural resources. Sustainable agriculture is ecologically sound, economically viable, socially just, humane and adaptable.

Farming system approach is therefore designated as a set of agricultural activities organized into functional unit(s) to profitably harness solar energy while preserving land productivity and environmental quality and maintaining desirable level of biological diversity and ecological stability. The emphasis is more on a system rather than gross output. In other words "farming system" is a resource management strategy to achieve economic and sustained production to meet diverse requirement of farm household while preserving resource base and maintaining a high-level environment quality (Lal and Miller 1990).

Farming system is a complex interrelated matrix of soil, plants, animals, implements, power, labour, capital and other inputs controlled in part by families and influenced to varying degrees by political, economic, institutional and social forces that operate at many levels.

The components of IFS include agriculture, fish farming, horticulture, duck rearing, forestry, pigeon rearing, mushroom cultivation, sericulture, *Azolla* farming, dairy, kitchen gardening, poultry, fodder production, goat rearing, nursery, sheep rearing, seed production, piggery and vermiculture. The different IFS are (i) crop-livestock-forestry farming system, (ii) crop-fish-poultry farming system, (iii) crop-livestock-poultry-fishery farming system and (iv) labour-intensive system for small area.

3 Country Scenario

India is the second most populous country in the world. With the population growth rate of over 2%, the population is increasing year after year. Therefore, the challenge of meeting the basic requirements of this increasing population is a major concern. The green revolution, a follow-up of modern agriculture, has helped to solve the problem to a great extent. But the exploitative tendency of modern agriculture aimed at higher production per unit area, by using chemical fertilizers and pesticides, is in fact rapidly degrading the basic production capacity of the ecosystem. During the past few years, in many situations, over-exploitation of natural resources initially did have short-term gains but has resulted in degradation of soil fertility and productivity and ecological imbalance. It has created serious problems of salinity, water logging, soil erosion, host of complex pests and diseases, menace and degradation of environment including air and water pollution. Therefore, to maintain the productivity of the land even after harvesting higher crop yields, on a sustainable basis and also to increase the productivity of the land year after year, a sound integrated farming system has to be developed. Such an integrated farming system should have the objective to produce enough to meet the genuine demand and needs of the society, be profitable to the farmer, conserve natural resource base and provide healthy and safe environment in the long run. Considering the large variations that exist in soils, climatic situations in different locations and economic and social conditions of the farmers, alternate integrated farming systems are to be developed (Nadagouda 2000).

The future of Indian agriculture depends heavily on the development of appropriate farming system as applicable to resource-poor farm families and as suited to different agroecological zones. The endowment of abundant sunshine, long growing season, responsive soil types and combination of surface water, groundwater and seasonal rains and above all a progressive peasantry offer vast scope for an intensive farming system through multiple cropping and diversified farming including animal husbandry, forestry, sericulture, fisheries and the like (Patil et al. 2008).

4 Integrated Farming Systems: Profitability and Sustainability

Integrated systems are about bringing crops and livestock into an interactive relationship with the expectation that together, as opposed to alone, they will generate positive effects on outcomes of interest, such as profitability, overall productivity and conservation of non-renewable resources. It is, however, much more than this. The "system" includes the environment, soil characteristics, landscape positions, genetics and ecology of plant and animals. It involves management practices, goals and lifestyles of humans, social constraints, economic opportunities, marketing strategies and externalities including energy supplies and costs and impacts of farm policies. Systems also reflect natural resources available and the impact on their use, wildlife issues, target and nontarget plant and animal species, micro-organisms and indeed all of the definable and indefinable factors that ultimately interact to result in an outcome that is never constant (Allen et al. 2007).

Integration of various farm enterprises in a farm ensures growth and stability in overall productivity and profitability. It also ensures recycling of residues, optimization of resources, minimization of risk and generation of employment. Various enterprises that could be included in farming system are crops, vegetables, fruits, flower cultivation, dairy, poultry, fish, goat, pig, sericulture, mushroom cultivation, agroforestry, bee keeping, silviculture, agro-based industries and food processing. A judicious mix of enterprises complementary to cropping and suited to the given farm situation and farmer's preference would bring overall prosperity.

Under farming system, the farm is viewed in a holistic manner. Farmers are subjected to many socio-economic, biophysical, institutional, administrative and technological constraints. Farming system conceptually is a set of elements or components that are interrelated which interact among themselves. At the centre of the interaction is the farmer exercising control and choice regarding the type and results of interaction (Lal and Miller 1990).

5 Role of Integrated Farming System

IFS approach as a biophysical and socio-economic capsule has immense potential to address instability of income, food and nutritional insecurity, unemployment, vulnerability and poverty of farmers as well as landless labourers. The urgency for addressing issues of livelihood security mainstreaming rural poor to the benefits of production technologies and development process in an integrated manner in the target districts needs no emphasis. This calls for establishing synergy complimentary to both production system levels and beyond production and farm activities to nonfarm activities. IFS for optimizing economic returns from resource allocation under different agroecological scenarios such as totally rainfed farming, rainfed farming supported by WHS and rainfed farming supported by underground water extraction are the need of the hour. In this action research study, the IFS will be used to pave the way for growth and sustainable development for larger uptake.

6 Objectives of Integrated Farming Systems

(i) To identify existing farming systems in a specific area and assess their relative viability
(ii) To formulate farming system models involving main and allied enterprises for different farming situations
(iii) To ensure optional utilization and conservation of available resources and effective recycling of farm residues within a system
(iv) To maintain sustainable production system without damaging resource base environment
(v) To raise overall profitability of farm household by complementing main allied enterprises with each other
(vi) Analysis of existing farming systems in terms of resource use efficiency, production and productivity, income, employment and sustainability across different agroclimatic zones
(vii) Evaluation and identification of farming system through participatory approach that ensures threshold level of income for the livelihood security
(viii) Development and evaluation of synergic effects and their actions associated with different farming systems
(ix) Developing appropriate institutional and market linkage including value addition for enhancing sustainability
(x) Imparting training and capacity building of various stakeholders on IFS

7 Goal of IFS

The goals of IFS are maximization of yield of all component enterprises to provide steady and stable income rejuvenation/amelioration of system's productivity and achieve agroecological equilibrium. A void build-up of insect pests, diseases and weed population through natural cropping system management keeps them at low level of intensity. Reducing the use of chemicals (fertilizers and pesticides to provide chemical I free healthy produce and environment to the society.

8 Components of IFS

The components of IFS include agriculture, fish farming, horticulture, duck rearing, forestry, pigeon rearing, mushroom cultivation, sericulture, *Azolla* farming, dairy, kitchen gardening, poultry, fodder production, goat rearing, nursery, sheep rearing, seed production, piggery and vermiculture. The different IFS are (i) crop-livestock-forestry farming system, (ii) crop-fish-poultry farming system, (iii) crop-livestock-poultry-fishery farming system and (iv) labour-intensive farming system for small area (Patil et al. 2008).

9 Food and Fodder Security Through Integrated Farming System Approach

Sustaining household food has been an issue of prime importance to majority of the farmers belonging to the categories of small and marginal holdings. Farmers under these categories are economically poor, work in diverse, location specific, risk-prone environment invite attention to develop technologies for interdisciplinary enterprises as farming system research. It envisages development of technologies by integration of allied enterprises with existing crop components for harnessing the productivity and profitability of each enterprise by taking into account the farmer priorities, socio-economic conditions and resource available with them. Thus, farming system is the appropriate combination of farm enterprises, viz. crops, livestock, fishery, poultry, mushroom, bee keeping and the means available to the farms which serve as valuable manures for recycling within the system. It further ensures interaction with the environment without dislocation of the ecological and socio-economic balance on one hand and attempts to meet the national goals on the other. Integration of economically viable, socially accepted and environmentally safe enterprises, along with existing farming systems, has been found in enterprising states including Tamil Nadu (Jayanthi 2006), Punjab (Gill et al. 2009) and Uttar Pradesh (Singh et al. 2006). While assessing the need of farming system approach in Indian context, Swaminathan (1990) described that having attained food security for the nation, the future strategy would necessitate a change in priorities through diversification to encompass farm-level horticulture, agroforestry, animal production and fisheries, etc. into the subsistence-level farming avocation. This calls for holistic configuration of different farm enterprises (Singh et al. 2008).

Of late farming community in India has been facing new challenges of food and nutrition security, human health and structural adjustments to comply WTO stipulations on the one hand and the sustainable environment on the other. The slow growth of the agricultural sector is mainly due to stagnation in productivity growth and a grave concern for policy makers and planners. The key challenges to Indian agriculture lie in designing, developing and managing farming systems that enabled farmers to be efficient, equitable and sustainable in the bioeconomic, biophysical and sociocultural environment.

The existing resource base of large segment of farming community does not permit them to derive full benefit of modern technology. Hence the concept of strengthening resource base and optimum allocation of resources is of crucial importance. Therefore, it is imperative to evolve suitable strategy for undertaking more than one enterprise in the farm to increase productivity and supplement the income. This can be effectively achieved by determining optimum sustainable farming systems, resulting in increasing the farm income. "Farming" is the process of producing economic and animal products, and "system" implies a set of interrelated practices organized into the functional entity. "Farming system" is a resource management strategy to achieve economic and sustained agricultural production to meet diverse requirements of the farm household, preserving resource base for future generation and maintaining high environmental quality. Within an agroecological zone, several farming systems will typically be found with variations in resource endowments, preferences and socio-economic positions of the rural households.

Implementation of IFS for distinct protection systems with weak livelihood security, such as totally rainfed farming, rainfed farming with water harvesting structures (WHS), rainfed farming with WHS and groundwater irrigation and landless labourers household with livestock, will substantially improve the livelihood security on a sustainable basis. Infrastructural support to facilitate IFS and soil and water conservation (SWC) will enhance economic security and employment opportunities, reducing drudgery and enhancing livelihood security. Appropriate institutional linkages will improve market orientation, efficiency, value addition and livelihood security. Human resource development through capacity building in IFS and income generating activity (IGA) enables to counter the challenges of vulnerable areas in the state. Social capital formation builds collective action, cooperation, trust and social cohesion required for improving livelihood security through IFS.

9.1 Farming System Concept

In farming system, all the activities, decision, management, input/output, purchase/sale and resource(s) utilized make the matrix which interacts with socio-economic and biophysical environment. Farm activities interact with market forces (socio-economic) and ecosystem (biophysical) for purchasing inputs and disposing outputs by utilizing and degrading natural resources (land, water, air, sunshine, etc.). Sustainability is the objective of the farming system where production process is optimized through efficient utilization of inputs without infringing on the quality of environment with which it interacts.

The overall objective is to evolve technically feasible and economically viable farming system models by integrating cropping with allied enterprises for irrigated, rainfed, hilly and coastal areas with a view to generate income and employment from the farm.

9.2 Scope of Farming System

Farming enterprises include crop, livestock, poultry, fish, tree crops, plantation crops, sericulture, etc. A combination of one or more enterprises with cropping, when carefully chosen, planned and executed, gives greater dividends than a single enterprise, especially for small and marginal farmers. Farm as a unit is to be considered and planned for effective integration of the enterprises to be combined with crop production activity. Integration of farm enterprises depends on many factors such as:

1. Soil and climatic features of the selected area
2. Availability of the resources, land, labour and capital
3. Present level of utilization of resources
4. Economics of proposed integrated farming system
5. Managerial skill of the farmer

For any sustainable and economically viable farming system, in situ conservation of rainwater to the maximum possible extent (which automatically takes care of soil conservation) and build-up of soil fertility and productivity largely through the incorporation of organic manures but only supplemented by minimum extent of need-based chemical fertilizers/amendments are the primary considerations. In situ rainwater harvesting and conserving can be achieved by appropriate bunding followed by land levelling and strengthening of the bunds by planting trees and grasses on the bunds. Regular maintenance of bunds and interbund management practices such as carrying out all cultural operations across the slope, compartment bunding, mulching, application of organic manures to improve the infiltration rate of soils, etc. will help to conserve soil and water. Soil fertility build-up and its regular maintenance can be achieved on sustainable basis only through the application of organic manures such as FYM, compost, vermicompost, crop residues and green manures. Building up of soil fertility through organics is very important since with such a practice, stability of increasing trend in crop yields can be achieved as compared to sudden rise and fall of crop yields when chemical fertilizers are used. Regular use of organics encourages multiplication of beneficial soil microbial population, and the soilborne pathogens are controlled. Organics such as vermicompost also make the plant resistant to certain pests and diseases. Moreover, quality of the produce from such a crop will be superior as compared to the produce obtained when chemical fertilizers are used. In irrigated agriculture, after few years of irrigation, degradation of soil is very common. But, wherever green manuring is practised as a regular feature, degradation of soil has not taken place. At present, it has been the experience of most of the farmers that the farming under the existing situation is not a profitable enterprise. The major reason for such a situation is the high cost of cultivation which is increasing year after year. Therefore, any effort to reduce the cost of cultivation by avoiding/minimizing the use of chemicals and middlemen in agriculture will definitely bring in stability of agricultural production, and farming will be a profitable concern.

Crop management practices such as crop rotation, mixed cropping, INM and IPM need to be incorporated in all the farming situations. Crop rotation and mixed cropping have many advantages. Build-up of pests/diseases and depletion of plant nutrients from the soil can be avoided by following the simple practice of crop rotation. It will also help to build soil fertility when a legume crop is involved in the crop rotation. Mixed cropping when followed in an appropriate manner will increase total output under normal situations but will act as an insurance against failure of crops under abnormal conditions. When a legume is involved as one of the component crops, in mixed cropping, it will improve soil fertility (Balasubramaniyan and Palaniappan 2007). Mixed cropping is also advantageous in minimizing the spread of pests and diseases and is efficient in the utilization of soil moisture, nutrients and sunlight. Instead of depending solely on chemical fertilizers and amendments, a combination of organic manures, biofertilizers and chemical fertilizers is always better to maintain good soil health and at the same time to harvest higher crop yields. In case of plant protection, sole dependence on chemicals and their indiscriminate use has resulted in making the insect pests resistant to chemical pesticides, thereby increasing the dosage of pesticides for the control of pests year after year. Moreover specific pesticides are to be used for particular pest at its particular stage. Many farmers have very little knowledge of such specific plant protection measures which is again complicated by almost frequent changes in the plant protection schedules. Under such situations, IPM has definite advantages, but complete elimination of chemicals in a phased manner from the IPM schedule and developing plant protection methods with appropriate cultural practices, cropping systems and use of bioagents and plant products will be ideal.

Mixed farming involving allied enterprises which are complementary to agriculture, such as dairy, apiculture, sericulture, inland fisheries, sheep/goat rearing and poultry, will provide employment and additional income to the farmer throughout the year. It will also act as an insurance against crop failures due to climatic vagaries and will help to build soil fertility when animal wastes are properly utilized for preparation of manures.

9.3 Integrated Agroforestry System

Agroforestry is a land use system that involves deliberate retention, introduction of tree or woody perennials in crop/animal production to benefit from the resultant ecological and economic intersections (Nair 1991). Agroforestry-based production systems are capable of yielding leaf fodder, wood and food at the same time besides conserving and rehabilitating the ecosystem. The woody perennial/tree is one of the components in such a system. Agroforestry is an important part of integrated farm production systems in four major agroecosystems, determining the system productivity and profitability. Area under agroforestry can be covered depending upon the land capability classification. Land capability classification is the systematic arrangement of different kinds of land according to properties which determine the ability of land to produce virtually on permanent basis. The classification is based on land feature and inherent potentials. Land classes I–III or IV are categorized for agricultural or arable land use, whereas land classes IV/V–VIII are categorized for nonagricultural or nonarable land use. It is being increasingly realized that in order to derive maximum benefits from the available resources and prevailing patterns of agriculture, a land use planning is essential to achieve a sustainable use of land, water and vegetation. In India 175 m ha low productive wasteland is the potential area for agroforestry developmental activities (Shrotriya et al. 2000).

Agroforestry/agri-horticulture is a very important component in an integrated farming system. Inclusion of tree component in the farming has many advantages. Some of the important advantages of the agroforestry system are:

(a) It meets the human needs of food, fuel, fodder, timber and pesticides (e.g. neem).
(b) It provides sustainable income with low cost of cultivation, and returns are higher as compared to any cropping system involving only annual crops.
(c) It controls soil erosion and improves soil fertility and productivity by regular leaf fall and tapping the nutrients from lower regions of the soil.
(d) It very well adjusts with any vagaries of nature. Efficient use of erratic rainfall is possible by trees.
(e) Trees act as resting place for birds, which are relatively beneficial for agriculture, since harm done by birds is more than compensated by their action for control of insect pests.
(f) Shade created by trees is beneficial in raising certain shade loving crops and horticulture nursery and for vermiculture.

Summing up, it can be said that an integrated farming system involving annual crops, tree crops (beneficial MPTS and fruit trees), dairy, poultry, inland fisheries, goat/sheep rearing, apiculture, sericulture and organic farming by minimizing the use of chemical fertilizers and pesticides is a profitable, sustainable and eco-friendly agriculture which needs to be practised by each and every farmer (Nadagouda 2000).

9.4 Integration of Enterprises

In agriculture, crop husbandry is the main activity. The income obtained from cropping is hardly sufficient to sustain the farm family throughout the year. Assured regular cash flow is possible when cropping is combined with other enterprises. Judicious combination of enterprises, keeping in view of the environmental conditions of a locality, will pay greater dividends. At the same time, it will also promote effective recycling of residues/wastes.

Choice of Enterprise The basic points that are to be considered while choosing appropriate enterprise in integrated farming system (IFS) are:

1. Soil and climatic features of an area/locality
2. Resource availability with the farmer
3. Present level of utilization of resources
4. Return/income from the existing farming system
5. Economics of proposed integrated farming system
6. Farmer's managerial skill
7. Social customs prevailing in the locality

9.5 Enterprise Integration

Livestock is the best complementary enterprise with cropping, especially during the adverse years. Installation of a biogas plant in crop-livestock system will make use of the wastes and at the same time provides valuable manure and gas for cooking and lighting. In a wetland farm, there are greater avenues for fishery, duck farming and buffalo rearing. Utilizing the rice straw, mushroom production can be started. Under irrigated conditions (garden lands), inclusion of sericulture, poultry and piggery along with arable crop production is an accepted practice. The poultry component in this system can make use of the grains produced in the farm as feed. Pigs are the unique components that can be reared with the wastes which are unfit for human consumption. In rainfed farming, sheep and goat rearing form an integral part of the landscape. Sericulture can be introduced in rainfed farming, provided the climatic conditions permit it. Agroforestries (silviculture and silvi-horticulture) are the other activities which can be included under dryland conditions. In the integrated system, selection of enterprise should be on the cardinal principle that there should be minimal competition and maximum complementary effect among the enterprises.

Integrated farming system meets spread-out demand for food, income and diverse requirements of food grains, vegetables, milk, egg, meat, etc., thereby improving the nutrition of small-scale farmers with limited resources. Integration of different agriculturally related enterprises with crops provides ways to recycle the products and by-products of one component as input to another and reduce the cost of production and increase the total income of the farm.

9.6 Choice of Crops and Allied Activities

Integrated farming system focuses around a few selected, interdependent, interrelated and often interlocking production systems. Normally, they are based on crops, livestock and related subsidiary professions. This integrated nature involves the utilization of primary and secondary produces of one system as basic input of the other systems, making them mutually integrated as one whole unit. This incidentally helps to reduce the dependence on procurement of inputs from open market, making the system sustainable on a long-term basis in the development of sustainable farming system models; the concepts of intensification, diversification and value addition must be kept in view. The interaction would also help to improve productivity in various activities.

9.7 IFS Expected Outcome

Enhanced Income IFS provide opportunity to make use of the produce of one component as input on another component at the least cost. By reductions of the cost of production, the profitability per rupee invested is enhanced by eliminating the interference of middleman in most of the inputs used. Small ruminants like goats and sheep form an important economic and ecological niche in Asian mixed farming systems. Approximately, 60% of goats and 20% of sheep population are in Asia. The sale of goats contributes 30% of the total farm income in India. The IFS research in Tamil Nadu provides interesting research results. In a small farm of 2 ha, goat component added an additional income of Rs. 12,000/– with six goats. In north-western and deltaic districts. In the rainfed black soil areas in Southern Tamil Nadu, tree legumes like *Leucaena leucocephala* (subabul), *Acacia senegal* (gum arabic tree) and *Prosopis cineraria* (khejri) and perennial fodder grass with inclusion of six goats yielded an additional income of Rs. 12,500 per year from a farm area of 1.6 ha. Reports state that for farm households, the average net income is shared between crops and livestock in the ratio of about 3:1.

Livestock keeping was more suited to small land holders to fetch additional income to the farm family. Reports show that there was an increase in the return to the tune of 148% due to the introduction of poultry-cum-dairy-based integrated farming systems over cropping alone.

9.8 Employment Generation

Gainful employment is one of the major considerations for evolving any farming system. IFS under dryland with sorghum + cowpea, *Leucaena leucocephala* + *Cenchrus ciliaris* (anjan grass) and *Acacia senegal* + grasses with goat rearing generated an additional employment I of 113 man-days/ha annually in a farm size of 1 ha. Maintenance of four milch cows with cropping could generate an additional employment of 274 man-days as against cropping alone in Thanjavur delta in Tamil Nadu. Integration of crop-dairy-biogas-silviculture I-spawn production could generate an additional employment of 562 man-days than cropping alone under lift-irrigated garden lands. Integration of duck-cum-fish culture and dairying could generate 396 and 702 man-days, respectively, as against 252 man-days with rice-based cropping alone in Cauvery delta region of I Tamil Nadu. Cropping with poultry + fish + mushroom generated the highest employment of 798 man-days. The allied enterprises added employment to the tune of 423 man-days providing opportunity for 1.16 family members to be employed per day round the year.

IFS study at Bhubaneshwar for a period of 2 years comprising of field and horticultural crops, fishery, poultry, duckery, apiary, mushroom, dairy and agroforestry generated an additional employment of 573 man-days on a small piece of land of 1.25 ha. At Kasaragod, 1 ha of coconut gardening required 150 man-days, and it increased to 1000 man-days on introduction of dairy-based integrated farming. Cropping alone generated 400 man-days as against 904 in integrated farming systems with six buffalo. Cattle and buffalo rearing involved intensive use of family labour and offered significant employment opportunity for small and marginal farmers. A herd of 200 goats under integrated farming systems provided full-time employment for two persons throughout the year. Labour utilization was found to increase by 182% in integrated farming systems by the introduction of 270 poultry birds in a crop-cum-poultry enterprise. Cropping alone generated I 245 man-days, and integrated farming systems with sericulture in one hectare generated 598 man-days in a year.

9.9 Nutrient Recycling

Replenishment of soil fertility status through substantial improvement in the post-harvest available NPK nutrients could be achieved even with higher removal of nutrients through crop uptake by the application of recycled or composted pigeon and poultry manure combined with inorganic fertilizer. Application of 50% nitrogen through fertilizer and 50% through goat manure enhanced the soil fertility status and provided better opportunity for recycling of manure to the crops. Continuous dairy-based farming system increases organic carbon and available status of nutrients.

9.10 Alternate Land Use Options

IFS provides alternate land use systems which are more appropriate in areas where subsistence farming is practised in fragile ecosystems, and it possesses more potentiality and flexibility in land use than the traditional crop production systems.

9.11 Agri-silvicultural System: Lesser Risk

Agroforestry is an integrated self-sustained land management system, which involves woody perennials with agricultural crops including pasture/livestock simultaneously or sequentially on the same unit of land and meeting ecological as well as socio-economic needs of the people. Due to low initial cost and ensured seasonal income through intercropping and supply of different kinds of raw materials to support cottage industries, tree farming could certainly offset the risky farming especially under dryland conditions.

9.12 Agrihorticultural System: Higher Income

Fruit-based cropping systems are not only known for their economic viability but also generate employment and give assurance against crop failure during drought years. Maize, sorghum and cowpea are compatible with trees like *Psidium guajava*, *Eugenia jambolana* and *Annona squamosa*. Under rainfed conditions in alfisol, agrihorticulture systems give the highest benefit-cost ratio compared to annual cropping.

9.13 Silvi–/Hortipastural System: Improved Sustainability

Horticulture is one of the agroforestry systems which involve integration of fruit trees with pasture. *Cenchrus ciliaris* and *Cenchrus glaucus* are grasses, and *Prosopis cineraria* and *Acacia senegal* are the trees suited for the system. *Stylo* and *Cenchrus* are compatible fodder crops with guava, custard apple and mango. In the southern zone of Tamil Nadu, the gross income and B:C ratio obtained from sorghum + tamarind, sorghum + neem, black gram + neem and black gram + tamarind were found sustainable.

9.14 IFS: Solution to Energy and Fodder Crisis

Solve Energy Crisis It is expected that the entire world is going to suffer for want of fossil fuel from 2030 AD. So it becomes inevitable to identify an alternative source to solve our energy crisis within a span of three to four decades. In IFS, by way of effective recycling techniques, the organic wastes available in the system can be utilized to generate biogas. Though this may not be a source for complete supplementation, to a certain extent, the energy crisis can be solved.

Solve Fodder Crisis In IFS, each and every piece of land area is effectively utilized. Growing of perennial fodder trees in the borders and water courses only helps in supplementing legume fodder but also enriches soil nutrients by fixing the atmospheric nitrogen. In the cropped land, IFS envisages intensification of cropping by including legume fodder like cowpea either as second tier or as third tier in the system. These practices relieve the crisis of non-availability of quality fodder to the animal component linked.

Solve Fuel and Timber Crisis The national demand of fuelwood in 2020 AD is 400 million m^3, whereas the current production is only 20 million m^3. Similarly, the requirement of industrial wood in 2020 AD is 64.4 million m^3, and the current production level is just 11 million m^3. The present level of production should be increased to 20-fold in case of fuelwood and 6-fold in industrial wood. This could be possible to a certain extent by afforestation programme in the shrub jungles and sparse forest areas. In IFS by linking agroforestry appropriately, the production level of fuelwood and industrial wood can be enhanced without detrimental effect on crop activity in the field level.

Avoid Degradation of Forests There is a vast gap between the demand and production level as far as fuelwood and timber are concerned. This naturally induces the users to encroach on the forests nearby illegally to bridge the gap. Right now our forest area is lesser (22%) than the prescribed norm of 33%, to the geographical area. Even the forest area at present has more than two thirds of sparse vegetation. By linking agroforestry in IFS, the degrading of forest area could be minimized to a certain extent by supplementation of fuel and timber wood.

9.15 Land Reclamation and Integrated Farming System

It is estimated that about seven million hectare of land in this country is affected by varying degrees of soil salinity and sodicity. Such problematic soils are inhibiting agricultural production in the affected areas. The farmers, generally unaware of the magnitude of the problem, continue to grow crops without soil reclamation measures and harvest very poor crop yields. In the spread of saline sodic soils, West Bengal ranks third, after UP and Gujarat, having an area of 0.85 million hectare of salt-affected soil distributed mainly in 24 Parganas and Midnapore districts. Sundarbans areas of 24 Parganas district alone constitute about 94% salt-affected area of the state.

9.16 Constraints

There are certain constraints like heavy investment in the initial stage, especially for the procurement of enterprises; involvement of multidisciplinary activities like animal husbandry, fishery, sericulture, horticulture, forestry, agricultural engineering, etc.; non-availability of improved cultivars/varieties/breeds of livestock at farm site; lack of know-how especially on the constituents of feed and the possibility of supplementing from their own produces with cheaper rate; and lack of marketing for the produces from different enterprises at village level which are anticipated in the progress of this technology.

10 Conclusion

Farming system models at different situations could enhance the productivity of the farm as a whole, improve the profitability in terms of additional net return and continuous flow of income to the farmer and sustain the soil health through residue addition, and improving the major and micronutrient supply effective in recycling of crop residues and livestock waste results in environmentally safe disposal. The enterprise linkage provides good opportunity for regular and gainful on-farm employment for farm family members with equitemporal distribution. This also helps for nutrition security through optimized carbohydrate, protein, fat and energy supply by integrating allied enterprises (Jayanthi and Vennila 2008).

References

Allen VG, Baker MT, Segarra E, Brown CP (2007) Integrated irrigated crop livestock systems in dry climates. Agron J 99:346–360. American Society of Agronomy

Balasubramaniyan P, Palaniappan SP (2007) Principles and practices of agronomy, 2nd edn. Agrobios, Jodhpur, pp 462–463

Jayanthi C (2006) Integrated farming system: A path to sustainable agriculture, 2nd edn. Published by department of Agronomy, Directorate of Soil and Crop management studies, Tamilnadu Agricultural University, Coimbatore, p 1

Jayanthi C, Vennila C (2008) Integrated farming systems and sustainability for livelihood security. National symposium on new paradigms in agronomic research, pp 281–284

Gill MS, Singh JP, Gangwar KS (2009) Integrated farming and system agriculture sustainability. Indian J Agron 54(2):128–139

Korikantimath VS, Manjunath BL (2008) Integrated farming systems for sustainability in agricultural production. In: Proceedings of national symposium on "New paradigms in agronomic research". Indian Society of Agronomy, Navsari Agriculture University, Gujarat, pp 279–281

Lal R, Miller FP (1990) Sustainable farming for tropics. In: Singh RP (ed) Sustainable agriculture. Issues and prospective, vol 1. Indian Society of Agronomy, IARI, New Delhi, pp 69–89

Nadagouda VB (2000) What next in Agronomy. Prospects of integrated farming systems. In: proceedings of the national seminar. Agronomy Club UAS Dharwad Karnataka, pp 33–34

Nair PKR (1991) State of the art of Agroforestry systems. In: Jarvis PG (ed) Agroforestry: principles and practices. Elsevier, Amsterdam, pp 5–29

Norman (1993) FAO farm systems management series, vol 10. FAO of United Nations, Rome, p 227

Patil HM, Todmal SM, Pastil VS, Wani AG, Gaikwad CB (2008) Farming systems approach for achieving sustainable agriculture. In: Proceedings of national symposium on "New paradigms in agronomic research", Indian Society of Agronomy, Navsari Agriculture University, Gujrat, pp 292–293

Shrotriya GC, Kaora SV, Wankhade KG (2000) Agricultural productivity improvement through farming systems approach. Fertil News 46(11):53–58

Singh K, Bohra JS, Singh JP (2006) Development of farming system models for the northeastern plain zone of Uttar Pradesh. Indian Farming 56(7):5–11

Singh JP, Gill MS, Tripathi D, Pandey DK (2008) Household food and fodder security through integrated farming system approach. In: National symposium on new paradigms in agronomic research, pp 281–293

Swaminathan MS (1990) Agriculture and food systems. In: Jager J, Ferguson HL (eds) Climate change: science, impacts and policy. Proc. Second World Conf. Cambridge: Cambridge University Press, pp 265–277

Resource-Conserving Technologies for Enhancing Resource Use Efficiency and Crop Productivity

Hari Ram, Balwinder Kumar, Navneet Aggarwal, and Jashanjot Kaur

Abstract

Declining soil, water and human resources in rice-wheat and other irrigated cropping systems need the resource-conserving technologies (RCTs) for agricultural sustainability. After the era of green revolution, the use of soil and water resources is overexploited. The income from irrigated agroecosystem especially from rice-wheat system is declining in many areas as cultivation costs are rising faster than crop prices. The various RCTs can save soil, water and other natural resources. Zero tillage with or without mulch/happy seeder sowing technology not only saves the money spent on inputs but also gives similar or higher grain yields in various crops by conserving soil moisture but also reducing the weed density and environmental pollution. Furrow-irrigated bed planting/permanent raised-bed technologies can be practices for sustainability of major cropping systems. Hydrogel (water-adsorbing material) can effectively be used under rainfed/dryland or limited water conditions for increasing the water use efficiency. Direct seeding of rice with seed-cum-fertilizer drill under well-levelled conditions holds the promise of saving water and human resources. Human resources are decreasing due to industrialization, so direct seeding or mechanical transplanting of rice is the best solution. Laser land levelling can save huge amount of the

H. Ram (✉) · J. Kaur
Department of Plant Breeding & Genetics, Punjab Agricultural University, Ludhiana, Punjab, India
e-mail: hari_ram@pau.edu

B. Kumar
Krishi Vigyan Kendra, Tarn Taran, Punjab, India

N. Aggarwal
South Australian Research and Development Institute, Adelaide, SA, Australia
e-mail: Navneet.Aggarwal@sa.gov.au

R. S. Sengar, A. Singh (eds.), *Eco-friendly Agro-biological Techniques for Enhancing Crop Productivity*, https://doi.org/10.1007/978-981-10-6934-5_7

irrigation water for getting similar productivity as in conventionally levelled fields. As the machinery required for adoption of these RCTs are heavy and costly, these machines should be used on co-operative basis.

Keywords
Soil · Water · Resource-conserving technologies · Zero tillage · Happy seeder · Mechanical transplanter · Laser land leveller · Furrow-irrigated raised-bed system · Grain yield

1 Introduction

Rice-wheat cropping system covers about 13.5 million hectares on the Indo-Gangetic plains of South Asia including India, providing livelihood and food to millions of people in the world. Irrigated rice-wheat system uses major proportion of the region's natural soil and water resources. Increasing competition for water from industrial and domestic users in the present era, questions are being raised about the productivity of soil and water used in agriculture. Adverse environmental effects related to soil and irrigation water are also increasing as over-exploitation of groundwater and poor water management lead to falling water table in some areas (Hira 2009) and increased water logging and salinity in others. Meanwhile, the profitability of rice and wheat production is declining in many areas as production costs rise faster than crop prices. Various research organizations have developed and promoted a number of technologies that increase farm productivity, conserve the scarce natural resources, and limit adverse environmental impacts. These resource-conserving technologies (RCTs), which form the basis for conservation agriculture, have been actively promoted in the Indo-Gangetic plains for many years. "Conservation agriculture" is the term used for a diverse array of crop management practices that involve minimal disturbance of the soil, retention of residue mulch on the soil surface and use of crop rotations to control pests and diseases.

2 The Various Technologies Which Can Be Enlisted as RCTs

1. Zero tillage
2. Mulching
3. Furrow-irrigated raised-bed system
4. Use of polymers/hydrogels
5. Direct-seeded rice
6. Mechanical transplanting of rice
7. Laser land levelling

2.1 Zero Tillage (ZT)

Among the various RCTs that have received the most attention is zero-tillage planting of wheat after rice. In ZT, the tractor-drawn ZT seed drill allows wheat seed to be sown directly into unploughed fields with a single pass of the tractor, whereas conventional tillage practices for wheat involve multiple passes of the tractor to accomplish as many as four to six or more separate ploughing, harrowing and seeding practices. ZT has a number of advantages like higher or similar crop yields, reduced tractor costs, reduced irrigation time for timely planting and reduced environmental pollutions. To obtain a good seedbed in conventional tillage, the repeated ploughings not only involve high expenditure but also consume time which many a times delay the sowing of the crops resulting in low yields in intensive cropping system where the gap between the harvest of one crop and sowing of the next crop is very short. In the past weeds were the most culprit to reduce the yields in zero tillage, but pre- and post-emergence herbicides made it possible to control these in zero tillage. Zero tillage has a number of advantages like saving fuel, energy and time and timely sowing of crops. Zero-tillage technique is an ecological approach for soil surface management and seedbed preparation resulting in economics of cultivation, less weed problem, better crop residue management, higher or equal yield and quality of produce, etc. Zero tillage has a number of advantages for which it should be adopted at the farmers' field to the resources. The advantages of zero tillage are given below:

(a) Zero tillage has potential of reduction in cost of production. This is one of the important factors which can play a major role in the rapid adoption of zero-till technology. Saving diesel, inputs and labour are other important parameters in reduction in production costs. Studies revealed that the cost of wheat production is reduced by INR 2000 to 3000 per hectare (Malik et al. 2004), INR 3470/ha in maize-wheat (Ram et al. 2012) and INR 3000/ha in soybean systems (Ram et al. 2013). In addition to providing higher crop yields, ZT can reduce production costs by 15–20% (Landers et al. 2001), by reducing four to eight tillage operations practised under conventional tillage.
(b) Zero tillage helps in saving of water and nutrients. The minimum and zero tillage has been advocated in the world for purpose of controlling erosion losses, run-off component, increasing water use efficiency of summer crops and improving crop productivity (Hangen et al. 2002). De Vita et al. (2007) stated that higher soil water content under zero till than under conventional tillage indicated the reduced water evaporation during the preceding period. They also found that across growing seasons, soil water content under zero till was about 20% greater than under conventional tillage. Choudhary (2015) and Kumar et al. (2002) reported that soil moisture content was higher with zero tillage than conventional tillage in pea crop. Bachmann and Freidrick (2002) reported that direct seeding significantly reduced wind and water erosion, increased soil moisture retention and improved soil structure and fertility. Thus, it may be

concluded that crops under zero till were less susceptible to drought than under plough till method of seedbed preparation.

(c) Zero tillage helps to reduce the incidence of weeds. Conventional tillage for preparation of seedbed in autumn-winter cereal can improve the soil conditions for germination of weeds (Mirsky et al. 2010; Morris et al. 2010), which will increase the population density of the weeds in the crop (Brautigam and Tebrugge 1997). However in zero tillage, the least disturbance to the soil and soil aggregates, there is little disturbance in the sowing furrows only which only favour the germination of crop seeds and weeds in the furrows only. Singh et al. (2015) reported that zero tillage helps to reduced incidence of the major weed like *Phalaris minor* in wheat.

(d) In properly managed zero-till planted wheat, yields were invariably higher compared to traditionally prepared fields for comparable planting dates. CA has been reported to enhance the yield level of crops due to associated effects like prevention of soil degradation, improved soil fertility, improved soil moisture regime (due to increased rain water infiltration, water holding capacity and reduced evaporation loss) and crop rotational benefits. Yield increases as high as 200–500 kg/ha are found with zero-till wheat compared to conventional wheat under a rice-wheat system in the Indo-Gangetic plains. Review of the available literature on CA provides mixed indications of the effects of CA on crop productivity. While some studies claim that CA results in higher and more stable crop yields, on the other hand there are also numerous examples of no yield benefits and even yield reductions particularly during the initial years of CA adoption.

    Mehla et al. (2000) and Srivastava et al. (2002) reported significantly higher yield of wheat under zero tillage than under conventional tillage system. Brar et al. (2004) reported from long-term experiment that after 25 years wheat yield in maize-wheat sequence remained the same in zero-tillage and conventional tillage treatments.

(e) Zero tillage plays an important role in reducing environmental pollution. Zero till is an excellent opportunity to eliminate burning of crop residue which contributes to large amounts of greenhouse gases like CO_2, CH_4 and N_2O. The environmental impact of burning diesel will be minimized due to less use in the zero-tillage technology.

    Due to various advantages of zero-tillage technology, it will not only save money and give similar grain yields in various crops but also reduce the pollution.

2.2 Mulching/Happy Seeder Technology

Tillage systems are integral part of crop production affecting numerous factors important to crop growth. Recently a shift towards conservation tillage has occurred for variety of reasons including soil surface conservation, fuel energy savings and erosion control. Covering the soil surface with crop residue could help in checking soil loss due to erosion and change its energy balance, which affects soil environment through its influence on hydrothermal regime. Burning crop residue would be an effective management tool, and additionally the time saved by burning could be important since the turnaround time between winter and summer crops would be only few days. However recent concern about environment pollution due to smoke from burning motivated the study of alternative tillage besides the planting and straw management practices. Although yields of crops like wheat, soybean, maize and other crops were not affected by burning but for undesirable effects in the soil physico-chemical properties, this practice should be avoided.

Why mulching in no tillage or conventional tillage should be followed?

Soil management has an important effect on the availability and the use efficiency of water in different crops (Hatfield et al. 2001). The raised-bed cultivation systems have involved tillage operations to prepare the raised beds before planting. However, in Mexico and some areas in China have essentially no tillage, the crop residues being mulched on the surface of the raised beds (Sayre, 2004). Raised-bed planting reduces compaction by confining traffic to the furrow bottoms. To improve the sustainability of raised-bed planting as resource conservation technology, reduce tillage and manage crop residues on the surface, reshaping the beds only as needed between cropping system, by following the even distribution of the previous crop residues.

The grain yield of maize-wheat in zero tillage or permanent bed planting along with mulch increased the grain yield as compared to conventional tillage in Mexico (Govaerts et al. 2005). Govaerts et al. (2005) reported 25–30% improvement in grain yield through adoption of zero tillage, appropriate rotation and retention of sufficient crop residue as compared to heavy tillage before seeding, monocropping and crop residue removal. Ram et al. (2012, 2013) reported that application of mulch on zero tillage and permanent bed planting numerically increased the grain yield of maize-wheat and soybean-wheat cropping systems. Upadhyay et al. (2016) reported that the maize grain yield was the highest in permanent beds along with mulch (4626 kg/ha) followed by zero tillage with mulch (3838 kg/ha). Whereas total water applied during the crop period were the highest in conventional tillage without mulch followed by conventional tillage with mulch. Under Sri Lankan conditions, the permanent bed with mulch increased the yield and reduced the water requirement of spring maize in Chitwan. Sharma et al. (2010) from the northwestern Himalayan regions of India reported that mulching is useful for soil moisture conservation which results in increased productivity and improved soil properties for the maize (*Zea mays* L.)-wheat cropping system. Recently, a machine called happy seeder had been developed, which cuts rice straw in front of the sowing tines and spreads it uniformly over the sown area (Sidhu et al. 2007). In the recent past, Sidhu et al. (2015) reported the advanced version of happy seeder, the 9-row turbo happy seeder having a weight of 506 kg, which can be operated by a 33.6 kW tractor with a work rate of 0.3 ha/h. Various on-farm trials showed that yield of wheat sown into rice residues with the 9-row turbo happy seeder (4.50 t/ha) is similar to or higher than yield with straw burning and conventional tillage (4.36 t/ha) prior to sowing while providing many benefits to the farmer (Fig. 1).

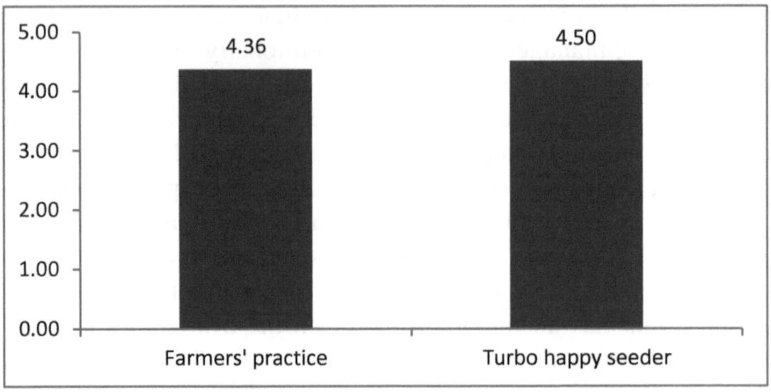

Fig. 1 Wheat grain yield (t/ha) in different sowing methods (Adapted from Sidhu et al. 2015)

Balwinder-Singh et al. (2016) reported that mulching in early sown crop showed more response than late sown crop. However, they further reported that if sowing was delayed to mid (sandy loam) or late (clay loam) November, the probability of yield loss with mulch increased. The retention of rice residue as a surface mulch could be beneficial for enhancing soil water status and moderating soil temperature thereby increasing root growth, plant canopy, wheat yield and water productivity (Balwinder-Singh et al. 2011; Chakraborty et al. 2010). Ram et al. (2013) reported that straw mulching decreased soil temperature and reduced the weed dry matter, increased yield attributes and yield in wheat. They further reported that mulching decreased the total water use in wheat, while water use efficiency was increased as the level of rice straw mulch increased. Limon et al. (2001) noticed that soil strength decreased with application of crop residue, whereas soil microbial biomass carbon increased. Retention of crop residue on the soil reduced the bulk density, enhanced the organic carbon and EC but reduced the pH of the soil (Ram et al. 2013). Yi et al. (2011) reported that as mulching changes the hydrothermal properties of the soil. They observed that over the whole season, the average topsoil water content was significantly higher and soil temperature significantly lower under the mulched conditions as compared to un-mulched treatment. Ram et al. (2012, 2013) also reported higher organic carbon in the soils after 5 years of field experiments. Zero tillage and straw mulching improved wheat tillering ability and soil nutrient content during the growth stages and improved some quality parameters, while year-round tillage practice had less effect on grain quality (Tang et al. 2013; Ram et al. 2013). Singh et al. (2014b) identified the weak link in zero tillage with mulch which demonstrated higher damage by pink stem borer. Some of the farmers also reported the damage of wheat by pink stem borer. However, lower termite damage and root aphid incidence in zero tillage and ZT + mulch sown crops are the positive effects of conservation tillage.

It can be concluded that mulching on no tilled flat as well as permanent raised bed has advantage of saving in irrigation water, reduced weed population, moderation of soil temperature and similar or higher grain yield and better returns. But some of the disadvantages like heavy machinery with low efficiency, mindset of the farmers and researchers, less number of demonstrations, shift in weed flora and fauna and non-availability of decomposing organisms do not allow this technology for its wider adoption in rice-wheat cropping system.

2.3 Furrow-Irrigated Raised-Bed System (FIRBS)

Furrow-irrigated raised-bed system is a method of planting a crop on top of raised beds with a definite number of rows often one or two per bed. Soil moved from the furrows is added to the crop zone, raising its surface level. The furrows serve as irrigation channels, drains and traffic lanes. The FIRB systems have a number of advantages like better irrigation management, increased availability of nutrient to crop roots, better crop establishment, better weed management through interbed cultivation and less soil compaction. FIRB system facilitates better light penetration within the crop canopy, improves water and fertilizer use efficiency, reduces crop lodging and allows lower seed rates. For such potential benefits and favourable effects, wheat could be sown by this technique. Beds could greatly improve drainage for crops sown after rice, improving their establishment and yield and increasing cropping flexibility. In bed planting, fertilizer can be managed for higher efficiency by placing the fertilizer in the soil rather than broadcasting on the surface in the traditional system. Moreover, losses of nitrogen fertilizer are minimal in beds as irrigation is applied in channels. The bed planting has already been recommended in wheat in Indian Punjab conditions. There is possibility of extending this crop-raising method to the *kharif* crops like maize and soybean. Further the permanent bed planting allows the bed to be reused for succeeding crop and thus has the potential to minimize the cost of cultivation similar to zero tillage. This technology may allow making large reductions in turnaround time between crops for maize-wheat and soybean-wheat sequence.

Majeed et al. (2015) from Punjab province of Pakistan reported that wheat planting on FIRB and nitrogen application at 120 kg ha^{-1} produced higher wheat grain yield by 15.06% than flat planting at the same nitrogen rate. They reported 29.0% higher economic gain in FIRB planting as compared to flat planting, when nitrogen was applied at 120 kg/ha. They further reported 25.04, 15.02, 14.59 and 29.83% higher nitrogen uptake, nitrogen use, and agronomic and recovery efficiencies, respectively, were recorded for bed compared to flat planting. Bed planting in wheat with a nitrogen application of 80 kg/ha gave a yield similar to that of flat planted crop with 120 kg/ha nitrogen.

Ram et al. (2012) from Indian Punjab reported that during the wheat emergence, raised beds recorded 1.3 °C higher soil temperature compared to the flat treatments. Maize and wheat yields in FIRB system were similar to flat planted crop. However, maize and wheat planted on raised beds recorded about 7.8% and 22.7% higher water use efficiency than under flat planting, respectively (Table 1). The net returns from the maize-wheat system were higher in permanent raised beds than with conventional tillage. They further reported in 2013 that soybean and wheat yields in a sequence were similar under raised beds and flat planting. However, soybean and

Table 1 Effect of planting methods on different parameters of maize and wheat

	Maize		Wheat	
Parameter	Flat planting	FIRBS	Flat planting	FIRBS
Grain yield (t/ha)	5.54	5.57	5.34	5.33
Water use (cm)	70.6	65.9	57.6	46.6
Water use efficiency (kg ha-cm^{-1})	80.8	86.8	93.0	114.6

Adapted from Ram et al. (2012)

wheat planted on raised beds recorded about 17% and 23% higher WUE, respectively, than in flat planting. They reported that net returns from SW system were greater in permanent raised beds than with conventional tillage.

Ram et al. (2013) reported the highest photosynthetically active radiation interception (PARI) and net returns of soybean were recorded in raised-bed planting. The seed yield was recorded in raised-bed sowing, which was 6.70 and 5.29% higher than ridge-furrow and flat sowing methods, respectively. The highest water productivity was also recorded in raised broad bed sowing which was significantly higher than flat sowing and ridge-furrow sowing methods but statistically on par with raised-bed sowing. Dhindwal et al. (2006) reported that FIRBS resulted in 22% higher grain yield of pigeon pea, 8% of cluster bean and 15% of green gram compared with flat-sown crops. In FIRBS 21.6, 18.8 and 14.4 cm irrigation water was applied in the three respective crops, with 25% saving compared with flat sown. Irrigation-water productivity (WPirri) of pigeon pea, cluster bean and green gram under FIRBS was 76, 44 and 54% more than that of flat-sown crops. Ram et al. (2016) recorded that pigeon pea biomass and grain yield recorded under furrow-irrigated raised-bed planting was 9.3–15.65 and 13.0–16.4% higher as compared to the flatbed system during different years of study. They also reported that gross returns, net returns and B:C ratio were also higher in raised-bed planting.

It is concluded that furrow-irrigated bed planting and permanent raised-bed technologies can be adopted for sustainable crop production in different crops.

2.4 Use of Polymers/Hydrogels

Water absorbing materials have been reported to be effective tools in increasing water holding capacity. Water conservation is a key step to attain sustainable agriculture growth, development and productivity. Reducing the frequency of irrigation during crop production is essential for minimizing post production crop loss in retail outlets. Hydrogels are the water-loving synthetic polymers which provide growth media with increased water and nutrient holding capacity. Hydrogel is one of the most popular, having also been used to reduce water run-off and increase infiltration rates in field agriculture, in addition to increasing water holding capacity for agricultural applications (Sharma 2004). The hydrogel has a range of environmental benefits like decreased erosion, reduced sediment and nutrient losses and absorbed nutrients which are gradually released. The influence of hydrogels depends on soil structure, concentration of salts and fertilizers, and the type of plant cultivated (Narjary et al. 2012). The use of hydrophilic polymers to improving soil water and fertilizer retention properties and thus crop productivity is attracting considerable interest. Hydrogel polymers are used for increasing water holding capacity and for retaining more water and fertilizer near it. Dar and Ram (2015) reported that with the application of hydrogel at the rate of 5 kg/ha, nutrient application can be reduced by 25% without decreasing the yield of wheat crop. The effect of hydrogel was more pronounced under rainfed or limited irrigation conditions as compared to irrigated conditions. Dar and Ram (2016) again reported that instead of applying four irrigations in wheat, three irrigations are sufficient if hydrogel used 2.5–5.0 kg/ha without any penalty on crop yields. Consumptive water used was reduced with hydrogel as compared to no hydrogel application; hence, water use efficiency was increased from 11.6 kg grain/ha/mm in no hydrogel application to 13.7 kg grain/ha/mm in 5 kg/ha hydrogel used at the time of sowing of the crop. Choudhary et al. (1995) while working on calcareous sand and loam soils reported that increase in polymer applied (0.0, 0.2, 0.4 and 0.6% on dry weight basis) increased the water holding capacity and decreased evaporation and as a result increased the amount of water conserved in both the soils. The hydrogel-amended soils require less frequent irrigations for crop production, especially early in plant establishment. Yangyuoru et al. (2006) found that the water storage at different tensions improved significantly in sandy soil treated with gel conditioner. The amended soil with natural and synthetic polymers improved maize yields by 36% and 31% and improved dry matter yields by 92% and 81%, respectively, than those of the control. Hydrogels increased the maize yields and water holding capacity of sandy soils (Allahbadi et al. 2005). Al-Rahim et al. (2007) concluded that hydrogels increase the water holding capacity for agricultural applications. They further reported that application of 0.6% hydrogel concentration prolonged the time of water loss from the soil by about 66%, and the seedlings grown in 0.6% hydrogel-mixed soil survived three times as long as those grown in the control soil, however, statistically at par with 0.4% hydrogel concentration. Bhardwaj et al. (2007) used water absorbents in sandy soils from arid and semiarid regions and reported it as an important tool in increasing water use efficiency and crop production. They reported that water retained by the absorbents alone ranged between 200 and 500 kg/kg of polymer, whereas water retained by the polymers when mixed with sand ranged between 40 and 140 kg/ kg. Yezdani et al. (2007) obtained more grain yield of soybean by hydrogel application in drought-prone soils, and the yield obtained was 1.77, 3.47, 4.98 and 6.41 q/ha with application of superabsorbent polymer at 0, 75, 150 and 225 kg/ ha, respectively. Narjary et al. (2012) conducted an experiment in PVC columns in a laboratory at the IARI (New Delhi) on four different soils, red sandy loam, sandy, black clay and alluvial sandy loam, and the results showed that hydrogel was unsuitable for black soils and hydrogel at 0.7% and was found most suitable than 0.5% hydrogel for growing crops on other soils because the water available to plants grown in gel-treated soils increased by 1.5–2 times over the water available to plants grown in non-gel-treated soils.

Moghadam et al. (2013) from Iran reported that the hydrogel at 7% concentration was able to reduce the destructive effect of water deficiency, by absorbing and preserving water and improving several agronomic characters. They also recorded increased yield and its components and less plant water requirement in six oilseed rape genotypes.

2.5 Direct-Seeded Rice (DSR)

Rice is the major rainy season crop of India covering about 43 million ha with production of 85.7 million tonnes. The conventional system of rice production in this region is basically water, labour and energy intensive, adversely affecting the environment. Therefore, to sustain the long-term production of rice, more efficient alternative methods of rice productions are needed. For this, DSR is the technology which is water, labour and energy efficient along with eco-friendly characteristics and can be a potential alternative to conventional system (Kumar and Ladha 2011). DSR is being followed in uplands where there is less possibility for water stagnation. This system of cultivation is followed in many parts of the world, however, mainly confined to tracts that don't have adequate irrigation facilities. Bhatt and Kukal (2014) reported that zero tillage in wheat followed by direct seeded rice retained higher amount of moisture content for a longer period of time than any other sowing method which may be helpful for taking any crop. Kumar et al. (2015) observed that DSR saves labour by 97% as compared to manual puddled transplanted rice. They reported that rice yield in manual puddled transplanted rice was significantly higher (10–12%) than DSR. However, when compared to manual puddled transplanted rice, a labour saving of 7–8% (overall) is recorded. The study showed that manually puddled transplanting establishment method could be replaced with DSR to save water and labour.

It is not being picked up at farmer's level due to infestation of weeds and non-availability of DSR drill. Although the weeds can easily be controlled with the use of pendimethalin + bispyribac Na +, one-hand weeding is the best integrated weed management strategy to control weeds and to increase yield, water productivity and profitability (Mahajan and Timsina 2011). Bhullar et al. (2016) reported that tank mixture of fenoxaprop and ethoxysulfuron gave good control of mixed weed flora in dry-seeded rice. Although the tank mixture of fenoxaprop with bispyribac herbicides has the potential to effectively control aerobic and aquatic grass weeds in dry-seeded rice, but the integration of one-hand weeding prevented the crop yield loss from weeds that escaped herbicide treatments within 6 weeks of sowing. Gangwar et al. (2008) reported that grain yield of hybrid rice was higher (8.52 t/ha) with drum seeding and statistically at par with direct seeding and mechanical transplanting (puddled) compared with manual transplanting (puddled) and mechanical transplanting (unpuddled). They further reported that direct seeding (dry bed, unpuddled) adopted in the previous rice crop gave higher mean yield of the succeeding wheat, chickpea and Indian mustard.

The adoption of DSR is being picked in the parts of NW India but it has different disadvantages which should be addressed by the researchers like high weed infestation, lack of technical knowhow, bird damage after sowing is common etc.. If the farmers do not follow the proper practices for direct-seeded rice, it may result in poor yield and gross income although the B:C ratio was higher in DSR (Sidhu et al. 2014). Ekta et al. (2013) reported other associated problems with DSR are increased incidences of blast disease, crop lodging, impaired kernel quality, increased panicle sterility and stagnant yields across the years. The farmers can get the higher grain yield by following proper weed management and other agronomic practices in direct-seeded rice. The technology can be used effectively with laser land levelling which is the prerequisite for the technology.

2.6 Mechanical Transplanting of Rice

Manual transplanting, a common method of crop establishment in SE Asia, is highly labour intensive (Pandey and Velasco 2005; Gangwar et al. 2008) and requires up to 240 man-hours ha^{-1} (Singh et al. 2014a). Nowadays, labour scarcity is a big problem and rice transplanting activity requires about 1/3 of total labour requirement for rice (Kumar et al. 2012). Implementation of the government's policies (Anonymous, 2005) has created a labour scarcity in northwest India as rice transplanting in this region is dependent on migrant labourers from the eastern states of India (Mahajan et al. 2012). Moreover, the rural labour is migrating to industrial sector where requirement is relatively stable across the year. As delay in transplanting from its scheduled time causes decline in rice yield (Chahal et al. 2007), therefore, timely transplanting with limited labour availability is a major challenge to sustain rice productivity.

The resource conservation technologies like mechanical transplanting help immensely in reducing cultivation costs, energy and labour requirements, sustaining the agricultural productivity and increasing the profitability of the farmers (Bijay-Singh et al. 2006). The mechanical transplanting enables more timely transplanting as it requires only 7% of time for transplanting and saves 66% cost over the manual transplanting (Sharma et al. 2002). Moreover, plant population in manual transplanting at the farmers' fields is generally low (18–20 plants/m^2) compared with the recommended 35–40 plants/m^2 (Malik and Yadav 2008) which results in reduced grain yields, whereas mechanical transplanting easily enables near-optimum density and results in higher productivity (Aggarwal and Singh 2015).

From a series of 571 on-farm trials conducted over a 5-year period from 2006 to 2010 in Haryana (India), Kamboj et al. (2013) observed that mechanical transplanting resulted in 3–11% yield increase over manual transplanting of rice in different years. In a study conducted in North India, Aggarwal et al. (2015a) reported that the use of machine for transplanting resulted in lowest consumption of energy in terms of human resources (287.5 MJ/ha), thus a step forward towards sustaining rice productivity with limited labour resources. In another study, Aggarwal et al. (2015b) observed that mechanical transplanting of rice in conventional tillage without puddling gave similar grain yields (6.5 t/ha) as that of puddled transplanted rice (6.4 t/ha). Mechanical transplanting of rice can save the labour used in manual transplanting along with getting higher or similar rice productivity.

2.7 Laser Land Levelling

As per previous studies, a significant (20–25%) amount of irrigation water is lost during its application at the farm due to poor farm designing and unevenness of the fields. This problem is more pronounced in the case of rice fields. Fields that are not level have uneven crop stands, increased weed burden and uneven maturing of crops. All these factors lead to reduced yield poor grain quality.

Laser land levelling is levelling the field within a certain degree of desired slope using a guided laser beam throughout the field. Unevenness of the soil surface has a significant impact on the germination, stand and yield of crops. Farmers also recognize this and therefore devote considerable time resources in levelling their fields properly. However, traditional methods of levelling land are cumbersome, time consuming as well as expensive. The advantages of laser levelling include better distribution of water, water saving, improvement in nutrient use efficiencies, option for precision farming, higher crop productivity and increased farming area and improved operational efficiency. Abdullaev et al. (2007) reported that that laser land levelling can reduce the water application rate as compared to the unlevelled field in cotton. They also found 92% higher gross and 22% higher net income and from the laser field than that from the control field (Table 2).

Researchers in Punjab found that laser levelling increases crop yields by around 11% and results in water saving of around 25%, holding constant other inputs like fertilizers and seed quality. These experiments have also demonstrated that levelling reduces weeds by up to 40% and labour time spent weeding by up to 75% (Bhatt and Sharma 2009). Naresh et al. (2014) reported from UP (India) that with laser levelling, farmers could save irrigation water by 21% and energy by 31% and obtained higher yields of 6.6, 5.4 and 10.9% in rice, wheat and sugarcane. They also reported that total irrigation duration and applied water depth were reduced to 10.9 and 14.7% in rice, 13.7 and 13.3% in wheat and 13.5, 20.3% in sugarcane as compared to traditional levelled fields which resulted in the highest water use efficiency (WUE), which was 48, 47 and 49% higher in precisely levelled field than control (unlevelled) and 22, 19 and 20% higher than traditionally levelling fields, respectively.

Table 2 Effect of laser land levelling on cotton yield, water productivity and duration of irrigation in cotton

Parameter	Control field	Laser land levelled
Cotton yield (t/ha)	2.50	3.39
Water productivity (kg/m^3)	0.32	0.47
Duration of irrigation (h)	365	344

Adapted from Abdullaev et al. (2007)

It was concluded from above studies that the use of laser land levelling increases yield and saves irrigation water as compared to traditional method of levelling in different cropping system. Aryal et al. (2015) reported that laser levelling in rice fields reduced time of irrigation time by 47–69 h/ha/season and improved 7% yield as compared to traditionally levelled fields. In wheat, reduction in irrigation time was 10–12 h/ha/season and yield increased was 7–9% in laser levelled fields. They also reported an additional benefit of USD 143.5/ha/year from rice-wheat systems as compared to traditionally levelled fields.

The main constraint in adoption was initial capital of farmers and scattered land location. However, the use of laser land levelling on corporate and cooperated basis was the success story in northwestern plain of India.

3 Conclusions

Rice-wheat cropping system occupies 13.5 million hectares land on the Indo-Gangetic plains of South Asia including India. Rice is the main water guzzler crop that needs 1300–1500 mm of water. The green revolution not only degraded soil but also overexploited the groundwater. The net returns from rice and wheat are declining in many areas as production costs rise faster than crop prices. These resource-conserving technologies (RCTs), which form the basis for conservation agriculture, has the only solution to save soil, water and other natural resources.

Zero tillage and happy seeder sowing technology in wheat, maize and soybean not only save money and give similar or higher grain yields in various crops but also reduce the weed problem and environmental pollution. But care should be taken for shift in weed and insect pest in zero tillage. Furrow-irrigated bed planting and permanent raised-bed technologies can be adopted for sustainable crop production in wheat, soybean, maize and pulse crops. It saves the irrigation water and also protects the rainy season crops from the water logging. Water absorbing materials (hydrogel) are effective tools in increasing water holding capacity of the soils. Under rainfed or limited water availability, hydrogels can effectively be used for increasing the water use efficiency. Direct seeding of rice can be adopted under well-levelled conditions with proper management of weeds through herbicides in the early stages of the crop. As human resources are decreasing day by day due to industrialization, rice sowing should be shifted either to direct seeding or through mechanical transplanters.

The machinery required for zero tillage/happy seeder/paddy transplanters are heavy and costly, so use of these machines should be used on co-operative basis. The laser land leveller technology which saves lot of water in different cropping systems has been widely adopted in North India and can easily be extended to the other similar agroclimatic conditions in the world.

References

Abdullaev I, Hassan MIU, Jumaboev K (2007) Water saving and economic impacts of land leveling: the case study of cotton production in Tajikistan. Irrig Drainage Syst 21:251–263

Aggarwal N, Singh A (2015) Crop performance, nutrient uptake vis-a-vis weed suppressive ability of mechanically transplanted rice (Oryza sativa) as influenced by age of seedling and planting density. Indian J Agron 60(2):255–260

Aggarwal N, Singh A, Kang JS, Singh SP (2015a) Energy utilization in production of mechanically transplanted rice as affected by age of seed lings and planting densities in the Indo-Gangetic plains. Prog Res 10:5–8

Aggarwal N, Singh A, Singh SP (2015b) Heat utilization vis-à-vis crop performance of mechanically transplanted rice (*Oryza sativa* L.) as affected by tillage systems and nitrogen levels. J Agromet 17:84–89

Allahbadi I, Moazzen GB, Akbari GA, Zohorianfar M J (2005) Investigating the different rates superabsorbent polymer (Superab A200) and irrigation on the growth and yield of *Zea mays*, 3rd specialized training course and seminar on the application of Super absorbent hydrogels in Agriculture IPPI, Iran, pp 52–56

Al-Rahim HAA, Hegazy EA, El-Mohdy HLA (2007) Radiation synthesis of hyhrogels to enhance sandy soils water retention and increase plant performance. J Appl Polym Sci 93:1360–1371

Anonymous (2005) GOI (Government of India), "The Mahatma Gandhi National Rural Employment Guarantee Act 2005," Government of India, Ministry of Rural Development, 2011. http://nrega.nic.in/netnrega/home.aspx

Aryal JP, Mehrotra MB, Jat ML, Sidhu HS (2015) Impacts of laser land leveling in rice–wheat systems of the north–western Indo-Gangetic plains of India. Food Sec 7:725–738

Bachmann T L and Freidrick T (2002) Conservation agriculture in Mangolia. In Proceedings of the international symposium on conservation agriculture for sustainable wheat production in rotation with cotton in limited water resource area, October 14–18. Institute of Irrigation and Agricultural Mechanization Engineers, Tashkent, Uzbekistan, pp 91–96

Balwinder-Singh, Humphreys E, Eberbach PL, Katupitiya A, Yadvinder-Singh, Kukkal SS (2011) Growth yield and water productivity of zero till wheat as affected by rice straw mulch and irrigation schedule. Field Crops Res 121:209–225

Balwinder-Singh, Humphreys E, Gaydond DS, Eberbach PL (2016) Evaluation of the effects of mulch on optimum sowing date and irrigation management of zero till wheat in central Punjab, India using APSIM. Field Crops Res 197:83–96

Bhardwaj AK, Shainberg I, Goldstein D, Warrington DN, Levy GJ (2007) Water retention and hydraulic conductivity of cross linked poly acrylamides in sandy soils. J America Soil Sci Soc 71:406–412

Bhatt R and Sharma M (2009) Laser leveller for precision land levelling for judicious use of water in Punjab. Krishi Vigyan Kendra, Kapurthala, Punjab Agricultural University, Ludhiana, Extension Bulletin

Bhatt R, Kukal SS (2014) Moisture retention trends during the intervening period of differently established rice-wheat cropping pattern in sandy loam soil. Int J Farm Sci 4(2):7–14

Bhullar MS, Kumar S, Kaur S, Kaur T, Singh J, Yadav R, Chauhan BS, Gill G (2016) Management of complex weed flora in dry-seeded rice. Crop Protect 83:20–26

Bijay-Singh, Gupta RK, Yadvinder-Singh, Gupta SK, Jagmohan-Singh, Bains JS, Vashishta M (2006) Need-based nitrogen management using leaf color chart in wet direct-seeded rice in northwestern India. J New Seeds 8:35–47

Brar SS, Kler DS, Singh A, Virk HK (2004) Zero Tillage – A resource conservation technology: Research results and adoption in Punjab. *Proc National Symp on Resource Conservation and Agricultural Productivity*, November 22–25. Punjab Agricultural University Ludhiana, India, pp 492–493

Bräutigam V, Tebrügge F (1997) Influence of long-termed no-tillage on soil borne pathogens and weeds. In: Tebrügge F, Böhrnsen A (eds) Experience with the applicability of no-tillage crop production in the West-European countries: proceedings of the EC-Workshoop – III -Évora, pp 17–29

Chahal GBS, Sood A, Jalota SK, Choudhury BU, Sharma PK (2007) Yield, evapotranspiration and water productivity of rice (*Oryza sativa* L.)-wheat (*Triticum aestivum* L.) system in Punjab (India) as influenced by transplanting date of rice and weather parameters. Agric Water Manag 88:14–22

Chakraborty D, Garg RN, Tomar RK, Singh R, Sharma SK, Singh RK, Trivedi SM, Mittal RB, Sharma PK, Kamble KH (2010) Synthetic and organic mulching and nitrogen effect on winter wheat (*Triticum aestivum*) in a semi-arid environment. Agric Water Manag 97:738–748

Choudhary VK (2015) Tillage and mulch effects on productivity and water use of pea and soil carbon stocks. Arch Agron Soil Sci 61:1013–1027

Choudhary MI, Shalaby AA, Al-Ornran AM (1995) Water holding capacity and evaporation of calcareous soils as affected by four synthetic polymers. Comm Soil Sci Plant Anal 26:2205–2215

Dar SB, Ram H (2015) Growth analysis, micro-climatic parameters and productivity of wheat (*Triticum aestivum* L.) in relation to hydrogel under different irrigation regimes and nutrient levels. J Wheat Res 8:59–62

Dar SB, Ram H (2016) Grain yield, nutrient uptake and water-use efficiency of wheat (*Triticum aestivum*) under different moisture regimes, nutrient and hydrogel levels. Indian J Agron 61:58–61

De Vita P, Di Paolo E, Fecondo G, Di Fonzo N, Pisante M (2007) No-tillage and conventional tillage effects on durum wheat yield, grain quality, and soil moisture content in southern Italy. Soil Tillage Res 92:69–78

Dhindwal AS, Hooda IS, Malik RK, Kumar S (2006) Water productivity of furrow-irrigated rainy season pulses planted on raised beds. Indian J Agron 51:49–53

Ekta J, Dinesh K, Lal B, Nepalia V, Priyanka G, Vyas AK (2013) Management of direct seeded rice for enhanced resource - use efficiency. Plant Knowl J 2(3):119–134

Gangwar KS, Gill MS, Tomar OK, Pandey DK (2008) Effect of crop establishment methods on growth, productivity and soil fertility of rice (*Oryza sativa*)-based cropping systems. Indian J Agron 52:102–106

Govaerts B, Sayre KD, Jozef D (2005) Stable high yields with zero tillage and permanent bed planting. Field Crops Res 94:33–42

Hangen E, Buezko U, Bens O, Brunotte J, Huttl RF (2002) Infiltration patterns into two soils under conventional and conservation tillage. Influence of the spatial distribution of plant root structures and soil animal activity. Soil Tillage Res 62:181–186

Hatfield J, Sauer T, Prueger J (2001) Managing soils to achieve greater water use efficiency: a review. Agron J 93:271–280

Hira GS (2009) Water management in northern states and the food security of India. J Crop Improv 23:136–157

Kamboj BR, Yadav DB, Yadav A, Goel NK, Gill G, Malik RK, Chauhan BS (2013) Mechanized transplanting of rice (*Oryza sativa* L.) in non puddled and no-till conditions in the rice-wheat cropping system in Haryana, India. American. J Plant Sci 4:2409–2413

Kumar V, Ladha JK (2011) Direct seeded rice: recent development & future research needs. Adv Agron 111:297–413

Kumar V, Singh S, Yadav A, Malik RK, Hobbs PR (2002) Studies on the effect of zero tillage in wheat on physico-chemical properties of soil in comparison to conventional tillage. In: Proceedings of the internaltional symposium on herbicide resistance management and zero tillage in rice-wheat cropping system, March 4–6. CCS Haryana Agricultural University, Hisar, India, pp 110–112

Kumar S, Singh SS, Sundaram PK, Shivani S, Bhatt BP (2012) Agronomic management and production technology of unpuddled mechanical transplanted rice. Technical bulletin no. R-37/PAT-24, ICAR Research Complex for Eastern Region, Patna (India)

Kumar A, Kumar S, Dahiya K, Kumar S, Kumar M (2015) Productivity and economics of direct seeded rice (*Oryza sativa* L.) J Appl Nat Sci 7:410–416

Landers JN, Saturnino HM, de Freitas PL (2001) Organizational and policy consideration in zero tillage. In: Saturnino HM, Landers JN (eds) The environment and zero tillage. Brazil, pp 13–24

Limon OA, Sayre KD, Drijber RA, Francis CA (2001) Soil attributes in a furrow-irrigated bed planting system in northwest Mexico. Soil Tillage Res 63:123–132

Mahajan G, Timsina J (2011) Effect of nitrogen rates and weed control methods on weeds abundance and yield of direct-seeded rice. Arch Agron Soil Sci 57:239–250

Mahajan G, Chauhan BS, Timsina J, Singh PP, Singh K (2012) Crop performance and water- and nitrogen-use efficiencies in dry-seeded rice in response to irrigation and fertilizer amounts in northwest India. Field Crops Res 134:59–70

Majeed A, Muhmood A, Niaz A, Javid S, Ahmad ZA, Shah SSH, Shah AH (2015) Bed planting of wheat (*Triticum aestivum* L.) improves nitrogen use efficiency and grain yield compared to flat planting. The Crops J 3:118–124

Malik RK, Yadav A (2008) Direct seeded rice in the indo-Gangetic Plains: progress, problems and opportunities. ACIAR Proc Ser 127:133–143

Malik RK, Yadav A, Singh S (2004) Resource conservation technologies for maintaining health of natural resources in rice-wheat cropping system. *Proc Workshop on Sustaining Agriculture: Problems and* Prospects, November 9–11. Punjab Agricultural University Ludhiana, India

Mehla RS, Verma JK, Gupta RK, Hobbs PR (2000) Stagnation in the productivity of wheat in the Indo-Gangetic plains: zero tillage seed-cum-fertilizer drill as an integrated solution: *Rice-wheat consortium series*. Rice-wheat consortium for Indogangetic plains, New Delhi, India, p 12

Mirsky SB, Gallandt ER, Mortensen DA, Curran WS, Shumway DL (2010) Reducing the germinable weed seedbank with soil disturbance and cover crops. Weed Res 50(4):341–352

Moghadam T, Shirani-Rad HR, Mohammadi AHN, Habibi G, Modarres-S MSAM, Dolatabadian MA (2013) Response of six oilseed rape genotypes to water stress and hydrogel application. Pesquisa Agro Tropical 39:243–250

Morris NL, Miller PCH, Orson JH, Froud-Williams RJ (2010) The adoption of non-inversion tillage systems in the United Kingdom and the agronomic impact on soil, crops and the environment-a review. Soil Till Res:10(1–2): 1–15

Naresh RK, Singh SP, Misra AK, Tomar SS, Kumar P, Kumar K, Kumar S (2014) Evaluation of the laser leveled land leveling technology on crop yield and water use productivity in Western Uttar Pradesh. African J Agric Res 9:473–478

Narjary B, Aggarwal P, Singh A, Chakraborty D, Singh R (2012) Water availability in different soils in relation to hydrogel application. Geoderma 187:94–101

Pandey S, Velasco L (2005) Trends in crop establishment methods in Asia and research issues. In: Heong KL, Hardy B, Toriyama, K (eds) Rice is life: scientific perspectives for the 21st century. International Rice Research Institute, Los Baños, Philippines and Japan Inter-national Research Center for Agricultural Sciences, Tsu-kuba, pp 178–181

Ram H, Yadvinder-Singh, Saini KS, Kler DS, Timsina J, Humphreys EJ (2012) Agronomic and economic evaluation of permanent raised beds, no tillage and straw mulching for an irrigated maize-wheat system in northwest India. Expl Agric 48:21–38

Ram H, Yadvinder-Singh, Saini KS, Kler DS, Timsina J (2013) Tillage and planting methods effects on yield, water use efficiency and profitability of soybean–wheat system on a loamy sand soil. Expl Agric 49:524–542

Ram H, Singh G, Aggarwal N (2016) Pigeonpea (*Cajanas cajan* l.) performance in relation to sowing methods, fertilizer application methods and rates. Prog Res 11:66–68

Sayre KD (2004) Raised-bed cultivation In: Lal R (ed) Encyclopedia of soil science. Marcel Dekker Inc. (eBook Site Online publication 04/03/2004)

Sharma J (2004) Establishment of perennials in hydrophilic polymer- amended soil. SNA Res Conf 42:530–532

Sharma SN, Bohra JS, Singh PK, Srivastava RK (2002) Effect of tillage and mechanization on production potential of rice (*Oryza sativa*)-wheat (*Triticum aestivum*) cropping system. Indian J Agron 47:305–310

Sharma AR, Singh R, Dhyani SK, Dube RK (2010) Moisture conservation and nitrogen recycling through legume mulching in rainfed maize (*Zea mays*)–wheat (*Triticum aestivum*) cropping system. Nutr Cycling Agroecosys 87:187–197

Sidhu HS, Manpreet-Singh, Humphreys E, Yadvinder-Singh, Balwinder-Singh, Dhillon SS, Blackwell J, Bector V, Malkeet-Singh, Sarbjeet-Singh (2007) The Happy Seeder enables direct drilling of wheat into rice stubble. Aust J Exp Agric 47:844–854

Sidhu AS, Kooner R, Verma A (2014) On-farm assessment of direct-seeded rice production system under central Punjab conditions. J Crop Weed 10:56–60

Sidhu HS, Singh M, Yadvinder-Singh, Blackwell J, Lohan SK, Humphreys E, Jat ML, Vicky-Singh, Sarbjeet-Singh (2015) Development and evaluation of the Turbo Happy Seeder for sowingwheat into heavy rice residues in NW India. Field Crops Res 184:201–212

Singh A, Kumar R, Kang JS (2014a) Crop residues and nitrogen improve the productivity of direct seeded rice and transplanted rice. Curr Agric Res J 2:14–29

Singh B, Kular JS, Ram H, Mahal MS (2014b) Relative abundance and damage of some insect pests of wheat under different tillage practices in rice-wheat cropping system. Crop Protec 61:16–22

Singh AP, Bhullar MS, Yadav R, Chowdhury T (2015) Weed management in zero-till wheat. Indian J Weed Sci 41:161–166

Srivastava RK, Sah D, Singh R (2002) Studies on varying mode of tillage operations, seeding rates and fertility levels on yield of wheat var. HUW-234 under puddle rice-wheat sequence. Res Crops 3:332–334

Tang O, Wu X, Li C, Wu C, Ma X, Huang G (2013) Long-term effect of year-round tillage patterns on yield and grain quality of wheat. Plant Prod Sci 16:365–373

Upadhyay IP, Jha SK, Karki TB, Yadav J, Bhandar B (2016) Tillage methods and mulch on water saving and yield of spring maize in Chitwan. J Maize Res Dev 2:74–82

Yangyuoru M, Boateng E, Adiku SK, Acquah D, Adjadeh TA, Mawunya F (2006) Effects of natural and synthetic soil conditioners on soil moisture retention and maize yield. J App Eco 9:1–8

Yezdani F, Allahdadi I, Akbari GAB (2007) Impact of the super absorbent polymer on growth of soybean under drought stress condition. Pak J Bio Sci 10:4190–4196

Yi L, Yufang S, Shenjiao Y, Shiqing L, Fang C (2011) Effect of mulch and irrigation practices on soil water, soil temperature and the grain yield of maize (*Zea mays L*) in Loess Plateau, China. Afr J Agric Res 6:2175–2182

Biotechnological Tools for Enhancing Abiotic Stress Tolerance in Plant

Amit Kumar, Rakesh Singh Sengar, Ashu Singh, Rekha Dixit, and Raj Singh

Abstract

Environmental abiotic stresses, for example, high temperatures, low water accessibility, mineral insufficiency, high salt levels, and lethality, are serious threats to the harvest survival which someway influence the harvest yield. A few traditional strategies are used for sustainable harvest efficiency; however, with the expanding abiotic stress because of changing climatic conditions and enhancing pressure of populace, the conventional procedures of overcoming abiotic stress are not ready to meet the demands. Biotechnology is the best ways by which the productivity of crops can be improved by enhancing their ability to resist or tolerate biotic and abiotic stresses. In biotechnology different strategies are involved for the improvement of harvest yield and quality. This chapter concentrates on the traditional and new enhanced biotechnological strategies for the betterment of abiotic stress tolerance in plants.

Keywords

Abiotic stress · Candidate gene · HSPs · MAS · QTL mapping

A. Kumar (✉) · R. Dixit
Department of Biotechnology, Faculty of Science, Swami Vivekanand Subharti University, Meerut, Uttar Pradesh, India

R. S. Sengar · A. Singh
Department of Agricultural Biotechnology, College of Agriculture, S.V.P University of Agriculture & Technology, Meerut, Uttar Pradesh, India

R. Singh
Department of Botany, Faculty of Science, Swami Vivekanand Subharti University, Meerut, Uttar Pradesh, India

R. S. Sengar, A. Singh (eds.), *Eco-friendly Agro-biological Techniques for Enhancing Crop Productivity*, https://doi.org/10.1007/978-981-10-6934-5_8

1 Introduction

Biotechnology is the best way by which the productivity of crops can be improved by enhancing their ability to resist or tolerate biotic and abiotic stresses. In biotechnology different strategies are involved for the improvement of crop yield and quality. In this chapter, we will focus on the impact of environmental conditions such as extreme temperature, salinity on crop plants, and water accessibility. Figure 1 shows a number of physical factors that may impose an abiotic stress on plants and adversely affect their quality and yield. These stresses also affect physiological, biochemical, and developmental processes of crop (Amit kumar and Sengar 2013). These include a number that can be gathered together as temperature stresses (heat, chilling, and freezing) which in turn belong to a bigger subgroup that can be arranged as stresses that result in water shortfall. The figure likewise accentuates the point that most abiotic stresses straightforwardly or indirectly lead to the generation of free radicals, creating oxidative stress and reactive oxygen species.

The impact of abiotic stresses on crop yield compared with biotic stresses (weed, pest, and disease effects) is shown in Table 1. One of the first things to notice is the large difference between the average yields of crops and the record yields. It is clear from these data that the major difference between record yield and average yield is accounted for by abiotic stress. Thus, the variation in environmental conditions from 1 year to the next produces such a variation in yield for wheat in which the average yield is only 13% of the maximum. In contrast, the control of biotic stresses in industrialized farming is such that they tend to reduce the annual yield by a fairly

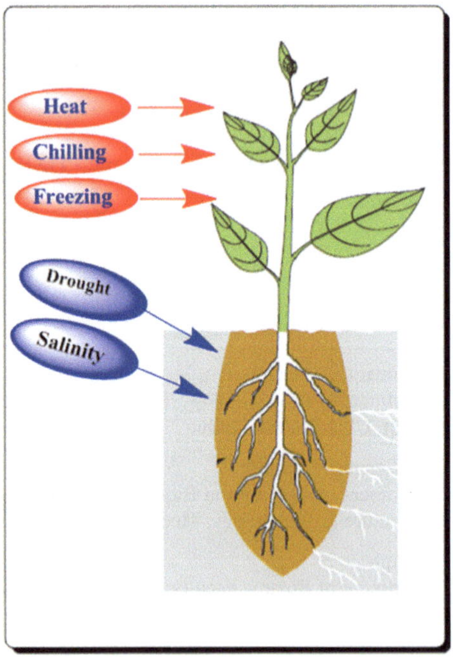

Fig. 1 Types of environmental stress

Table 1 Average and record yields of some major crops

Crop	Record yield (kg/ha)	Average yield (kg/ha)	Average yield (% of record yield)	Average losses (% of record yield)	
				Biotic	Abiotic
Wheat	14,500	1880	13.0	5.0	82.1
Barley	11,400	2050	18.0	6.7	75.4
Soybean	7390	1610	21.8	9.0	69.3
Corn	19,300	4600	23.8	10.1	65.8

Source: Bray et al. (2000)

stable proportion, which is generally less than the most adverse abiotic stresses. Improving the tolerance of crops to abiotic stresses could therefore enable them to maintain growth and development during the normal fluctuations of adverse conditions and consequently buffer crops against the large swings in yield experienced from 1 year to the next.

In the long time, the predicted depletion of the ozone layer and climate changes related with a worldwide temperature alteration are probably going to add to the burden of environmental stresses on harvest plants and enhance the imperative to prepare stress-tolerant varieties. Moreover, there is increasing pressure to extend the area of harvest cultivation to environments that are not ideal for the development of significant harvests (high-salt condition). The preparation of stress-tolerant plant is therefore a major aim of agricultural biotechnology and one that is likely to become increasingly important.

2 Nature of Abiotic Stress

When discussing the subject of stress tolerance, it is necessary first to try to define stress in relation to plant physiology. Plants are subject to many types of fluctuation in the physical environment. Many of the strategies used by animals to avoid the effects of these fluctuations are not available to plants, because of the sessile nature of their growth habit. Plants therefore depend largely upon internal mechanisms for tolerating variations in the external environment. Not all such fluctuations present a stress to plants, since they can adapt to typical variety by virtue of their plasticity. Consequently, plants are adjusted to work in a fluctuating environment, and normal outside changes are countered by inward change without detriment to development and advancement. It is only acute or chronic extremes of environmental condition that lead to environmental stress that has the potential to cause physical damage to the plant.

Abiotic stresses, for example, high temperatures, low water accessibility, high salt levels, mineral insufficiency, and toxicity, are regularly skirmish by plants in both natural and agriculture frameworks. By and large, a few classes of abiotic stress challenge plants in blend; for instance, high temperature and shortage of water can be exacerbated by mineral toxicities that constrain root development. Over a range of cropping frameworks around the globe, abiotic stresses are estimated to shorten yields not as much as a half of that possible under perfect development conditions (Boyer 1982).

Abiotic stress responses, particularly to water inadequacy and high level of salts, are unpredictable physiological and morphological phenomena in plants (Wang et al. 2003). At the cellular level, alterations in extracellular solute concentrations cause osmotic stress and finally inadequate water availability. This water deficiency causes an abatement in turgor pressure and an expansion in concentration of intracellular solutes, which gives a strain on layers and macromolecules.

Abiotic stresses exhibit a big challenge in our journey for sustainable food generation as these may lessen the potential yields by 70% in cultivates. Intense water insufficiency hinders photosynthesis (Gallagher et al. 1975). Changes in the worldwide atmosphere, remarkably in territorial spatial and temporal temperature designs, are anticipated to have vital outcomes for crop production (Parry and Duinder 1990); both plant development and improvement are influenced by temperature (Porter and Moot 1998).

Given the range of abiotic stresses to which plants are exposed, it might be thought that a wide range of different strategies would be required to engineer particular types of stress. This chapter will concentrate on the variety of different stresses. These damages result from water deficit caused by various distinctive environmental conditions, including temperature, cold, drought, and salinity.

3 Tolerance to Drought Stress

Agriculture is a major client of water assets in numerous regions of the world. With expanding aridity and a developing populace, water will turn significantly scarcer in the future. Suboptimal accessibility of water for unlimited plant development and transpiration, i.e., drought, is a noteworthy constraint to agriculture yield (Boyer 1982 and Delmer 2005). Drought is a standout among the most well-known environmental stresses that influence development and advancement of plants through shifts in metabolism and expression of gene. It is changeless to agriculture creation in many developing nations and periodic reason for losses of agriculture production in created ones (Ceccarelli and Grando 1996). In India, 29 percent of the total cultivable area faces drought condition, out of which 10 percent is under severe drought (Anonymous 2003).

Harvest plants developed under dry spell conditions are presented to a mix of stresses that are attributable from high temperatures, over the top irradiance, and soil resistance to root infiltration and low water potential. Drought is one of the main factors adding to a few yield losses of wheat developed in minimal land and to essentially diminish yields in temperate area (Morris et al. 1991; Trethowan et al. 2001). Drought is the most genuine abiotic stress restricting wheat production in various parts of the world (Chaves et al. 2003).

In spite of the fact that it is not a crop plant, *Arabidopsis* has played an important role in the explanation of the essential procedures underlying stress tolerance, and the information acquired has been transferred to a specific degree to imperative food

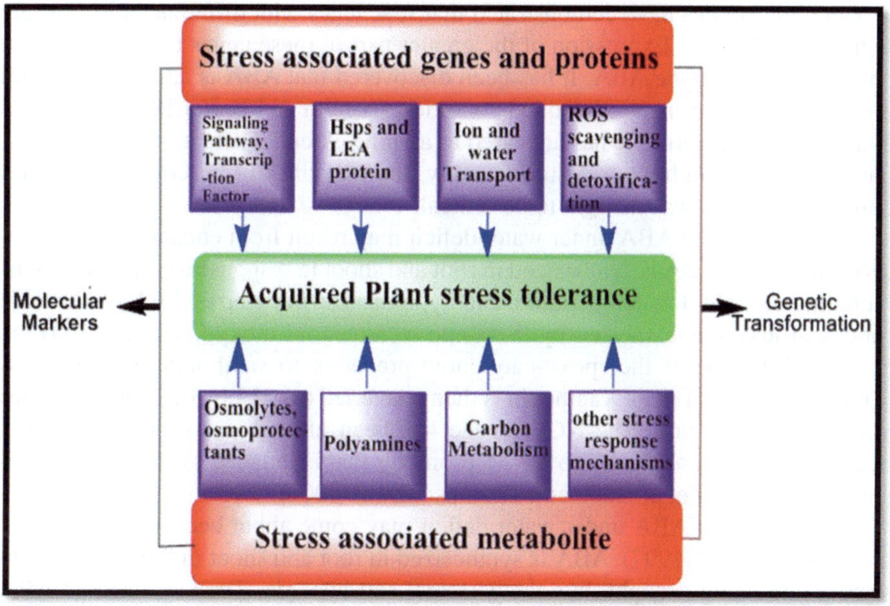

Fig. 2 Approaches to develop stress tolerance plant

plants. Different types of the gene known to be required in stress resistance have been isolated at first from *Arabidopsis*. Two general procedures for the metabolic engineering of abiotic stress resistance have been proposed which expanded production of particular specific desired compounds, or the decrease in the enzymatic step is typically managed by the tendency of cell frameworks to reestablish homeostasis, thus limiting the potential of this approach (Fig. 2). The transgenic approach is a helpful innovation to beat reproductive isolation among species and use useful exotic genes.

Numerous loci for gene that control tolerance to abiotic stress in plants have been recognized by hereditary investigation. However, numerous genes that control agronomically imperative characteristics remain to be recognized and changed to create new varieties with desirable traits. There is a confirmation that transgenic plants in which the expression of a solitary gene has been changed have improved resistance to abiotic stress.

3.1 Trait Associated with Drought Stress

Plant reaction to drought stress is a complex procedure molded by various segment reactions that both interact and vary in their individual reactions to the intensity and span of water deficiencies. It is a standout among the most widely recognized environmental stresses that influence development and advancement of plants through modifications in metabolism and gene expression.

Genomic studies have demonstrated that few genes with different capacities are induced by drought. Expression of the greater part of these genes is induced by the collection of hormone ABA. It has been discovered that ABA is amassed 40 times more in drought over the controlled condition (Ingram and Bartel 1996; Shinozaki et al. 2003). The hormone has appeared to assume essential role in imparting resistance against drought stress. Subsequently, the part of ABA in controlling genes under water stress was thought to be critical.

Accumulation of ABA under water deficit may result from enhanced biosynthesis (Bray 1997). ABA is synthesized in root and shoot in response to various stresses including drought, low temperature, and hypoxia. The key step of ABA synthesis in roots and leaves is catalyzed by 9-cis-epoxycarotenoid dioxygenase (NCED), an enzyme that converts the epoxy-carotenoid precursor to xanthonin in the plastids (Qin and Zeevaart 1999). Xanthonin is then converted to ABA by cytosolic enzymes via abscisic aldehyde. The NCED gene has been cloned in cowpeas, tomatoes, beans, and maize and shown to have a promoter that is induced by drought in roots and leaves (Taylor et al. 2000 and Thompson et al. 2000).

Aggregation of ABA under water deficit may come about because of increased biosynthesis (Bray 1997). ABA is synthesized in root and shoot in light of different stresses including drought, low temperature, and hypoxia. The key stride of ABA synthesis in roots and leaves is catalyzed by 9-cis-epoxycarotenoid dioxygenase (NCED), a protein that changes over the epoxy-carotenoid precursor to xanthonin in the plastids (Qin and Zeevaart 1999). Xanthonin is then changed over to ABA by cytosolic proteins by means of abscisic aldehyde. The NCED gene has been cloned in tomatoes, cowpeas, maize, and beans and appeared to have a promoter that is induced by dry season in roots and leaves (Taylor et al. 2000; Thompson et al. 2000).

Availability of *Arabidopsis* mutants for defined traits opened paths to dissect biochemical processes at genetic level. When ABA-deficient and ABA-insensitive (abi) mutants were subjected to drought stress, several genes were induced. This suggested that expression of these genes followed an ABA-independent pathway. Indeed, alterations in poly(A) RNA population were observed prior to accumulation of ABA in water-stressed pea leaves.

While working with *rd* clones in *Arabidopsis thaliana*, Yamaguchi-Shinozaki et al. (1992) reported that *rd29* cDNA was induced very quickly and strongly by desiccation stress. This observation suggested that *rd29* could have no less than two cis-acting components. One was involved in ABA-associated slow response and the other functioned independently of ABA. Genomic clone investigation demonstrated that rd29 were situated in tandem in an 8048-bp-long region of *Arabidopsis* genome.

3.2 Marker-Assisted Selection for Drought Stress

Marker-assisted selection (MAS) is the breeding technique in which determination for a gene depends on molecular markers firmly connected to the gene of interest instead of the gene itself, and the markers are utilized to screen the joining of the desirable allele from the contributor source. The correct route in which hereditary diagnostics are applied in plant breeding will be impacted by numerous considerations, both biological and economic, yet the accessibility of cost-effective diagnostic technology that is good with the goals will be the main factor.

Molecular breeding requires the following innovation: hereditary maps, molecular markers connected to agronomic attributes, high throughput, automated diagnostic procedure, and an adjustment in breeding practice that takes full favorable position of the information given by such indicative examines. The prescient estimation of molecular markers utilized as a part of MAS relies upon their inherent repeatability, position of map, and linkage with economically imperative qualitative and quantitative attributes. The presence of a tight linkage between qualitative characteristics and molecular marker might be helpful in MAS to increase gain from selection. MAS may have potential in populace and inbreed line advancement. Whenever QTLs and single gene are satisfactorily mapped, they can be isolated by map-based cloning methodologies. The viability of any MAS will rely upon the precision of the phenotypic order of attribute expression and the level of linkage between the markers and traits of interest.

Molecular marker is utilized to develop firmly separated genetic maps of a specific genome utilizing an appropriate mapping populace of plants (Sehgal et al. 2008; Kumar 1999). Molecular marker mapping has reinforced our realization that, in a few taxonomic groups of yield plants, e.g., the Gramineae, which share the similar fundamental chromosome number, the linkage group and the individual chromosome maps look fundamentally the same, i.e., they are syntenic, especially when we take out the repetitive DNA and analyze the maps for single duplicate sequences. This implies that even between crops as differing as wheat and rice, the genes we are occupied with are fundamentally the same in both species, and their areas in the hereditary maps are very much alike. The greater part of these maps has been made utilizing RFLP (restriction fragment length polymorphism) markers, and more recently, different PCR-based markers are utilized for the development of genetic map (Gupta and Rustogi 2004; Sehgal et al. 2008), for example, SSR (simple sequence repeats), RAPD (random amplified polymorphic DNA), AFLP (amplified fragment length polymorphism), STS (Sequence Tagged Sites), and so on permit loci controlling quality related to stress tolerance to be recognized and mapped in the genome.

3.3 Candidate Gene Related to Drought Stress

Conventional assessment of plant reactions to drought has depended chiefly on leaf to atmosphere gas exchange, development investigation, and in vitro assurance of enzymes, metabolites, and gene expression patterns. The main molecular way to help breeders in their endeavors to expand drought tolerance has been with molecular marker, genomics, and post-genomic techniques (Nguyen and Blum 2004; Lancers et al. 2004). Candidate gene can be characterized as the gene demonstrating molecular polymorphisms and hereditarily related with QTL or statistically connected with drought tolerance trait. The candidate genes can be partitioned into two classes on the premise of their functions: in the first category, gene qualities are

Table 2 List of candidate gene for drought tolerance

Candidate gene	Gene product	Function	References
DREB	Drought-responsive element (DRE) binding factor	Transcription factor	Oh et al. (2005) and Ito et al. (2006)
ERA1	Farnesyltransferase	Negative regulator of ABA sensing	Wang et al. (2005)
Mn-SOD	Mn-superoxide dismutase	ROS-scavenging proteins	McKersie et al. (1996)
AVP1	Vacuolar H^+ pyrophosphatase	Ion transport	Park et al. (2005)
HVA1	LEA proteins	Protective proteins	Xiao et al. (2007)
P5CS	Pyrroline-5-carboxylate synthetase	Proline synthesis	Zhu et al. (1998)
ABF3	ABA-responsive element	Transcription factor	Kang et al. (2002)
PARP	Poly ADP-ribose polymerase	ROS (reactive oxygen species)-scavenging mechanism	De Block et al. (2005)
LTP	Lipid transfer protein	Membrane biogenesis	Vinod et al. (2006)
NADP-Me	NADP-malic enzyme	Carbon metabolism	Laporte et al. (2002)
MAPK	Mitogen-activated protein kinase	Signaling factor	Shou et al. (2004)
HRD	Hardy protein	Transcription factor	Karaba et al. (2007)
AtNF-YB1	NF-YB protein	Transcription factor	Nelson et al. (2007)
LFY	Leafy protein	Transcription factor	Tuyen and Prasad (2008)
SRK2C	Protein kinase	Signaling factor	Umezawa et al. (2004)

required in cell assurance during stress, for example, osmotic, and, second, genes are included in the direction of different genes for the stress responses.

Various such genes have been recognized, separated, cloned, and expressed in plants, which are potential sources or resistant to drought stress. In Table 2 a portion of the candidate genes have been recognized for drought tolerance accumulated from literature survey.

Chen et al. (2002) recognized groups of transcription components directed by the abiotic stress in *Arabidopsis*. The expression patterns of these transcription components are extremely complex which propose that stress tolerance and resistance are controlled at the transcriptional level by an amazingly multifaceted gene regulatory system. Genome-wide transcriptome investigation has distinguished several candidate genes encoding transcription components which are induced or repressed by ecological stresses (Chen et al. 2007a, b).

In *Arabidopsis* plants, transcript profiling (1300 genes) for overexpressing the gene encoding for dehydration-responsive element binding protein (DREB 1a) distinguished 12 gene as cold and drought season target gene having a place with DREB 1 transcription factor family (Seki et al. 2000). The gene enacted by drought incorporates those included in mechanism to maintain a strategic distance from water misfortune, secure the cellular machinery, and repair damage (Shinozaki and Yamaguchi-Shinozaki 2007). The DREB 2 genes, incited by drought stress, have been accounted for to give drought tolerance in a few plant species, for example, *Arabidopsis* (Sakuma et al. 2002), rice (Dubozet et al. 2003), wheat (Egawa et al. 2006), maize (Qin et al. 2007), and barley DRFs (Xue and Loveridge 2004).

3.4 QTL Mapping for Drought Tolerance

Quantitative trait is represented by polygenes and is markedly influenced by the environment. Thus, it demonstrates a constant variation rather than the discrete variation that is characteristic of qualitative traits. Polygenes are those genes that have little however combined impact on the concerned traits, and a few polygenes influence a single trait. A quantitative trait locus is a position in a chromosome that contains at least one polygenes involved in the assurance of a quantitative trait.

Many of DNA marker maps are not adequately thick to accomplish satisfactory QTL mapping, since inadequate marker maps extremely confine the power of QTL mapping. The mapping populace must be relatively huge in order to identify QTLs having minor impacts, and the biological relevance of the revealed QTLs relies upon the cutoff decided for statistical significance. In QTL mapping, natural variables and genetic background have a marked effect on the outcomes; as a result, some QTLs may be detectable in some but not in other environments. A standout among the most capable applications of QTL mapping is to analyze gene x gene and gene x environment interactions, yet this requires some extensive, tedious trials to allow a thorough investigation of a system.

QTL mapping involved testing DNA markers all through the genome for the probability that they are related with a QTL. Individuals in a suitable mapping populace are investigated regarding DNA marker genotypes and the phenotypes and the phenotype of interest. For every DNA marker, the individuals are split into classes as indicated by marker genotypes. These markers are being utilized to distinguish drought-related quantitative trait loci and their productive transfer into economically developed crop varieties of rice, wheat, maize, and millet. Great hereditary maps in view of molecular marker technologies are presently accessible for large cereal species (Snape et al. 2005; Langridge et al. 2006). In cultivated species with huge, complex genomes, QTL investigation is a vital tool in the recognizable proof of genetic markers to help breeding efforts. This approach is complicated in wheat on account of the polyploidy nature of the genome. Furthermore, the low levels of polymorphism, however, is straightforward in rice, barley, and maize (Snape et al. 2005)

Studies on the abiotic stress resistance of cereals incorporate the broad investigation of QTLs linked to the field assessment of stress tolerance (Langridge et al. 2006). Approach and improvement of molecular markers in quantitative genetics greatly incredibly encourage the investigation of complex quantitatively acquired inherited traits by the development of high-density genome linkage maps for harvests, for example, wheat (Xiao et al. 1996). This infers the utilization of molecular markers, and interim mapping is an intense approach, which allows the distinguishing proof and genetic mapping of loci controlling complex traits like grain yield and its contributing traits having extraordinary significance in plant breeding (Broman and Speed 1999). Lacking accessibility of water during drought may constrain the size of the plant and furthermore influence the improvement of different plant parts. Recent advances in plant genomics have prompted the identification of a vast number of potentially beneficial water stress-related gene, in addition to innovations for gene overexpression or silencing. Also, these can be brought into transgenic plants under the control of proper promoters and are transmitted to subsequent generations (Delmer 2005; Ma and Bohnert 2007).

Recently, Tuberosa and Saliva (2007) detailed that genomic-based methodologies give access to agronomically attractive alleles present at quantitative trait loci (QTLs) that influence such responses, accordingly enabling us to enhance the drought resistance and yield of crops in submerged restricted conditions more effectively. QTLs for drought tolerance, one each situated on chromosome 7AL (Quarrie et al. 2005), and three QTLs for heat resistance, one each situated on chromosome arms 2DL (Mohammadi et al. 2008a), IBS, and 5BL (Mohammadi et al. 2008b) and their linked SSR markers, were utilized for validation for their inevitable exploitation during amid MAS.

4 Tolerance to Heat Stress

For many years, it has been known that heat stress applied to a wide range of organisms induces a specific set of heat shock proteins (HSPs); they fall into five classes, four of which are highly conserved in prokaryotes and eukaryotes. These four are categorized according to size as the HSP 100, HSP 90, HSP 70, and HSP 60 classes whose members appear to function as molecular chaperones. Some of them are expressed constitutively and are involved in normal protein synthesis and folding. Those induced by heat appear to be involved in countering the effects of heat stress by protecting or refolding denatured proteins. Their expression is induced by heat treatment and, in some cases, can be correlated with the acquisition of thermotolerance. The fifth group of several classes of small HSPs is particularly abundant in plants, but their function is not yet clear.

In a way analogous to strategies for engineering cold tolerance, individual HSPs have been transformed into plants to enhance heat tolerance. However, it is also known that the rapid heat shock response is coordinated by a heat shock factor. This protein is expressed constitutively but in normal conditions exists as a monomer bound to one of the HSP70 proteins. Heat binds to a heat shock element common to the promoters of HSP genes. Sequence NGAAN: five to seven of these repeats occur in the promoter close to the TATA box.

When the AtHSF 1 gene was overexpressed in *Arabidopsis*, the translation factor was not dynamic, and there was no impact on thermotolerance. Nonetheless, combination of AtHSF1 to the N or C end of the gus A reporter gene created a fusion protein that could trimerize without heat. Transformation of this fusion protein into *Arabidopsis* created transgenic plants that expressed HSPs constitutively and showed upgraded thermotolerance without requiring earlier heat treatment.

4.1 Marker-Assisted Selection for Heat Stress

Molecular marker have been applied in quantification of genetic diversity, genotype identification, mapping and tagging of utilize genes, and MAS in cereals for biotic stresses, abiotic stresses, and quality traits. Many genes for those traits have been mapped, tagged, and cloned, and linked markers have been created. Those have been successfully used in marker-assisted breeding program to develop genotypes with resistance.

This approach involves the use of molecular markers associated with important agronomic traits for selection of desirable plants in the segregating generations. This is particularly desirable for traits, for which selection through conventional method of plant breeding is either difficult or cost/time ineffective. This has assumed significance in recent years due to the realization that improvement of traits like water use efficiency and nutrient use efficiency along with tolerance to a number of other abiotic stresses would be necessary to augment the productivity of all major food crops, as is required to meet the future demands of food and nutritional security.

In order to meet the demands of molecular breeding, one needs to have the knowledge either about the marker trait association, as determined in case of linkage-based QTL interval mapping and LD-based association mapping, or about the genomic estimated breeding values of individual markers as worked out in case of genomic selection. The merits and limitation of these three different approaches have been widely discussed, and improvements in the basic proposed initially have been suggested.

MAS and hereditary engineering are two most basic molecular methodologies for enhancing stress resistance in plants. With the appearance of molecular markers, MAS has turned into a fundamental component of new discipline named as molecular breeding with the assistance of which allelic variation among the genes' basic traits can be accurately and proficiently detected (Mohammadi et al. 2008b). Various markers, for example, RAPDs, SSRs, RFLPs, and AFLPs, have been accounted for different abiotic stresses for accomplishing mapping of the QTL involved in stress tolerance (Korzun et al. 1999). Relatively, however, limited research has been directed to recognize genetic markers related with heat resistance in various plant species.

4.2 Candidate Gene Related to Heat Stress

The enhancement of heat tolerance is expected to be much higher if information is generated about the presence of genetic variation for the traits related with heat resistance in crops, and tagging and mapping of these traits with an appropriate marker can be utilized in marker-assisted breeding. It has been observed that marker-assisted selection (MAS) approaches have contributed enormously to a superior comprehension of the hereditary bases of plant stress tolerance, which prompted the improvement of plants with enhanced resistance of abiotic stress.

A few transgenic approaches have been utilized for the advancement of heat tolerance crops (Rodriguez et al. 2005). Many heat shock proteins (HSPs) have been utilized for the advancement of thermotolerance crops. Many scientists studied the expression level of HSPs by making an improvement in the heat shock factor (HSF) which is responsible for the production of high-temperature stress-tolerant *Arabidopsis*. Various such genes have been recognized, isolated, cloned, and expressed in plants, which are potential sources or resistant to drought stress. In Table 3 some of the candidate genes have been identified for heat tolerance gathered from literature survey. Candidate gene can be defined as the gene indicating molecular polymorphisms and genetically related with QTL or measurably associated with thermoresistance trait.

Table 3 List of candidate gene for heat tolerance

Candidate gene	Function of candidate gene	References
HSP 101	Synthesis of heat shock protein for temperature tolerance	Queitsch et al. (2000)
APX 1	H2O2 detoxification and conferred heat tolerance	Shi et al. (2001)
HSP 70	Synthesis of heat shock protein for temperature tolerance	Montero-Barrientos et al. (2010)
Cod A	Glycine betaine synthesis for tolerance to temperature tolerance	Alia et al. (1998)
HSP 17.7	Synthesis of sHSP	Malik et al. (1999)
FAD 7	Enhanced the level of unsaturated fatty acid and provide temperature tolerance	Sohn and Back (2007)
AtPARP2	Increased tolerance to heat stress	Vanderauwera et al. (2007)
TLHS1	Synthesis of class I HSP	Park and Hong (2002)
AtHSF 1	HSF fused with β-glucuronidase and such modification will increase HSP production	Lee et al. (1995)
ANP 1	H2O2-responsive MAPK kinase kinase (MAPKKK) production to protect against the lethality in HT	Kovtun et al. (2000)
MT-sHSP	Molecular chaperone	Sanmiya et al. (2004)
Dnak 1	Tolerance to temperature	Ono et al. (2001)
BADH	Overproduction of GB osmolyte that will enhance the heat tolerance	Salvucci and Crafts-Brandner(2004)

4.3 QTL Mapping for Heat Tolerance

One of the aims of molecular mapping is to produce a sufficiently fine-scale map to pinpoint the location of genes that play a role in determining important agronomic traits. Many of these traits are described as quantitative; that is, they are impacted by various hereditary and environmental factors. However, it is possible to map genes that have majorly affected quantitative traits by assessing the correlation between estimation of the quantitative trait and the allelic states at linked hereditary markers. A QTL (quantitative trait locus) is therefore a chromosomal location where there is considered to be a reasonable probability that practically unique alleles segregate and cause huge impacts on a quantitative trait. QTL mapping requires a statistical analysis of molecular marker and phenotypic information from an extensive segregating populace to determine those markers where allelic polymorphism correlates with the quantitative trait phenotype. This primary QTL mapping is coarse and locates the gene within a chromosome region of approximately 10–30 cm, which could contain several hundred genes. To identify the gene involved in the quantitative trait, two methods are available: positional cloning and association mapping.

Positional cloning requires further steps to the QTL to a much finer resolution and relates this map position to the DNA sequence. One strategy is to cross nearly isogenic lines in which the only allelic variation occurs in the short region of the coarse-mapped QTL. In the resulting populations, the QTL is described as a mendelized and fine map with more precise centimeter distance between the QTLs, and adjacent molecular markers can be produced. This is relatively straightforward when there are many polymorphic markers mapped in the region, but this is currently possible only for those plants whose genome has already been, or is in the process of being, sequenced. At this stage, the markers closest to the QTL are used to anchor the genetic map to the physical map, and it may then be possible to determine the gene responsible from candidate gene in the location by identifying the mutation responsible for the QTL effect. Alternatively, it may be necessary to test each predicted coding sequence in the region functionally, by overexpressing or downregulating the gene.

To date, the level of accomplishment in distinguishing hereditary markers related with high-temperature tolerance in wheat, and to be sure other crop species, has been limited. Dependable marker trait associations are essential for a viable marker-assisted breeding program (Kato et al. 2000), and these are most successfully settled by means of quantitative trait locus mapping (Patterson 1998). QTLs for heat resistance have been reported by various researchers (Mohammadi et al. 2008a; Mason et al. 2010; Pinto et al. 2010; Barakat et al. 2011; Tiwari et al. 2013). At present, the good breeders' technique is to challenge the material by sowing late, in the information that this will expose the plants to high temperatures amid grain filling; after this, determination is regularly based on yield performance. Recent outcomes demonstrate that synchronous improvement of yield potential and thermotolerance is possible and that the CIMMYT strategy to grow high-yielding early developing wheat lines is promising for South Asia and Mexico (Mondal et al. 2013). The recognizable proof of QTLs for heat tolerance has given a chance to deploy MAS for the improvement of the high-temperature resistance wheat (Paliwal et al. 2012).

5 Tolerance to Salt Stress

Salt resistance might be characterized as a differential impact on different life processes of a similar tissue concentration of salt in different genotypes of a species. There is considerable evidence that genotypes differ in resistance to a similar amount of salt in their tissues. However, the enzyme and cellular processes of halophytes are as sensitive to salt as those of glycophytes. Most of the crops grow under the saline environment by which yields are generally poor in the initial 3–4 years of the reclamation of saline lands through application of gypsum. Therefore, improving salt tolerance of crops may help to a great extent in increasing food production of the growing population of India. Therefore, cultivation of salt-tolerant varieties can play important role in rehabilitation of such lands besides reducing the requirement of input in the form of chemical amendment.

5.1 Salt Tolerance Mechanism at Physiological Levels

Salt tolerance is increasingly becoming a major target for crop improvement as substantial areas of irrigated land are damaged by the accumulation of salt. Furthermore, the pressure for land has made it necessary to consider the possibility of growing crops in more saline conditions, with poorer quality water. As described above, saline conditions lead to osmotic stress by preventing water uptake by the roots and water efflux from the cells. However, the accumulation of Na^+ and Cl^- ions in the cytoplasm may also have direct toxic effects by inhibiting protein synthesis, photosynthesis, and susceptible enzymes (Estan et al. 2005). Thus, strategies for engineering water stress tolerance via the production of compatible solutes may provide protection against the osmotic effect of saline conditions, but not against ion toxicity. Additional approaches to minimize the toxic effects of specific ions may also be required. Strategies for engineering salt tolerance have been developed by comparing the response to salt stress of plants that are sensitive to high-salt conditions (glycophytes) with that of plants that can tolerate high-salt conditions (halophytes). As a general rule, glycophytes respond to salt stress by accumulating osmoprotectants, whereas halophytes employ specific mechanisms to avoid the toxic effects of Na^+ and Cl^- ions. Some halophytes actually excrete salt via specialized glands on their leaf surfaces, but it is more common for plants to regulate Na^+ ions out of the cell, whereas other cells with large vacuoles may act as sinks for the accumulation of excess sodium by transport into the vacuole.

Resistance to salt stress is an unpredictable phenomenon at both the entire plant level and the cellular level, and serious research endeavors have concentrated on understanding the physiological basis of resistance in higher plants (Cuartero et al. 2008). Different physiological parameters which are highly specific for saline condition were identified and used regularly for the selection of high-tolerant varieties. Some of these parameters were Na^+ transport to shoot, preferential accumulation of Na in older leaves, high Cl^- uptake, lower K^+ uptake, and low Na^+/K^+ ratio (Sharma and Goyal 2003)

One approach to enhancing salt tolerance would be to copy the strategies used by halophytes to transport Na^+ ions out of the cytoplasm. In order to put this into practice, it is necessary to consider the mechanisms of ion transport out of the cytoplasm. The first transgenic experiments considered transport of Na^+ ions into the vacuole. Since this transport is working against a concentration gradient, it requires the input of energy. This is achieved by coupling the transport protein to a proton pump, transporting H^+ ions in the opposite direction. The vacuolar Na^+/H^+ antiporter protein AtNHX of *Arabidopsis* has been studied extensively and is known to be coupled to proton pumps such as AVP1, a vacuolar H^+-translocating pyrophosphatase. An analogy that has been used is to compare AtNHX1 with a revolving door and AVP1 as providing the energy for the door to spin. To increase the traffic through the membrane, one could therefore either increase the number of doors or provide more energy for the existing doors to spin faster.

The first approach was effectively used to engineer salt resistance in tomato plants by transformation with the *Arabidopsis* AtNHX1 antiporter protein gene (Yokoi et al. 2002). AtNHX1 enduring state transcript levels were expanded in response to NaCl, recommending that AtNHX1 transcript upregulation is not particular to ionic stress but rather is regular to osmotic stress (Zhu 2001). High expression of NHX antiporters has been utilized to enhance salt resistance (Wu et al. 2004). The *Arabidopsis* gene has also been transformed into maize and wheat plants, prompting to increased salt resistance, and improved grain yields in salt tolerance have been investigated. For example, rice changed with the OsNHX1 gene demonstrated enhanced development and accumulation of biomass under salt stress conditions (Wu et al. 2005, Chen et al. 2007a, b).

More recently, the effects of introducing a plasma membrane Na^+/H^+ antiporter to extrude Na^+ ions from the plant cells have been investigated. Overexpression of the *Arabidopsis* plasma membrane Na^+/H^+ antiporter AtSOS1 gene in transgenic *Arabidopsis* improved the salt tolerance of callus cultures and reduced the levels of Na^+ ions in the xylem of transgenic plants (Shi et al. 2003). Expression of the fission yeast SOD2 gene in *Arabidopsis* improved seed germination and seedling salt tolerance (Zhao et al. 2006). Transformation of rice with the *E.coli nhaA* plasma membrane Na^+/H^+ antiporter improved germination rate, growth, and yield under conditions of both salt and drought stress (Wu et al. 2005). Interestingly, expression of *nhaA* also increased the levels of proline in the transgenic rice, suggesting that enhanced osmoregulation might in plat be due to activation of proline biosynthesis.

Another approach to enhancing vacuolar transport has been to overexpress the gene encoding *AVP1*, initially in *Arabidopsis*, to increase the proton pumping potential of the vacuole and hence its ability to transport sodium (Gaxiola et al. 2001). This has improved not only the salt tolerance of these experimental plants but also the drought tolerance, since the altered ion balance has enabled the plants to retain more water. Similar results have been obtained by overexpressing the homologues from *Thellungiella halophila* (*TsVP*) in tobacco (Gao et al. 2006) and *Triticum aestivum* (*TVP1*) also in *Arabidopsis* (Brini et al. 2007).

5.2 QTL Mapping for Salt Tolerance

QTL is therefore a chromosomal location where there is considered to be a reasonable probability that functionally different alleles segregate and cause significant effect on a quantitative trait. QTL mapping requires a statistical analysis of molecular marker and phenotypic data from a large segregating population to determine those markers and phenotypic data from a large segregating population to determine those markers where allelic polymorphism correlates with the quantitative trait phenotype.

Molecular technology is a new technology for analyzed the quantitative trait such as salt tolerance and detect the chromosomal location, which is associated with such character, known as Quantitative trait loci. Few scientists have been reported that the QTL for salinity damage at early embryo stage in rice (Prasad et al. 2000).

6 Tolerance to Cold Stress

Diverse plants differ enormously in their capacity to withstand cold and freezing temperatures (Fig. 3). Most tropical plants have essentially no ability to survive chilling conditions. Furthermore, plants can survive a scope of chilling temperatures from −5 to −30 °C relying on the species. Plants from colder districts routinely withstand temperatures even lower than this. It is realized that plants are better ready to withstand colder or chilling stress in the event that they initially experience a time of chilly acclimation, at a low however nonchilling temperature. For instance, wheat plants developed at ordinary warm temperature are killed by chilling at −5 °C; however, after a time of cold acclimation when the plant develops at temperatures underneath 10 °C, they can survive chilling temperature down to −20 °C.

Plants differ in their ability to withstand cold or freezing condition, and cold tolerance is one of the traits that plant breeders have selected for over many centuries. However, there has been little improvement in the cold tolerance of major crop species over the past two decades by conventional breeding, prompting the search for molecular solutions to this problem.

One approach has been to study the mechanisms of freezing resistance that exist in some plant species. Amid the time of acclimation, plants produce various chilly instigated proteins that are expected to assume a part in the ensuing cold tolerance. Around 50 frosty prompted proteins have been recognized in various plant species. These fall into few groups, but they all share the property of being to a high degree hydrophilic. A significant number of them additionally have relatively basic amino acid composition, with current motifs. Some of these groups had already been distinguished as late embryogenesis abundant (LEA) proteins, which seem to play a

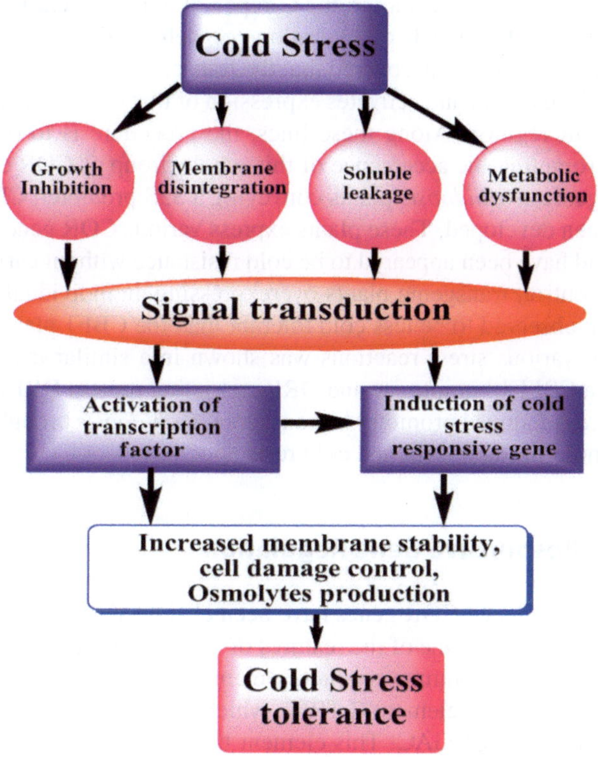

Fig. 3 Cold stress response in plant

defensive role amid seed desiccation. Different groups of proteins encoded a class of genes assigned as chill-responsive gene as per their patterns of expression. The exact function of these chill prompted genes is not yet known, but rather it has been hypothesized that they may contribute specifically to chilling tolerance by mitigating the possibly harming impacts of lack of hydration related with freezing. Overexpression of this chilly prompted protein could in this way be a possible route to be particular technology of cold or chilling stress resistance.

There is some case of the expression of chilly induced proteins in transgenic plants. For instance, constitutive expression of the little, hydrophilic, chloroplast-targeted COR protein COR 15a in *Arabidopsis* enhanced the chilling resistance of chloroplast frozen in situ or protoplast frozen in vitro. In any case, COR 15a expression has no perceptible impact on the survival of frozen plants. One clarification for this observation is that the cold-prompted proteins might be focused to various vulnerable cell segments and that they are altogether required to provide full protection to the cell. By implication, many COR genes would need to be changed into a transgenic crop to get in appreciable change in cold resistance.

One answer for the issue of engineering a multigene trait has risen after the recognition that few distinctive chill resistance-related genes contain a comparative regulatory element in their promoter: the C-repeat (CRT) element/low-temperature response element (LTRE)/lack of hydration- or dehydration-responsive element (DRE). Besides, it has been discovered that the transcription factor CBF1 ties to the CRT/DRE/LTRE element and activates expression of this gathering of genes, which contain the COR regulon. Along these lines, the procedure is to overexpress the CBF1 gene, prompting the acceptance of this whole group of COR cold resistance gene. In transgenic, *Arabidopsis* plants conveying a 35S promoter, CBF1 gene constructs have been developed. These plants express various COR genes without chill acclimation and have been appeared to be cold resistance without earlier cold acclimation. As a control, transgenic plants overexpressing an individual COR protein, COR15a, were observed to be less cold tolerant than the CBF1 plants. The interrelated nature of various stress reactions was shown in a similar investigation. The expression of a CBFF1 homologue and DRE restricting protein DREB1A under the control of a stress-induced promoter in transgenic *Arabidopsis* brought about plants that had enhanced drought, salt, and cold resistance.

6.1 Cold-Responsive Gene Regulation

Some of the cold-induced COR genes have been characterized and the sequence of their promoter compared. One of the features of several different COR genes is that their promoters share a common regulatory element termed the C-repeat (CRT) or low-temperature response element (LTRE), which is five nucleotide long and has a consensus sequence of CCGAC. This element had already been linked to drought resistance and termed the dehydration-responsive element (DRE). The CRT/LTRE/DRE is bound by a CBF transcription factor, CBF1. The structure of CBF1 is shown in Fig. 4, indicating the nuclear localization sequence, DNA-binding domain, and an acidic region that may be involved in interactions with other protein CBF1 expressions which are induced by cold acclimation and lead to the expression of the COR genes. This group of genes, sharing a common regulatory mechanism, has been termed the COR regulon.

CBF1 is an individual from a small gene family member; CBF2 and CBF3 and also transcription factors and expression of all three CBF genes are induced rapidly by low temperatures. In addition, CBF3 overexpression results in several biochemical changes related with cold acclimation, such as elevated levels of compatible osmolyte, proline, and soluble sugar.

Although low-temperature-induced gene expression, mediated by the CRT element, appears to be well conserved in plants, not all cold-induced genes have the CCGAC element in their promoters. Other pathways of low-temperature gene expression, not mediated through CRT/CBF, appear to be present in plants, and another sequence element, CCGAAA, has been identified as conferring low-temperature inducibility in some genes.

Genetic engineering and molecular studies have recognized numerous genes which are incited by many stresses. Several stress-inducible genes encoding useful protein have been utilized for the improvement of stress resistance. Most of the

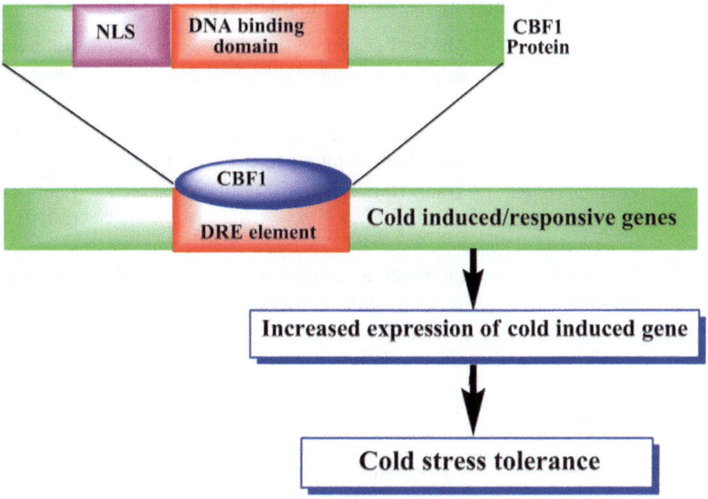

Fig. 4 Cold-responsive gene contains CBF1 transcription factor and drought-responsive element (DRE) by which increased the expression of cold-induced gene

reviews have been published on stress resistance (Christensen and Feldmann 2007; Umezawa et al. 2006; Valliyodan and Nguyen 2006). Numerous transcriptional activators, for example, DREB1/CBF (Kasuga et al. 1999; Liu et al. 1998; Kreps et al. 2002), which incite the stress-responsive gene, have been used to deliver the low-temperature-tolerant transgenic plants (Zhang 2003). The DREB/CBF qualities have been effectively used to engineer low-temperature stress resistance in different plant species, for example, rice (Dubouzet et al. 2003; Ito et al. 2006), pepper (Hwang et al. 2005), chickpea (Mantri et al. 2007), and potato (Rensink et al. 2005).

Screening for gene required in cold resistance is a critical introductory step. Vast quantities of studies have been accomplished for chilly stress resistance in plants. All the cold resistance-related genes contain a comparative regulatory element in their promoters, the C-repeat component and low-temperature-responsive component. A rundown of genes and transcription factor which are enhancing the chill resistance in various plants is given in Table 4.

7 Conclusion

In this chapter, we studied the effect of environmental stress, for example, temperature, water, cold, and salt, on cultivated plants and also highlighted the importance of abiotic stress in determining the large annual fluctuations in crop yield. Molecular genetics is used for the understanding of the underlying biology of plants under abiotic stress condition. Molecular mapping is to produce a sufficiently fine-scale map to pinpoint the location of genes that play a key role in determining important agronomic traits. Many of these traits are described as quantitative; that is, they are

Table 4 List of plant genes conferring increased cold tolerance

Genes	Plant	Mode of action	References
SICZFP1	*Arabidopsis*	Regulating cold-responsive gene	Zhang (Zhang 2010)
OrbHLH001	*Arabidopsis*	Involved in metabolic regulation or ionic homeostasis pathways in stress	Li et al. (2010)
Osmyb4	*Osteospermum ecklonis, apple, Arabidopsis*	Transcription factor	Laura et al. (2010)
OsLTP	*Phalaenopsis amabilis*	Increased accumulation of total soluble sugar, proline, antioxidant superoxide dismutase	Qin et al. (2011)
OsSPX1	*Arabidopsis and tobacco*	Accumulation of proline and sugar	Zhao et al. (2009)
OsDREB1D	*Arabidopsis*	Transcription factor	Zhang et al. (2009)
OsiSAP8	*Tobacco*	Cytoplasmic zinc finger protein that is involved in the signal transduction	Kanneganti and Gupta (2008)
OsDREB1F	*Arabidopsis*	Transcription factor	Wang et al. (2008)
Cat	*Wheat*	Use for the catalase	Matsumura et al. (2002)
GS2	*Rice*	Glutamine synthase	Hoshida et al. (2000)
P35S-ZFP245	*Rice*	Accumulation of proline, activation of the pyrroline-5-carboxylate synthetase and proline transporter genes, and enhancement of the ROS-scavenging enzymes	Huang et al. (2009)
OsP5CS2	*Rice*	Accumulation of proline	Hur et al. (2004)

influenced by different heredity and environmental factors. However, it is possible to map genes that have a large effect on quantitative traits by evaluating the correlation between the value of quantitative trait and the allelic states at linked heredity markers. QTL mapping is essential for the development of recombinant inbred line (RIL) from parents, which is wildly related to resistance to abiotic stress. These populace will help to detect the QTL in chromosome region.

Although crops experience a number of different abiotic stresses, several of these cause the same two major problems: water-deficit and oxidative stress. Two general strategies for engineering tolerance to abiotic stresses in plants are therefore possible. Some measure of tolerance to water-deficit stress can be provided by the synthesis of compatible solutes. On the other hand, expressing enzymes involved in protection against ROS can combat oxidative stress.

In some cases, single-gene mechanisms for tolerating specific stresses can be deployed (e.g., salt stress and cold stress). However, the overriding theme from this chapter is that abiotic stresses induce complex reaction from plants and that optimal protection may well involve several genes.

References

Alia HH, Sakamoto A, Murata N (1998) Enhancement of the tolerance of *Arabidopsis* to high temperatures by genetic engineering of the synthesis of glycinebetaine. Plant J 16:155–161

Anonymous (2003) Agriculture statistics, Ministry of Jihad – e – Agriculture, Te'hran, Iran

Barakat MN, Al-Doss AA, Elshafei AA, Moustafa KA (2011) Identification of new microsatellite marker linked to the grain filling rate as indicator for heat tolerance genes in F2 wheat population. Aust J Crop Sci 5:104–110

Boyer JS (1982) Plant productivity and environment. Science 218:443–448

Bray EA (1997) Plant responses to water deficit. Trends Plant Sci 2:48–54

Bray EA, Bailey-Serres J, Weretilnyk E (2000) Responses to abiotic stresses. In: Biochemisty and molecular biology of plants. American Society of Plant Physiologists, Rockville, pp 1158–1203

Brini F, Hanin M, Mezghani I (2007) Overexpression of wheat Na^+/H^+ antiporter TNHX1 and H^+ pyrophosphatase TVP1 improve salt and drought stress tolerance in *Arabidopsis thaliana* plant. J Exp Bot 58:301–308

Broman KW, Speed TP (1999) A review of methods for identifying QTLs in experimental crosses. In: Seiller-Moiseiwitsch F (ed) Statistics in molecular biology and genetics, IMS lecture notes-monograph series 33, pp 114–142

Ceccarelli S, Grando S (1996) Drought as a challenge for the breeder. Plant Growth Regul 20:149–155

Chaves MM, Marco TJ, Pereira SJ (2003) Understanding plant responses to drought from genes to the whole plant. Funt Plant Biol 30:239–264

Chen H, An R, Tang JH et al (2007a) Over expression of a vacuolar Na^+/H^+ antiporter gene improves salt tolerance in an upland rice. Mol Breed 19:215–225

Chen M, Wang QY, Cheng XG, Xu ZS, Li LC, Ye XG, Xia LQ, Ma YZ (2007b) Gm DREB2, a soybean DRE binding transcription factor, conferred drought and high salt tolerance in transgenic plants. Biochem Biophys Res Comun 353:299–305

Chen X, Moore M, Milus EA, Long DL, Line RF, Marshall D, Jackson L (2002) Wheat stripe rust epidemics and races of Puccinia striiformis f. sp. tritici in the United States. Plant Diseas 86:39–46

Christensen CA, Feldmann KA (2007) Biotechnology approaches to engineering drought tolerant crops. Springer, Dordrecht, pp 333–357

Cuartero J, Bolarin MC, Moreno V et al (2008) Toleranica a la salinidad. In: Moreno MT, Cubero JI, Atienza S et al (eds) La Adaptacion al Ambiente Y los Estreses Abioticos en la Mejora Vegetal, Sevilla, Spain

De Block M, Verduyn C, De Brouwer D, Cornelissen M (2005) Poly (ADP-ribose) polymerase in plants affects energy homeostasis, cell death and stress tolerance. Plant J 41:95–106

Delmer DP (2005) Agriculture in the developing world: connecting innovation in plant research to downstream applications. PNAS 102:15739–15746

Dubouzet JG, Sakuma Y, Ito Y, Kasuga M, Dubouzet EG, Miura S, Seki M, Shinozaki K, Yamaguchi-shinozaki K (2003) OsDREB genes in rice, *Oryza sativa* L, encode transcription activators that function in drought-high salt and cold responsive gene expression. Plant J 33:751–763

Egawa CF, Kobayshi M, Ishibashi T, Nakamura C, Nakamura T, Takaum S (2006) Differential regulation of transcript accumulation and alternative splicing of a DREB2 homolog under abiotic stress conditions in common wheat. Genes Genet Syst 81:77–91

Estan MT, Martinez Rodriguez MM, Perez-Alfocea F et al (2005) Grafting raises the salt tolerance of tomato through limiting the transport of sodium and chloride to the shoot. J Exp Bot 56:703–712

Gallagher J, Biscoe P, Scott R (1975) Barley and its environment V. stability of grain weight. J Appl Ecol 12:319–336

Gao F, Gao Q, Duan XG (2006) Cloning of an H^+ PPase gene from Thellungiella halophila and its heterologous expression to improve tobacco salt tolerance. J Exp Bot 57:3259–3270

Gaxiola RA, Li JL, Undurraga S (2001) Drought and salt tolerant Plants result from overexpression of the AVP1 H^+ pump. Proc Natl Acad Sci USA 98:11444–11449

Gupta PK, Rustogi S (2004) Molecular markers from the transcribed/expressed region of the genome in higher plants. Func Integr Genomics 4:139–162

Hoshida H, Tanaka Y, Hibino T, Hayashi Y, Tanaka A, Takabe T, Takabe T (2000) Enhanced tolerance to salt stress in transgenic rice that overexpresses chloroplast glutamine synthetase. Plant Mol Biol 43:103–111

Huang J, Sun SJ, DQ X, Yang X, Bao YM, Wang ZF, Tag HJ, Zhang H (2009) Increased tolerance of rice to cold, drought and oxidative stresses mediated by the overexpression of a gene that encodes the zinc finger proteins ZFP245. Biochem Biophys Res Commun 389:556–561

Hur J, Jung KH, Lee CH, An G (2004) Stress inducible OsP5CS2 gene is essential for salt and cold tolerance in rice. Plant Sci 167:417–426

Hwang EW, Kim KA, Park SC, et al. (2005) Expression profiles of hot pepper (Capsicum annuum genes under cold stress conditions J Biosci 30: 657–667

Ingram J, Bartels D (1996) The molecular basis of dehydration tolerance in plants. Ann Rev Plant Physiol Plant Mol Biol 47:377–403

Ito Y, Katsura K, Maruyama K et al (2006) Functional analysis of rice DREB1/CBF-type transcription factor involved in cold responsive gene expression in transgenic rice. Plant Cell Physiol 147:141–153

Kang JY, Choi HI, Im MY et al (2002) Arabidopsis basic leucine zipper proteins that mediate stress-responsive abscisic acid signalling. Plant Cell 14:343–357

Kanneganti V, Gupta AK (2008) Overexpression of OsiSAP8, a member of stress associated protein (SAP) gene family of rice confers tolerance to salt, drought and cold stress in transgenic tobacco and rice. Plant Mol Biol 66:445–462

Karaba A, Dixit S, Greco R et al (2007) Improvement of water use efficiency in rice by expression of HARDY, an *Arabidopsis* drought and salt tolerance gene. Proc Natl Acad Sci U S A 104:15270–15275

Kasuga M, Liu Q, Miura S et al (1999) Improving plant drought, salt and freezing tolerance by gene transfer of a single stress-inducible transcription factor. Nat Biotechnol 17:287–291

Kato K, Miura H, Sawada S (2000) Mapping QTLs controlling grain yield and its components on chromosome 5A of wheat. Theor Appl Genet 101:1114–1121

Korzun V, Roder MS, Wendehake K, Pasqulone A, Lotti C, Ganal MW, Blanco A (1999) Integration of dinucleotide microsatellites from hexaploid bread wheat into a genetic linkage map of durum wheat. Theor Appl Genet 98:1202–1207

Kovtun Y, Chiu WL, Tena G, Sheen J (2000) Functional analysis of oxidative stress-activated mitogen-activated protein kinase cascade in plants. Proc Natl Acad Sci U S A 97:2940–2945

Kreps JA, Wu Y, Chang HS, Zhu T, Wang X, Harper JF (2002) Transcriptome changes for Arabidopsis in response to salt, osmotic, and cold stress. Plant Physiol 130(4):2129–2141

Kumar A, Sengar RS (2013) Effect of delayed sowing on yield and proline content of different wheat cultivars. Res On crop 14(2):409–415

Kumar LS (1999) DNA markers in plant improvement: an overview. Biotech Adv 17:143–182

Lanceras JC, Pantuwan G, Jongdee B, Toojinda T (2004) Quantitative trait loci associated with drought tolerance at reproductive stage in rice. Plant Physiol 135:344–399

Langridge P, Paltridge N, Fincher G (2006) Functional genomics of abiotic stress tolerance in cereals. Briefings Funct Genomics Proteomics 4:343–362

Laporte MM, Shen B, Tarczynski MC (2002) Engineering for drought avoidance: expression of maize NADP-malic enzyme in tobacco results in altered stomatal function. J Exp Bot 53:699–705

Laura M et al (2010) Metabolic response to cold and freezing of Oseospermum ecklonis overexpressing OsmyB4. Plant Physiol Biochem 48(9):764–771

Lee JH, Hubel A, Schoffl F (1995) Derepression of the activity of genetically engineered heat shock factor causes constitutive synthesis of heat shock proteins and increased thermotolerance in transgenic *Arabidopsis*. Plant J 8:603–612

Li F, Guo S, Zhao Y, Chen D, Chong K, Xu Y (2010) Overexpression of a homopeptide repeat containing bHLH protein gene (OrbHLH001) from Dongxiang wild rice confers freezing and salt tolerance in transgenic. Arabidopsis Plant Cell Rep 29(9):977–986

Liu Q, Kasuga M, Sakuma Y, Abe H, Miura S, Yamaguchi-Shinozaki K, Shinozaki K (1998) Two transcription factors, DREB1 and DREB2, with an EREBP/AP2 DNA binding domain separate two cellular signal transduction pathways in drought- and low-temperature-responsive gene expression, respectively, in Arabidopsis. Plant Cell 10(8):1391–1406

Ma S, Bohnert HJ (2007) Integration of Arabidopsis thaliana stress related transcript profiles, promoter structures, and cell-specific expression. Genome Biol 8:R49. doi:https://doi.org/10.1186/gb-2007-8-4

Malik MK, Slovin JP, Hwang CH, Zimmerman JL (1999) Modified expression of a carrot small heat shock protein gene, *Hsp17.7*, results in increased or decreased thermotolerance. Plant J 20:89–99

Mantra NL, Ford R, Coram TE et al (2007) Transcriptional profiling of chickpea genes differentially regulated in response to high salinity, cold and drought. BMC Genomics 8:303

Mason RE, Mondal S, Beecher FW, Pacheco A, Jampala B, Ibrahim AMH, hays DB (2010) QTL associated with heat susceptibility index in wheat (*Triticum aestivum* L.) under short term reproduction stage heat stress. Euphytica 174:423–436

Matsumura T, Tabayashi N, Kamagata Y, Souma C, Saruyama H (2002) Wheat catalase expressed in transgenic rice can improve tolerance against low temperature stress. Physiol Plant 116:317–327

McKersie BD, Bowley SR, Harjanto E (1996) Water-deficit tolerance and field performance of transgenic alfalfa overexpressing superoxide dismutase. Plant Physiol 111:1177–1181

Mohammadi V, Modarraesi M, and Byrne P (2008a) Detection of QTLs for heat tolerance in wheat measured by grain filling duration. In: Appels R, Eastwood R, Lagudah E, Langridge P, Lynne MM (eds) Proceedings of 11th international wheat genetics symposium, Brisbane, Australia, pp 1000–1001

Mohammadi V, Zali AA, Bihamta MR (2008b) Mapping QTLs for heat tolerance in wheat. J Agri Sci and Tech 10:261–267

Mondal S, singh RP, Crossa J, Huerta-Espino J, Sharma I, Chatrath R, Singh GP, sohu VS, mavi GS, sukuru VSP, Kalappanavar IK, Mishra VK, Hussain M, Gautam NR, Uddin J, Barma NCD, Hakim A, Joshi AK (2013) Earliness in wheat: a key to adaptation under terminal and continual high temperature stress in South Asia. Field Crop Res 151:19–26

Montero-Barrientos M, Hermosa R, Cardoza RE, Gutierrez S, Nicolás C, Monte E (2010) Transgenic expression of the *Trichoderma harzianum hsp70* gene increases *Arabidopsis* resistance to heat and other abiotic stresses. J Plant Physiol 167:659–665

Morris ML, Belaid A and Byerlee D (1991) Wheat and barley production in rainfed marginal environments of the developing world. Plart 1 of 1990–91 CIMMT world wheat factors and trends. Wheat and barley production in rainfed marginal environments of the developing world. CIMMYT, Mexico, D.F

Nelson DE, Repetti PP, Adams TR et al (2007) Plant nuclear factor Y (NY-Y) B subunits confer drought tolerance and lead to improved corn yields on water-limited acres. Proc Natl Acad Sci U S A 104:16450–16455

Nguyen HT, Blum A (2004) Physiology and biotechnology integration for plant breeding: epilogue. Marcel Dekker, New York

Oh SJ, Song SI, Kim YS et al (2005) Arabidopsis CBF3/DREB1A and ABF3 in transgenic rice increased tolerance to abiotic stress without stunting growth. Plant Physiol 138:341–351

Ono K, Hibino T, Kohinata T, Suzuki S, Tanaka Y, Nakamura T, Takabe T (2001) Overexpression of DnaK from a halotolerant cyanobacterium *Aphanothece halophytica* enhances the high-temperature tolerance of tobacco during germination and early growth. Plant Sci 160:455–461

Paliwal R, Roder MS, Kumar U, Srivastava JP, Joshi AK (2012) QTL mapping of terminal heat tolerance in hexaploid wheat (*T. aestivum L.*) Theor Appl Genet 125:561–575

Park S, Li J, Pittman JK et al (2005) Up-regulation of a H+-pyrophosphatase (H+ PPase) as a strategy to engineer drought-resistant crop plants. Proc Natl Acad Sci U S A 102:18830–18835

Park SM, Hong CB (2002) Class I small heat shock protein gives thermotolerance in tobacco. J Plant Physiol 159:25–30

Parry ML, Duinder PN (1990) The potential effects of climate change on agriculture. In: Intergovernmental Panel on Climate change the IPCC Impacts Assessment, WMO and UNEP, Geneva, Switzerland, pp 2-1- 2-45

Patterson AH (1998) Molecular dissection of complex traits. CRC Press, New York

Pinto RS, Reynolds MP, Mathews KL, mcintyre CL, Olivares-Villegas J, Champman SC (2010) Heat and drought adaptive QTL in a wheat population designed to minimize confounding agronomic effects. Theor Appl Genet 121:1001–1021

Porter JR, Moot DJ (1998) Research beyond the means: climatic variability and plant growth. In: Dalezios NR (ed) International symposium on applied agrometeorology and agroclimatology. Office for Official Publication of the European Commission, Luxembourg, pp 13–23

Prasad SR, Bagali PG, Shailaja H, Shashidhar HE, Hittalmani S (2000) Molecular mapping of quantitative trait loci associated with seedling tolerance to salt stress in rice (Oryza sativa L.) Curr Sci 78(2):162–164

Qin X, Zeevart JAD (1999) The 9-cis epoxycarotenoid cleavage reaction in the key regulatory step of abscisic acid biosynthesis in water stressed bean. PNAS USA 96:15354–15361

Qin X, Liu Y, Mao S, Li T, Wu H, Chu C, Wang Y (2011) Genetic transformation of lipid transfer protein encoding gene in *phalaenopsis amabilis* to enhance cold resistance. Euphytica 177(1):33–43

Qin F, Kakimoto M, Sakuma Y, Maruyama K, Ssakabe Y, Phan tran LS, Shinozaki K, Yamaguchi Shinozaki K (2007) Regulation and functional analysis of ZmDREB2A in response to drought and heat stresses in Zea mays L. Plant J 50:54–69

Quarrie SA, Steed A, Calestani A, Semikhodskii A, Lebreton C et al (2005) A high-density genetic map of hexaploid wheat(*Triticum aestivum L.)* from the cross Chinese Spring x SQI and its use to compare QTLs for grain yield across a range of environments. Theor Appl Genet 110:865–880

Queitsch C, Hong SW, Vierling E, Lindquist S (2000) Hsp101 plays a crucial role in thermotolerance in *Arabidopsis*. Plant Cell 12:479–492

Rensink WA, Lobst S, Hart A et al (2005) Gene expression profiling of potato responses to cold, heat and salt stress. Funct Integr Genom 5:201–207

Rodriguez M, Canales E, Borrás-Hidalgo O (2005) Molecular aspects of abiotic stress in plants. Biotechnol Appl 22:1–10

Sakuma Y, Liu Q, Dubouzet JG, Abe H, Shinozaki K, Yamaguchi Shinozaki K (2002) DNA binding specificity of the ERF/AP2 domain of Arabidopsis DREBs transcription factor involved in dehydration and cold inducible gene expression. Biochem Biophys Res Commun 290:998–1009

Salvucci ME, Crafts-Brandner SJ (2004) Inhibition of photosynthesis by heat stress: the activation state of rubisco as a limiting factor in photosynthesis. Physiol Plant 120:179–186

Sanmiya K, Suzuki K, Egawa Y, Shono M (2004) Mitochondrial small heat shock protein enhances thermotolerance in tobacco plants. FEBS Lett 557:265–268

Sehgal D, Bhat V, Raina SN (2008) Applicability of DNA markers for genome diagnostics of grain legumes. In Kirti PB (ed) Handbook of new technology for genetic improvement of grain legumes. CRC Press, New York, pp 497–557

Seki M, Nzrusaka M, Abe H, Kasuga M, Carninci KP, hayashizaki Y, Shinozaki K (2000) Monitoring the expression pattern of 1300 Arabidopsis gens under drought and cold stresses using a full length c-DNA microarray. Plant Cell 13:61–72

Sharma SK, Goyal SS (2003) Progress in salinity resistance researches: integration of physiological, genetic and breeding approaches. In Crop production in saline environments. Howorth Press, New York, pp 387–407

Shinozaki K, Yamaguchi-Shinozaki K (2007) Gene networks involved in drought stress response and tolerance. J Exp Bot 58:221–227

Shi WM, Muramoto Y, Ueda A, Takabe T (2001) Cloning of peroxisomal ascorbate peroxidase gene from barley and enhanced thermotolerance by overexpressing in *Arabidopsis thaliana*. Gene 273:23–27

Shi HZ, Lee BH, Wu SJ et al (2003) Overexpression of a plasma membrane Na^+/H^+ antiporter gene improves salt tolerance in *Arabidopsis thaliana*. Nat Biotechnol 21:81–85

Shinozaki K, Yamaguchi-Shinozaki K, Seki M (2003) Regulatory network of gene expression in the drought and cold stress. Curr Opin Plant Biol 6:410–417

Shou H, Bordallo P, Wang K (2004) Expression of the Nicotiana protein kinase (NPK1) enhanced drought tolerance in transgenic maize. J Exp Bot 55:1013–1019

Snape J, Fish L, Leader D, Bradburne R, Turner A (2005) The impact of genomics and genetics on wheat quality improvement. Turk J Agric For 29:97–103

Sohn SO, Back K (2007) Transgenic rice tolerant to high temperature with elevated contents of dienoic fatty acids. Biol Plant 51:340–342

Taylor IB, Burbidge A, Thompson AJ (2000) Control of abscisic acid synthesis. J Exp Bot 51:1563–1575

Thompson A, Jackson A, Rarker R, Morpeth D, Burbidge A, Taylor I (2000) Abscisic acid biosynthesis in tomato: regulation of zeaxanthin epoxidase and 9-cis epoxycarotenoid dioxygenase m-RNAs by light/dark cycles, water stress and abscisic acid. Plant Mol Biol 42:833–845

Tiwari C, Wallwork H, Kumar U, Dhari R, Arun B, Mishra VK, Reynolds MP, Joshi AK (2013) Molecular mapping of high temperature tolerance in breed wheat adapted to the Eastern Gangetic Plain region of India. Field Crops Res 154:201–210

Trethowan RM, Crossa J, van Ginkel M, Rajaram S (2001) Relationships among bread wheat international yield testing location in dry areas. Crop Sci 41:1461–1469

Tuberosa R, Salvi S (2007) Genomics-based approaches to improve drought tolerance of crops. Trends Plant Sci 11:405–412

Tuyen DD, Prasad DT (2008) Evaluating difference of yield trait among rice genotypes (*Oryza sativa* L.) under low moisture condition using candidate gene markers. Omonrice 16:24–33

Umezawa T, Fujita M, Fujita Y et al (2006) Engineering drought tolerance in plants: discovering and tailoring genes to unlock the future. Curr Opin Biotechnol 17:113–122

Umezawa T, Yoshida R, Maruyama K, Yamaguchi-Shinozaki K, Shinozaki K (2004) SRK2C, a SNF1- related protein kinase 2, improves drought tolerance by controlling stress-responsive gene expression in Arabidopsis thaliana. Proc Natl Acad Sci U S A 101:17306–17311

Valliyodan B, Nguyen HT (2006) Understanding regulatory networks and engineering for enhanced drought tolerance in plants. Curr Opin Plant Biol 9:189–195

Vanderauwera S, De Block M, Van de Steene N et al (2007) Silencing of poly(ADP-ribose) polymerase in plants alters abiotic stress signal transduction. Proc Natl Acad Sci U S A 104:15150–15155

Vinod MS, Shrama N, Manjunatha K et al (2006) Candidate genes for drought tolerance and improved productivity in rice. J Biosci 31:69–74

Wang Y, Ying J, Kuzma M et al (2005) Molecular tailoring of farnesylation for plant drought tolerance and yield protection. Plant J 43:413–424

Wang Q, Guan Y, Wu Y, Chen H, Chen F, Chu C (2008) Overexpression of a rice OsDREB1F gene increases salt, drought and low temperature tolerance in both Arabidopsis and rice. Plant Mol Biol 67:589–602

Wang W, Vinocur B, Altman A (2003) Plant response to drought, salinity and extreme temperatures; towards genetic engineering for stress tolerance. Planta 218:1–14

Wu CA, Yang GD, meng QW et al (2004) The cotton GhNHX1 gene encoding a novel putative tonoplast Na^+/H^+ antiporter plays an important role in salt stress. Plant Cell Physiol 45:600–607

Wu YY, chen Q, Chen M et al (2005) Salt tolerant transgenic perennial ryegrass (Lolium perenne L.) obtained by agrobacterium tumefaciens mediated transformed of the vacuolar Na^+/H^+ antiporter gene. Plant Sci 169:65–73

Xiao B, Huang Y, Tang N (2007) Over-expression of a LEA gene in rice improves drought resistance under the field conditions. Theor Appl Genet 115:35–46

Xiao J, Li J, Yuan L, Tanksley SD (1996) Identification of QTLs affecting traits of agronomic importance in a recombinant inbred population derived from a sub specific rice cross. Theoretical Appl Genet 92:230–244

Xue GP, Loveridge CW (2004) HvDRF1 is involved in abscisic acid mediated gene regulation in barley and produced two forms of AP2 transcriptional activators, interacting preferably with a CT-rich element. Plant J 37:326–339

Yamaguchi-Shinozaki K, Koizumi M, Urao S, Shinozaki K (1992) Molecular cloning and characterization of 9 cDNAs for genes that are responsive to desiccation in *Arabidopsis thaliana*: sequence analysis of one cDNA clone that encodes a putative transmembrane channel protein. Plant Cell Physiol 33:217–224

Yokoi S, Quintero FJ, Cubero B (2002) Differential expression and function of Arabidopsis thaliana NHX Na^+/H^+ antiporters in the salt stress response. Plant J 30:529–539

Zhang JZ (2003) Overexpression analysis of plant transcription factors. Curr Opin Plant Biol 6:430–440

Zhang X (2010) Overexpression of SicZfp1, a novel TFIIIA type zinc finger protein from tomato, confers enhanced cold tolerance in transgenic Arabidopsis and rice. Plant Mol Biol Rep:1–12

Zhang Y, Chen C, Jin XF, Xiong AS, Peng RH, Hong YH, Yao QH, Chen JM (2009) Expression of a rice DREB1 gene, OsDREB1D, enhances cold and high salt tolerance in transgenic Arabidopsis. BMB Rep 42(8):486–492

Zhao L, Liu F, Xu W, Di C, Zhou S, Xue Y, Yu J, Su Z (2009) Increased expression of OsSPX1 enhances cold/subfreezing tolerance in tobacco and Arabidopsis thaliana. Plant Biotechnol J 7(6):550–561

Zhao R, Guo S, Zhang H (2006) Expression of yeast SOD2 in transgenic rice results in increased salt tolerance. Plant Sci 170:216–224

Zhu BC, Su J, Chan MC et al (1998) Overexpression of a d-pyrroline-5-carboxylate synthetase gene and analysis of tolerance to water-stress and salt-stress in transgenic rice. Plant Sci 139:41–48

Zhu JK (2001) Plant salt tolerance. Trends Plant Sci 6:66–71

Bioengineering of *DREB* and *NAC* Transcriptional Factors for Enhanced Plant Tolerance Against Abiotic Stresses

Manoj K. Sharma, Ashu Singh, and Rakesh Singh Sengar

Abstract

Plants growing in their native environment face several types of abiotic stresses that intentionally affect their yields at significant levels. Plant reactions toward stress are complicated and involve various cellular, physio-biochemical, and molecular adaptations. Several recent studies show that under stress conditions, plants exhibit a series of several physiological and molecular responses as a part of their stress tolerance mechanisms. Such types of interactions among various stresses point to an in-between talks among their responsive pathways of cell signaling. This type of cross talk may be both synergistic and antagonistic and commiserate the defense system which combines the plant growth hormones, transcriptional factors, cascades such as kinase, and reactive oxygen species (ROS) as an aid. Such cross talk could lead to a cross-tolerance and the enhancement of plant's resistance levels against abiotic stresses. In recent years, transcriptional factors (TFs) have been reported to play important roles in crop improvement from the advent of agriculture. Transcriptional factors (TFs) have reported to be therefore good candidates for molecular genetics to enhance plant tolerance toward abiotic stress because of their major roles as regulators of the clusters of several genes. In this chapter, the current status of transgenic or genetically modified plants developed for enhanced tolerance against abiotic stresses by overexpressing *DREB* and *NAC* transcriptional factors has been discussed in detail. Therefore, the collective efforts and the results of several collaborative studies would definitely contribute toward the sustainable food production at global level and would also be helpful to prevent the large-scale environmental damages that result from the course of several abiotic stresses.

M. K. Sharma (✉) · A. Singh · R. S. Sengar
Department of Agricultural Biotechnology, College of Agriculture, S.V.P University of Agriculture & Technology, Meerut, Uttar Pradesh, India

R. S. Sengar, A. Singh (eds.), *Eco-friendly Agro-biological Techniques for Enhancing Crop Productivity*, https://doi.org/10.1007/978-981-10-6934-5_9

Keywords
Environmental stress · Gene/regulon · Transcriptional factors · Transgenics

1 Introduction

Abiotic stresses such as salinity, drought, and temperature/heat are location specific, displaying inner variation in occurrence, intensity, and duration, and generally cause reduced crop productivity. There is a necessary concern for food security in developing countries including India for the following reasons: (1) increasing food demand for the ever-growing population; (2) declining productivity in past high-productivity regions, often described as "'Green Revolution" fatigue; and (3) increasing susceptibility of agriculture to potential climate change. Climatic change is becoming a large concern toward humanity in the coming decades. Agriculture also produces significant effect on environmental changes as a potential contributor of several greenhouse gases and also is an important industry that is highly sensitive to environmental changes. Climate plays a pivotal role in productivity; it should be noted that it's not only a boon for us but a bane when it imposes its extreme threats. Thus there is an urgent need to correct this aberration. Of all the parameters, abiotic stresses contribute significant reduction in crop yields (Flowers and Yeo 1995). In recent years, water has been an increasingly limited resource; this asset is posing serious problem of scarcity, and we do know that Asia and Southeast Asia are abundant in fertile lands suited for cultivation but are left uncultivated due to this drought condition accompanied with sodic saline soil. Now dealing with this water-scarce condition, agronomic research mainly focuses on the augmentation of water-use efficiency for irrigation and on other sources of water. On the other hand, breeding for encountering this problem is very primitive approach of growing the varieties of desired traits and then crossing them to get the required progeny and evaluating their endurance under extreme environmental conditions. Biotechnologists, on the other hand, exploit engineering directly the genes responsible for this stress endurance/tolerance (Katerji et al. 2003, 2004; Manavalan et al. 2009; Umezawa et al. 2006). The genetic basis of stress tolerance is complex and is still an enigma. In order to manage environmental stresses, which have a very diverse timing of manifestation and varying levels of severity, plant activates its own molecular machinery to fight against it. It is the ability of plant to adapt to these conditions by fine-tuning its physiological processes, namely, photosynthesis, to sustain life, but it leads to adverse effect on growth and development, i.e., poor productivity (Chaves et al. 2011). The unique feature of plant adjustment is manifested by its activation of a wide array of genes responsible for abiotic stresses along with various responses collectively working in adaptation to abiotic stresses by intricate coordinations between various metabolic, cellular, and physiological pathways (Basu 2012). Abiotic stresses elicit complex cellular responses that have been elucidated by progresses made in exploring and understanding plant abiotic responses at the whole-plant, physiological, biochemical, cellular, and molecular levels (Grover et al.

2001). At subcellular level the response is manifested by successful induction of stress-related genes to induce either avoidance, tolerance, or resistance (Matsui et al. 2008), which is often conducted by the phytohormone abscisic acid (ABA). ABA is addressed as the plant stress hormone because not only does it play a prime role in developmental biology of the plant but also does directly control the responses to abiotic stress by regulating stomatal closure to optimize transpiration and by triggering the activation of many stress-related genes (Finkelstein et al. 2002; Cutler et al. 2010). Thus ABA hormones also function in reducing transpirational water loss and photosynthetic rate which also activates the water-use efficiency (WUE) of the plant. Numerous genes are accelerated in response to abiotic stresses at the transcriptional level, and their products are altered in different combinations to meet the goal of stress tolerance through altered metabolism and also in synchronizing the downstream genes (Kavar et al. 2007). Transcript profiling can be a significant tool for the characterization of stress-responsive genes. With the detailed analysis of the protein factors, it is now confirmed that they can be characterized into two groups (Bohnert et al. 2001; Seki et al. 2002; Fowler and Thomashow 2002); with this bifurcation the first genes in this family include proteins that regulate the accumulation of compatible solutes (enzymes for osmolyte biosynthesis like proline, betaine, sugars, etc.); passive and active transport systems across membranes (water channel proteins and membrane transporters); and protection and stabilization of cell structures from damage by ROS (the detoxification enzymes such as glutathione S-transferase, catalase, superoxide dismutase , ascorbate peroxidase, etc.); fatty acid metabolism enzymes, proteinase inhibitors, ferritin, and lipid-transfer proteins; and other proteins for the protection of macromolecules (LEA protein, osmotin, chaperons, etc.). The second of this group includes those factors regulating gene expression through delicate network of cell signaling. This includes various transcription factors (TFs) highlighting their role as transcriptional regulatory body in the stress signal transduction pathways; protein kinases (MAP kinase, CDP kinase, receptor protein kinase, etc.) and protein phosphatases and proteinases are some of the examples in this stress condition by regulation signaling and gene expression (Seki et al. 2003; Shinozaki and Yamaguchi-Shinozaki 2007) (Fig. 1).

The prime aim of attaining sustainable agriculture is to implement knowledge to obtain crops which are easy to breed as well as can be readily engineered to get desired agronomic trait (Lee et al. 2007). The crux of research in plants for abiotic stress is to answer this potent question as to how does a plant sense the stress and then how to withstand this stress situation using its indigenous memory. As we know this puzzle can be solved by novel techniques that allow modification/alteration of internal machinery of the crop as a whole to investigate the matter by using microarrays which has gained a lot of interest, while the other techniques such as whole-genome sequencing and chromatin immunoprecipitation have gained focus as they help us to track epigenetic changes induced by stress via mutagenesis, and metabolic profiling has gained interest as they decipher the metabolic pathways and their rescue operations during stress. As we know that the classical tools of genetics as well as plant breeding have already confirmed that abiotic stress tolerance in plants is multigenic and quantitative trait. This makes it a pretty nut job to

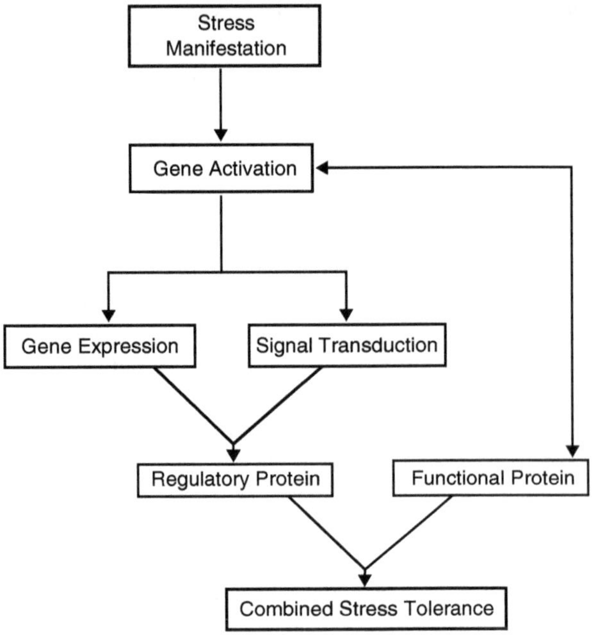

Fig. 1 Stress-induced proteins: classified into two groups, viz., regulatory and functional

manipulate these abiotic stress-related genes using modern molecular genetic techniques. Instigation of a single gene encoding for late embryogenesis abundant (LEA) proteins belongs to the diverse gene family that is present in almost all of the plant species. LEAs are suggested to play a role in many stress tolerance responses but would not confer a hundred percentage tolerance single-handedly, while antichill proteins, and molecular chaperons, would provide tolerance to some extent but would not give prolonged tolerance to most of the abiotic stresses. However, during the course of evolution, plant has equipped itself to sustain such adverse conditions through its unique molecular mechanism. So in continuation with this, another category of proteins is found called regulatory proteins. These proteins, encoded by stress-induced gene, act as on/off switch on sets of downstream stress-responsive genes. Annotations in recent development in functional and computational biotechnology have opened new avenues in the field of genomics and proteomics. For the first time, we can now obtain a blueprint of the cellular activities in real time using transcript profiling and quantitative PCR. Such a defined procedure gives reliable image of the metabolic activities of the cell not only in one genus, but also intrageneric species can be studied simultaneously having a comparative approach (Kitano 2002). A plant genome normally has around 7% of the coding sequences belonging to transcription factors (Udvardi et al. 2007), and these genes range from early- to immediate-responsive gene in abiotic stress conditions (Kilian et al. 2012). In the first manifestation of stress, be it drought, salinity, freeze, or temperature, the rescue method is conducted either by ABA-dependent or ABA-independent pathway which leads to subsequent activations of various physiological and metabolic responses with activation of DREB/NAC genes which leads to the production of DREB/NAC mRNA which after translation leads to the production of DREB/NAC TF protein. These TFs work in collaboration with the stress-induced rescue gene by binding to conserved "stress recognition sequences" which is located as *cis*-acting elements upstream the promoter of the gene concerned, the result of which is achieved stress tolerance. Thus in order to achieve multiple stress tolerance, the clue is in manipulation of those categories of gene that modulate the rescue operations, which in turn is possible by instigating the TFs. Furthermore, in order to attain multiple abiotic stress tolerance, there is an urgent need to test the potential NAC/DREB genes in model plant species, and field evaluation of the transgenic crops conferring stress tolerance must be established in order to attain the ultimate goal of any food research cult, i.e., productive success.

2 Structure and Functions of *DREB* Transcription Factors

At transcriptional level, all the inducible genes depending upon proteins for their expression are known as transcription factors. A typical transcription factor usually contains a DNA-binding domain, a transcription regulation domain, a dimerization site, and a nuclear localization domain. What will be the characteristic, location, and regulatory function of a transcription factor is defined by functional domain. The functional domain acts as a key which fits at a specific *cis*-acting element or interacts with other proteins and then activates or represses the transcription of target genes. In recent years, molecular biology has elucidated that all TFs have a DNA-binding domain containing a short peptide region, called the DNA-binding motif. The amino acid sequences of the DNA-binding domain are conserved within a family. Proteins belonging to *CBF/DREB* have highly conserved DNA-binding domain called *AP2/ERF* domain, consisting of 60 amino acids. This domain is considered plant-specific (Riechmann and Meyerowitz 1998). However, Magnani et al. (2004) have also reported the presence of homologues in some cyanobacteria, viruses, and bacteriophages, which uniquely bind poly(dG)/poly(dC) displaying evolutionary conservation of function with plant *AP2/ERF* proteins. The wide species of plant kingdom including both monocots and dicots have been reported to contain DREB/CBF genes containing AP2. On a three-dimensional analysis of *AP2/ERF* domain, the presence of three-stranded antiparallel β-sheets connected by loops and an α-helix was seen, packed approximately parallel to each other (Allen et al. 1998; Lata and Prasad 2011). This conserved sequence collectively forms the drought-responsive element (*DRE*). The presence of this DRE element is reported within the promoter region of many stress-induced genes (Thomashow 1999; Shinozaki and Yamaguchi-Shinozaki 2000). Thus for regulating cold-inducible promoters, there are similar *cis*-acting elements, named C-repeat (*CRT*) and low-temperature-responsive element (*LTRE*), which are present at the core of the *DRE* sequence, and both comprise of an A/GCCGAC motif (Stockinger et al. 1997; Thomashow 1999). Thus under abiotic stress condition, the two major *cis*-acting elements (i.e., *ABRE and DRE/CRT*) that can function either as ABA-dependent and ABA-independent elements for gene expression are activated. There are conserved sequences in *AP2*

DNA-binding domain within *DREB1/CBF* proteins, which distinctively binds to the *CRT/DRE* sequences, and thus the *CRT/DRE sequences* of the genes are transcribed cooperatively. In *Arabidopsis*, at chromosome 4 there are three *DREB1/CBF* proteins that are encoded by genes which lie in tandem order of *DREB1A/CBF3*, *DREB1B/CBF1*, and *DREB1C/CBF2* (Gilmour et al. 1998; Liu et al. 1998; Riechmann et al. 2000; Maruyama et al. 2004). *AP2/ERFs* belong to a class of protein that is unique to plants and plays an important role in biotic and abiotic stress responses (Agarwal et al. 2006). The *DREBs* (dehydration-responsive element binding) which are also referred as *CBF* (C-repeat binding factor) proteins belong to *ERF* subfamily and have received much attention due to its pivotal role in abiotic stress tolerance. *DREB protein* subfamily can be further classified into six small groups termed as A-1 to A-6; among these A-1 and A-2 embody the two largest groups (Sakuma et al. 2002). As we know that there is a continuous cross talk between the DREB1/CBF genes which belong to the A-1 subgroup and DREB2-like genes belonging to the A-2 subgroup and both these genes are mainly involved in cold and osmotic stress-responsive gene expression respectively. But still there remains a wide gap to be filled to elucidate the characteristics and functions of members of the other *DREB*. ERF proteins belong to the subfamily of APETLA2 (AP2)/ethylene-responsive element binding proteins (EREBP) and DREBs/CBFs, which are prime members of this subfamily belong to the ERF (ethylene-responsive element binding factors) family of transcription factors that are specific for each plant. The *ERF* family comprises two subfamilies, the *EREBP* (single *AP2* domain) and *AP2*/family (two copies of *AP2*). The *EREBP* subfamily is further divided into two classes, i.e., *ERFs* and *DREBs/CBFs*. *ERFs* bind to the GCC box found in the promoters of many pathogenesis-related (PR) genes conferring ethylene responsiveness (Gu et al. 2000). *DREBs/CBFs* bind to the dehydration-responsive element (*DRE/CRT*) in the promoters of cold- and dehydration-responsive LEA genes including *rd29A*, *rd17*, *cor6.6*, *cor15a*, *erd10*, and *kin1* (Yamaguchi-Shinozaki and Shinozaki 1994; Kasuga et al. 1999). From the promoter of a stress-inducible *rd29A* gene, a nine-base pair conserved sequence (TACCGACAT) was identified that is essential for *rd29A* induction under dehydration and cold stress in transgenic *Arabidopsis* (Yamaguchi-Shinozaki and Shinozaki 1994). The *DREBs/CBFs* are further divided into two subclasses, i.e., *DREB1/CBF* and *DREB2*, induced by cold and dehydration stress, respectively.

3 Bioengineering of *DREBs* for Enhancing Abiotic Stress Tolerance in Plants

Engineering plants for overexpression of genes encoding for transcriptional factors is identified as an effective method for generating abiotic stress-tolerant plants. Gene expression studies using constitutive promoters provide very little biological information compared with the use of inducible promoters or cell type-specific promoters. The choice of promoters can significantly affect the results from a transgenic manipulation; basic molecular biology research has established that expression of a given gene is governed by the promoter sequence present mostly at 5′ end of gene. The promoter sequences determine the strength of the expression (i.e., the extent to which the gene will be expressed) as well as the preciseness of the quadrants to which the gene will be expressed (temporal and/or spatial). For instance, regulation of any gene, for example, heat shock genes, is mediated by core DNA sequences termed as heat shock elements (HSE) that are located in the promoter region of *hs* genes which upon stimulation by stress condition transcribe certain proteins called transcription factors which act as *cis*-acting element in enhancing the promoter activity, thereby increasing the gene expression of the concerned gene. Thus abiotic stress-tolerant crops could be engineered by (1) successful fine-tuning of the stress response by engineering novel regulatory targets, (2) proper knowledge of posttranslational modifications which regulate plant growth performance under stress, (3) overexpression of miRNAs or their targets, (4) maintaining hormone homeostasis to avoid pleiotropic effects under stress, and (5) applying plant synthetic biology approaches to improve genetic engineering strategies (Abogadallah et al. 2011); thus this could be successfully achieved by changing the expression pattern of genes through altering the transcription factors, which is more or less like epigenetic modifications as there is no alteration of gene sequence. The tight regulation and fine-tuning of genes governing transcription factors (TFs) at the time of stress contribute toward the foundation of complicated signaling web. And the significant performances of DREB proteins are controlling gene expression through interacting with cis-elements, which are nothing but conserved sequences present upstream the promoter, which in turn is being induced by the respective TFs involved, whereas the cis-acting elements (like DRE/CRT, ABRE, and CE) and the signaling kinase that phosphorylate the concerned TFs carry out the gene expression of the abiotic stress-related genes. The development of transgenic plants consisting of single gene governing transcription factors is not of so much importance due to breaking of the durability of resistance level. Therefore, the efforts were also made to develop transgenic plants having multiple transcription factor regulatory genes to enhance resistance durability for longer time periods. It is possible that several stress-inducible genes can share similar transcription factors. Various transcriptional factors such as *DREB*, *NAC*, *MYB/MYC*, *bZIP*, *WRKY*, *AP2/ERF*, and Cys_2His_2 zinc fingers have known to be induced in many abiotic stresses. They have the capability to bind with the promoter or regulatory elements of stress-responsive genes during stress conditions. The members of different groups of transcriptional factors could be involved for single stress response or the members of similar groups also for diverse stress responses. In recent years, several transgenic plants resistant to various abiotic stresses have been generated through the genetic manipulation of genes governing transcription factors. The DRE/CRT have been reported as *cis*-acting regulatory elements in response to several abiotic stresses such as salt, drought, and low temperature in *A. thaliana* (Yamaguchi-Shinozaki and Shinozaki 1994). The introductory attempts for developing transgenic plants through the genetic transformation of *DREB* genes were started with *Arabidopsis thaliana*. The overexpression of *DREB1/CBF* transcription factor in *Arabidopsis thaliana* exhibits the expression of several genes responsible for enhanced tolerance to high-salinity, drought, and cold stress (Jaglo-Ottosen et al. 1998). Transcription factor *DREB1/CBF3* with CaMV 35S promoter also showed improved resistance to drought, cold, and salt stress (Kasuga et al. 1999; Gilmour et al. 2000). The constitutive expression of *DREB1A* gene with *rd 29A* promoter reduced the unfavorable effects on plant's growth and development (Kasuga et al. 1999). The transformation of wheat with *DREB1A* gene indicated increased resistance to water stress (Pellegrineschi et al. 2004). The overexpression of *OsDREB1A* in *A. thaliana* resulted in salinity, drought, and cold stress tolerance (Dubouzet et al. 2003). The ectopic expression of *AtDREB1A/CBF3* under the control of CaMV 35S promoter in transgenic *Arabidopsis* and rice plants showed increase in drought and salt stress tolerance (Dubouzet et al. 2003; Oh et al. 2005; Ito et al. 2006). In another experiment, the stress-inducible expression of maize transcription factor *ZmDREB2A* exhibited improved tolerance against drought and heat stress (Qin et al. 2007). Similar findings were also reported earlier in *Arabidopsis thaliana* plants showing the overexpression of constitutively active *DREB2A* gene for increased thermal and water stress tolerance (Sakuma et al. 2006a, b). Some transgenic events demonstrated 70% increase over wild-type JL 24; RD 2 had 40% more TE than WT JL 24 under dehydration situation (Bhatnagar-Mathur et al. 2007). Overexpression of *HvCBF4* gene isolated from barley in transgenic rice emanated in an increase in tolerance to a variety of stress such as low-/high-temperature salinity stresses without retarding growth. *HvCBF4* gene activated 15 rice genes in studies using the 60K Rice Whole Genome Microarrays, and in a comparative study with 12 target rice genes of *CBF3/DREB1A*, 5 genes were common to both *HvCBF4* and *CBF3/DREB1A*, while 10 and 7 genes were particular to *HvCBF4* and *CBF3/DREB1A*, respectively (Oh et al. 2007). *OsDREB1F* gene is activated not only by only abiotic stress but also on ABA application (Wang et al. 2008). An experiment with transgenic rice plants transformed with *OsDREB1F* gene displayed increased resistance to salt, drought, and cold stress. The expression of stress-inducible genes, such as *rd29B* and *RAB18* that have the *DRE/CRT* element in their promoter regions, is greatly increased by overexpression of *OsDREB1F* in *Arabidopsis*, which advocates the participation of *OsDREB1F* in the expression of osmotic stress-responsive genes via ABA-dependent signaling pathway. *OsDREB1G gene* overexpression also protects plant from drought stress (Chen et al. 2008). Transgenic *Arabidopsis* plant's *OsDREB2B* genes' functional form demonstrated increased expression of DREB2A target genes which in turn resulted in increased tolerance to drought and high-temperature stresses (Matsukura et al. 2010). These researches do suggest the role of *OsDREB2B* as a key gene that encodes the *DREB2*-type TF which plays a major role in stress-induced gene expression in rice. Similarly, overexpression of *DREB2A CA* in transgenic *Arabidopsis* greatly enhanced the overexpression of stress-protectant genes to provide an array of tolerance to all the abiotic stresses (Mizoi et al. 2011). The ubiquitin E3 ligase modulates protein stability and is an important factor for activation of *DREB2A* (Qin et al. 2011). Transgenic *Arabidopsis* plants overexpressing *AtDREB1A* also displayed increased tolerance to water scarcity and chilling stress (Liu et al. 2011a, b, c). Rice plants overexpressing *OsDREB2A* are relatively more tolerant to salt and water stress than untransformed plant (Mallikarjuna et al. 2011).

Zhou et al. (2012) also reported the overexpression of PeDREB2a in Arabidopsis and *Lotus corniculatus*, thus resulting tolerance against salinity and drought stress. While in Glycine max the overexpression of AtDREB2A-CA and OsDREB2A genes resulted in tolerance against salinity stresses and drought stress (Engels et al. 2013; Zhang et al. 2013). Overexpression of *SbDREB2A* in *Nicotiana tabacum* enhanced the tolerance to hyperionic and hyperosmotic stresses, while resistance to drought, salinity, cold, heat, and heavy metal stress is achieved via *EsDREB2B* gene (Gupta et al. 2014; Li et al. 2014). The endurance of transcription factors has resulted in plants better suited to harsh environment through elevated expression of stress-related genes, for example, *Nicotiana tabacum*, where *SsDREB* gene resulted better tolerance to heat and salt stress (Zhang et al. 2015), while *GmDREB2* and *FeDREB* show cold and drought stress tolerance, respectively (Tan et al. 2015; Fang et al. 2015a, b). The overexpression of *DREB* genes in *Arabidopsis thaliana* for tolerance against drought, salinity, and freezing stress (Chen et al. 2016; Hichri et al. 2016; Jiang et al. 2016; Peng et al. 2016; Yin et al. 2016; Zong et al. 2016; Kudo et al. 2017; Li et al. 2017; Wu et al. 2017); *Betula platyphylla* for salinity and osmotic stress (Zhang et al. 2016b); *Malus domestica* for drought stress (Liao et al. 2017); *Manihot esculenta* for drought and freezing stress (An et al. 2016); *Nicotiana tabacum* for Cold, heat and drought stress tolerance (Arroyo-Herrera et al. 2016); 2016); rice for drought, salinity, and freezing stress (Samara Reddy et al. 2016; Tang et al. 2017a, b; Tian et al. 2017); *Robinia pseudoacacia* for drought stress (Xiu et al. 2016); *Salvia miltiorrhiza* for drought stress (Wei et al. 2016b, 2017); *Solanum lycopersicum* for salinity and freezing stress (Hichri et al. 2016; Shah et al. 2016); and *Solanum tuberosum* for cadmium stress (Bouaziz et al. 2017; Charfeddine et al. 2017) has been reported in recent years, and the work is also in progress. For more information regarding the overexpressing *DREB* genes, kindly refer to Table 1.

Table 1 Transgenic plants developed for abiotic stress tolerance by the overexpression of *DREBs*

DREB genes	Targeted plant species	Targeted trait	References
AtDREB1A	*Arabidopsis thaliana*	Drought stress tolerance	Liu et al. (1998)
AtCBF1	*Arabidopsis thaliana*	Salinity stress tolerance	Jaglo-Ottosen et al. (1998)
AtCBF3	*Arabidopsis thaliana*	Cold stress tolerance	Gilmour et al. (2000)
AtCBF4	*Arabidopsis thaliana*	Cold stress tolerance	Haake et al. (2002)
OsDREB1A	*Arabidopsis thaliana*	Salinity, drought, and cold stress tolerance	Dubouzet et al. (2003)
ZmDREB1A	*Arabidopsis thaliana*	Drought and cold stress tolerance	Qin et al. (2004)
DREB1A	*Arabidopsis thaliana*	Drought, salinity, and cold stress tolerance	Maruyama et al. (2004)
GmDREB	*Arabidopsis thaliana*	Cold stress tolerance	Shiqing et al. (2005)
AtDREB2A	*Arabidopsis thaliana*	Drought and cold stress tolerance	Sakuma et al. (2006a)
AtDREB2A-CA	*Arabidopsis thaliana*	Drought, heat, and salinity stress tolerance	Sakuma et al. 2006b
AtDREB2C	*Arabidopsis thaliana*	Heat stress tolerance	Lim et al. (2007)

(continued)

Table 1 (continued)

DREB genes	Targeted plant species	Targeted trait	References
GmDREB2	*Arabidopsis thaliana*	Drought and salinity stress tolerance	Chen et al. (2007)
ZmDREB2A	*Arabidopsis thaliana*	Drought and heat stress tolerance	Qin et al. (2007)
TaERF1	*Arabidopsis thaliana*	Salinity and dehydration stress tolerance	Xu et al. (2007)
OsDREB1F	*Arabidopsis thaliana, Oryza sativa*	Drought, cold, and salinity stress tolerance	Wang et al. 2008
ZmDBP3	*Arabidopsis thaliana*	Drought and cold stress tolerance	Wang and Dong (2009)
OsDREB1D	*Arabidopsis thaliana*	Salinity and cold stress tolerance	Zhang et al. (2009)
DREB1A/CBF3, DREB2A	*Arabidopsis thaliana*	Drought and cold stress tolerance	Maruyama et al. (2009)
PeDREB2	*Arabidopsis thaliana*	Salinity stress tolerance	Chen et al. (2009b)
GmDREB3	*Arabidopsis thaliana*	Cold, drought, and salinity stress tolerance	Chen et al. (2009a)
HvDREB1	*Arabidopsis thaliana*	Drought stress tolerance	Xu et al. 2009
DgDREB1A	*Arabidopsis thaliana*	Drought and cold stress tolerance	Tong et al. (2009)
OsDREB2B	*Arabidopsis thaliana*	Drought and heat stress tolerance	Matsukura et al. (2010)
LcDREB3a	*Arabidopsis thaliana*	Drought and salinity stress tolerance	Xianjun et al. (2011)
ZmDBP2	*Arabidopsis thaliana*	Drought stress tolerance	Wang et al. (2011b)
ZmDBP4	*Arabidopsis thaliana*	Drought and cold stress tolerance	Wang et al. (2011c)
PeDREB2L	*Arabidopsis thaliana*	Drought and cold stress tolerance	Chen et al. (2011)
JcDREB	*Arabidopsis thaliana*	Salinity and freezing stress tolerance	Tang et al. (2011)
MbDREB1	*Arabidopsis thaliana*	Cold, drought, and salt stress tolerance	Yang et al. (2011)
PeDREB2a	*Arabidopsis thaliana, Lotus corniculatus*	Salinity and drought stress tolerance	Zhou et al. (2012)
MsDREBA5	*Arabidopsis thaliana*	Drought stress tolerance	Zhao et al. (2012)
GmDREB2A	*Arabidopsis thaliana*	Drought, heat, and cold stress tolerance	Mizoi et al. (2013)
VaCBF4	*Arabidopsis thaliana*	Cold, drought, and salinity stress tolerance	Li et al. 2013
VmCBF	*Arabidopsis thaliana*	Freezing stress tolerance	Oakenfull et al. (2013)
MsDREB2C	*Arabidopsis thaliana*	Drought, heat, and cold stress tolerance	Zhao et al. (2013)
SiARDP	*Arabidopsis thaliana, Setaria italica*	Drought and salinity stress tolerance	Cong et al. (2014)
VuDREB2A	*Arabidopsis thaliana*	Drought stress tolerance	Sadhukhan et al. (2014)

(continued)

Table 1 (continued)

DREB genes	Targeted plant species	Targeted trait	References
BpDREB2	*Arabidopsis thaliana*	Salinity and cold stress tolerance	Sun et al. (2014a)
LcERF080	*Arabidopsis thaliana*	Salt stress tolerance	Sun et al. (2014b)
MdCBF1	*Arabidopsis thaliana*	Cold stress tolerance	Xue et al. (2014)
NnDREB2	*Arabidopsis thaliana*	Salinity stress tolerance	Cheng et al. (2015)
FeDREB1	*Arabidopsis thaliana*	Drought and cold stress tolerance	Fang et al. (2015a, b)
GmDREB1s	*Arabidopsis thaliana*	Drought, salinity, heat, and cold stress tolerance	Kidokoro et al. (2015)
VrDREB2A	*Arabidopsis thaliana*	Drought and salinity stress tolerance	Chen et al. (2016)
SlDREB1	*Arabidopsis thaliana*	Drought stress tolerance	Jiang et al. (2016)
PmhCBFc	*Arabidopsis thaliana*	Freezing and oxidative stress tolerance	Peng et al. (2016)
CsCBF3	*Arabidopsis thaliana*	Cold stress tolerance	Yin et al. (2016)
AaDREB1	*Arabidopsis thaliana, Oryza sativa*	Drought, salinity, and cold stress tolerance	Zong et al. (2016)
DREB1A	*Arabidopsis thaliana*	Drought stress tolerance	Kudo et al. (2017)
SmDREB1	*Arabidopsis thaliana*	Drought stress tolerance	Li et al. (2017)
MrCBF2	*Arabidopsis thaliana*	Drought and freezing stress tolerance	Wu et al. (2017)
AtDREB1A	*Arachis hypogea*	Drought stress tolerance	Bhatnagar-Mathur et al. (2007)
AtDREB1	*Arachis hypogea*	Drought stress tolerance	Bhatnagar-Mathur et al. (2014)
AtCBF1	*Brassica napus*	Cold stress tolerance	Jaglo et al. (2001)
BNCBF5, BNCBF17	*Brassica napus*	Cold stress tolerance	Savitch et al. (2005)
FaDREB1	*Broussoneta papyrifera*	Drought and salinity stress tolerance	Meiru et al. (2012)
BpERF11	*Betula platyphylla*	Salinity and osmotic stress tolerance	Zhang et al. (2016a)
AtDREB1A	*Chrysanthemum morifolium*	Cold, drought, and salinity stress tolerance	Hong et al. (2006)
AtDREB1A	*Chrysanthemum morifolium*	Heat stress tolerance	Hong et al. (2009)
AtDREB1A	*Cicer arietinum*	Drought stress tolerance	Anbazhagan et al. (2015)
AtDREB1A/CBF3	*Festuca arundinacea*	Drought stress tolerance	Zhao et al. (2007)
AtDREB1A	*Glycine max*	Drought stress tolerance	Polizel et al. (2011)
AtDREB2A-CA	*Glycine max*	Drought stress tolerance	Engels et al. (2013)
OsDREB2A	*Glycine max*	Salinity stress tolerance	Zhang et al. (2013)
AtDREB1A	*Glycine max*	Drought stress tolerance	De Paiva Rolla et al. (2014)
DREB 1A	*Gossypium hirsutum*	Drought stress tolerance	Amudha et al. (2014)

(continued)

Table 1 (continued)

DREB genes	Targeted plant species	Targeted trait	References
TaDREB3	*Hordeum vulgare*	Frost stress tolerance	Kovalchuk et al. (2013)
AtDREB1A/ CBF3	*Lolium perenne*	Drought and cold stress tolerance	Li et al. (2011)
MsDREB6.2	*Malus domestica*	Drought stress tolerance	Liao et al. (2017)
AtCBF3	*Manihot esculenta*	Cold and drought stress tolerance	An et al. (2016)
GmDREB1	*Medicago sativa*	Salt stress tolerance	Jin et al. (2010)
MtDREB1C	*Medicago truncatula, Rosa chinensis*	Freezing stress tolerance	Chen et al. (2010)
AtDREB1A	*Nicotiana tabacum*	Drought and cold stress tolerance	Kasuga et al. (2004)
LeCBF1	*Nicotiana tabacum*	Cold stress tolerance	Zhang et al. (2004)
OPBP1	*Nicotiana tabacum*	Salt stress tolerance	Guo et al. (2004)
CAP2	*Nicotiana tabacum*	Dehydration and salinity stress tolerance	Shukla et al. (2006)
GhDREB1	*Nicotiana tabacum*	Drought, salinity, and cold stress tolerance	Shan et al. (2007)
PpDBF1	*Nicotiana tabacum*	Drought and salinity stress tolerance	Liu et al. (2007)
OsDREB1B	*Nicotiana tabacum*	Oxidative and cold stress tolerance	Gutha and Reddy (2008)
WDREB2	*Nicotiana tabacum*	Freezing and osmotic stress tolerance	Kobayashi et al. (2008)
BjDREB1B	*Nicotiana tabacum*	Drought and salinity stress tolerance	Cong et al. (2008)
CAP2	*Nicotiana tabacum*	Heat stress tolerance	Shukla et al. (2009)
PeDREB2	*Nicotiana tabacum*	Salinity stress tolerance	Chen et al. (2009c)
CkDBF	*Nicotiana tabacum*	Drought, salinity, and cold stress tolerance	Wang et al. (2010)
PgDREB2A	*Nicotiana tabacum*	Hyperionic and hyperosmotic stress tolerance	Agarwal et al. (2010)
CkDREB	*Nicotiana tabacum*	Salinity and osmotic stress tolerance	Wang et al. (2011a)
LbDREB	*Nicotiana tabacum*	Heavy metal stress tolerance	Ban et al. (2011)
GmERF7	*Nicotiana tabacum*	Salinity stress tolerance	Zhai et al. (2013)
SbDREB2A	*Nicotiana tabacum*	Hyperionic and hyperosmotic stress tolerance	Gupta et al. (2014)
EsDREB2B	*Nicotiana tabacum*	Drought, salinity, cold, heat, and heavy metal stress tolerance	Li et al. (2014)
SsDREB	*Nicotiana tabacum*	Salinity and drought stress tolerance	Zhang et al. (2015)
GmDREB2	*Nicotiana tabacum*	Drought stress tolerance	Tan et al. (2015)

(continued)

Table 1 (continued)

DREB genes	Targeted plant species	Targeted trait	References
FeDREB1	*Nicotiana tabacum*	Cold and drought stress tolerance	Fang et al. (2015b)
PsAP2	*Nicotiana tabacum*	Salinity and drought stress tolerance	Mishra et al. (2015)
CpDREB2	*Nicotiana tabacum*	Cold, heat and drought stress tolerance	Arroyo-Herrera et al. (2016)
CBF1/DREB1b	*Oryza sativa*	Cold stress tolerance	Lee et al. (2004)
AtDREB1A	*Oryza sativa*	Drought and salinity stress tolerance	Oh et al. (2005)
DREB1/CBF1	*Oryza sativa*	Cold stress tolerance	Ito et al. (2006)
OsDREB1G	*Oryza sativa*	Drought stress tolerance	Chen et al. (2008)
AP37	*Oryza sativa*	Drought stress tolerance	Kim and Kim (2009)
SbDREB2	*Oryza sativa*	Drought stress tolerance	Bihani et al. (2011)
ZmCBF3	*Oryza sativa*	Drought, salinity, and cold stress tolerance	Xu et al. (2011)
OsDREB2A	*Oryza sativa*	Dehydration and salinity stress tolerance	Mallikarjuna et al. (2011)
OsDREB2A	*Oryza sativa*	Drought stress tolerance	Cui et al. (2011)
AtDREB1A, *OsDREB1B*,	*Oryza sativa*	Dehydration and salinity stress tolerance	Datta et al. (2012)
OsERF922	*Oryza sativa*	Salinity stress tolerance	Liu et al. (2012)
AtDREB1A	*Oryza sativa*	Drought stress tolerance	Ravikumar et al. (2014)
AtDREB1A	*Oryza sativa*	Drought stress tolerance	Hussain et al. (2014)
OsDREB1BI	*Oryza sativa*	Cold stress tolerance	Zhuang et al. (2015)
DaCBF7	*Oryza sativa*	Cold stress tolerance	Byun et al. (2015)
AtDREB1A	*Oryza sativa*	Drought stress tolerance	Samara Reddy et al. (2016)
JcDREB2	*Oryza sativa*	Salinity stress tolerance	Tang et al. (2017b)
HsDREB1A	*Paspalum notatum*	Salinity and drought stress tolerance	James et al. 2008
PeCBF4	*Populus euphratica*	Drought, high-salt, and cold stress tolerance	Tian et al. (2017)
FpDREB2A	*Robinia pseudoacacia*	Drought stress tolerance	Xiu et al. (2016)
AtDREB2A CA	*Saccharum officinarum*	Drought stress tolerance	Reis et al. (2014)
EaDREB2	*Saccharum officinarum*	Drought and salinity stress tolerance	Augustine et al. (2015)
AtDREB1A	*Salvia miltiorrhiza*	Drought stress tolerance	Wei et al. (2016b)
AtDREB1B	*Salvia miltiorrhiza*	Drought stress tolerance	Wei et al. (2016a)
AtDREB1C	*Salvia miltiorrhiza*	Drought stress tolerance	Wei et al. (2017)
AtCBF1	*Solanum lycopersicum*	Drought and cold stress tolerance	Hsieh et al. (2002)
SlDREB2	*Solanum lycopersicum*, *Arabidopsis thaliana*	Salinity stress tolerance	Hichri et al. (2016)
AtDREB1A	*Solanum lycopersicum*	Cold stress tolerance	Shah et al. (2016)

(continued)

Table 1 (continued)

DREB genes	Targeted plant species	Targeted trait	References
DREB1A	*Solanum tuberosum*	Salinity stress tolerance	Celebi-Toprak et al. (2005)
AtDREB1A	*Solanum tuberosum*	Salinity stress tolerance	Behnam et al. (2006)
StEREBP1	*Solanum tuberosum*	Salinity and cold stress tolerance	Lee et al. (2007)
DREB1/CBF1	*Solanum tuberosum*	Drought stress tolerance	Waterer et al. (2010)
AtDREB1B	*Solanum tuberosum*	Drought and cold stress tolerance	Movahedi et al. (2012)
StDREB2	*Solanum tuberosum*	Salinity stress tolerance	Bouaziz et al. (2012)
StDREB1	*Solanum tuberosum*	Drought and salinity stress tolerance	Bouaziz et al. (2013)
StDREB1	*Solanum tuberosum*	Salinity stress tolerance	Bouaziz et al. (2015b)
StDREB1, StDREB2	*Solanum tuberosum*	Drought stress tolerance	Bouaziz et al. (2015a)
StERF3	*Solanum tuberosum*	Salinity stress tolerance	Tian et al. (2015)
StDREB1	*Solanum tuberosum*	WPG/cadmium stress tolerance	Bouaziz et al. (2017)
StDREB1, StDREB2	*Solanum tuberosum*	Cadmium stress tolerance	Charfeddine et al. (2017)
AhDREB1	*Triticum aestivum*	Drought and salinity stress tolerance	Shen et al. (2003)
AtDREB1A	*Triticum aestivum*	Drought stress tolerance	Pellegrinischi et al. (2004)
AtDREB	*Triticum aestivum*	Drought stress tolerance	Jun-Wei et al. (2006)
GhDREB	*Triticum aestivum*	Drought, salinity, and cold stress tolerance	Gao et al. (2009)
TaDREB2, TaDREB3	*Triticum aestivum, Hordeum vulgare*	Drought stress tolerance	Morran et al. (2011)
TaERF3	*Triticum aestivum*	Drought and salinity stress tolerance	Rong et al. (2014)
GmDREB1	*Triticum aestivum*	Salt stress tolerance	Jiang et al. (2014a, b)

4 *NAC* Transcription Factors and Their DNA-Binding Structure

Collectively we can term NAC as a mnemonic of three genes that were originally discovered containing peculiar domain (the NAC domain): NAM stands for no apical meristem, ATAF1, and ATAF2, and CUC2 stands for cup-shaped cotyledon (Souer et al. 1996; Aida et al. 1997). Another group of plant-specific transcription factors is YABBY, which functions as a key player in organ polarity (Bowman et al. 2002). Abaxial-adaxial polarity is carried out by YABBY in lateral organs (Goldshmidt et al. 2008); this YABBY family transcription factors in its amino-terminal region contain a zinc-finger domain, and a special YABBY domain in the carboxyl-terminal region is reported (Bowman 2000). Thus the presence of NAC proteins as well as their function as potent TFs is well documented in a wide number of plant species, and they comprise one of the largest families of plant-specific TFs. Around 117 NAC genes in *Arabidopsis*, 151 in rice, 79 in grape, 26 in citrus, 163 in poplar, and 152 each in soybean and tobacco have been identified through the use of sophisticated genomic scanning techniques as well as whole-genome sequences (Rushton et al. 2008; Hu et al. 2010; Nuruzzaman et al. 2010, 2012; Le et al. 2011). The vigorous analysis of promoter of the gene induced by drought in Arabidopsis indicates that the DNA-binding site of stress-inducible NAC TFs plays the key role. The NAC recognition sequence (NACRS) contains the CACG core motif which works on activation by early response to dehydration 1 (ERD1) gene; it encodes a protein with homology to the ATP-binding subunit of the Clp ATP-dependent protease from *Escherichia coli* (Simpson et al. 2003; Tran et al. 2004). Molecular analysis of the ERD1 promoter demonstrated that ERD1 expression during water scarcity stress mainly depends upon the integrity of both the “CAC TAA ATT GTC AC” ZFHDRS and the “ANN NNN TCN NNN NNN ACA CGC ATG T” NACRS sequences (Tran et al. 2004, 2007). The rice drought-responsive ONAC TFs bind to a similar NACRS found in rice genome, which clearly suggests that the binding of NACRS to the gene pre-induced by drought might be conserved across the plant kingdom (Hu et al. 2006; Fang et al. 2008). In addition, numerous other sequences have also been reported as NAC binding sites. The *Arabidopsis* NAC1 protein was examined to bind to a 21-bp segment (CTG ACG TAA GGG ATG ACG CAC) within the 35S-90 promoter fragment (Xie et al. 2000). Independently, Duval et al. (2002) also modulated that purified *AtNAM* recombinant protein salvages the region of the CaMV 35S promoter between -70 and -76, at the location 21-bp segment (AGGGATG). On the other hand, CBNAC TF found in *Arabidopsis* selectively binds to CBNACBS which was confirmed both in vitro and in vivo analysis. Thus the NAC TFs demonstrate a highly conserved N-terminal DNA-binding NAC domain and an inconstant transcriptional regulation C-terminal domain which varies its splicing at transcriptional levels to suit the variations of stress induced (Ernst et al. 2004; Olsen et al. 2005). Cloned NAM, ATAF, and CUC2 analyses revealed the presence of NAC domain in the NAC TFs which was confirmed by database homology searching (Ooka et al. 2003; Olsen et al. 2005). Typically, a number of NAC protein sequences, which is of ~150 amino acids in length, were determined, namely, crystal structure of the NAC domain of an *Arabidopsis* NAC protein and the ANAC (ABA-responsive NAC) (Ernst et al. 2004), which confirms their sequence homology conserved in evolutionary history of each haplotype of the species genome. The structure of NAC transcription factors has been depicted in Figs. 2 and 3.

Fig. 2 Transcriptional regulatory networks of *cis*-elements and DREB/NAC transcription factors involved in abiotic stress-induced gene expression (Adapted from Nuruzzaman et al. 2013)

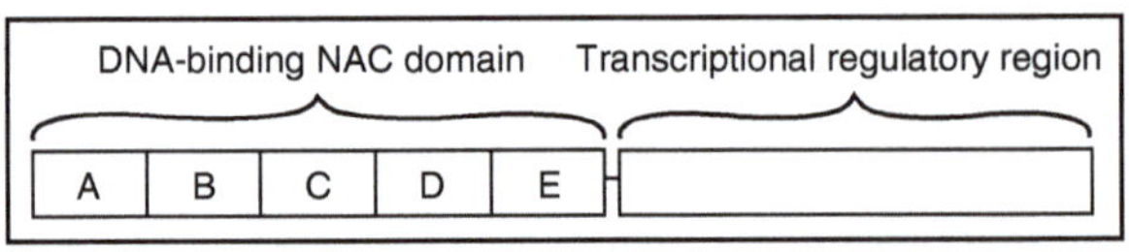

Fig. 3 Proposed structure of NAC transcription factor: the N-terminal DNA-binding domain of NAC consists of five subdomains (*A–E*). C-terminal acts as transcriptional regulatory sequence which interacts with stress-responsive sequences either as activators or repressors of the gene

5 Bioengineering of *NACs* for Enhancing Abiotic Stress Tolerance in Plants

NAC transcription factors play an important role in the growth of plants and its development and its reflexes in stress situations (Hao et al. 2011). The genes encoding NAC proteins are one of the largest families of plant-specific transcription factors and are uniquely present only in plant while it lacks other eukaryotes (Gao et al. 2009). There are approximately 75 and 105 putative *NAC* genes in rice and *Arabidopsis* genomes, respectively. For example, the modification of 33 *NAC* genes was reported specifically in *Arabidopsis* when treated with salt (Jiang and Deyholos 2006), while 38 *NAC* genes were reported in retaliation to drought in soybean (Le et al. 2011); 40 *NAC* genes combined in rice responded to both temperature and salinity (Fang et al. 2008), and in *Chrysanthemum lavandulifolium*, a cocktail of treatments lead to the activation of 32 *NAC* genes (Huang et al. 2015). Expression data from genome-wide transcriptome analyses suggests a significant proportion of *NAC* gene function in stress retaliation in many plants. *Arabidopsis* has been rendered as a model crop for elucidation of the role of NAC gene family in abiotic stresses, and this was a significant discovery; these NAC genes were persuaded under abiotic stresses. In *Arabidopsis*, overexpression of three genes *ANACO19*, *ANACO55*, and *ANACO72* was increased by water stress (Tran et al. 2004). In rice it demonstrated a sequence of multiple transcription factor activation at a very early stage when salinity stress was given (Chao et al. 2005). Multiple stresses such as cold, salt, drought, and abscisic acid (ABA) lead to the activation of OsNAC6, a member of ATAF family (Ohnishi et al. 2005). In transgenic rice, resistant to drought variety, there was overexpression of stress-responsive gene *SNAC1* (stress-responsive *NAC* 1) when exposed to severe drought (22–34% higher seed setting than control) in the field conditions at the reproductive stage, while there were no manifestations at phenotypic stage or yield penalty. The other NAC family *SNAC3*-OE transgenic plants unveiled considerably significant levels of H_2O_2, malondialdehyde (MDA), and relative electrolyte leakage when compared to wild-type control under temperature stress, clearly signifying that *SNAC3* may render its contribution of plant rescue at stress condition by harmonizing reactive oxygen species (ROS) homeostasis physiology. It was vividly observed that the *SNAC3*-RNAi transgenic plant's quantitative PCR experiments verified that the expression of a large number of ROS-scavenging genes was significantly increased in abiotic stress situations which were in contrast to wild types comparatively (Xiong et al. 2015). Very promising results were observed in transgenic rice which quantitatively showed a considerably better tolerance. The different NAC transcription families, bHLH, MYB, bZIP, WRKY, ZF, and AP2/ERF, demonstrate a higher degree of tolerance, suggesting that there is an interaction with the other regulatory proteins, as instantiated by the cross talk between the C-terminal domain of *HvNAC013* and radical-induced cell death1 (RCD1) (Kjaersgaard et al. 2011; Kragelund et al. 2012). The transgenic rice have demonstrated less sensitivity to abscisic acid and loss of water more steadily by closing more stomatal pores compared with wild type, while there is no significant amount of changes in photosynthetic rate. Through a study using DNA chip, an analysis demonstrated a large number of stress-related genes which were upregulated in the *SNAC1*-overexpressing rice plants. *SNAC1* holds a very bright future in improving drought and salinity tolerance in rice (Hu et al. 2006). Similarly a myriad of *NAC* genes are reported to be induced by drought, high-salinity, and cold stresses. In rice *OsNAC6/SNAC2* encodes a NAC transcription factor (Hu et al. 2008). Thus the expression of *OsNAC6*/SNAC2 is also induced by cold, drought, and high salinity. A very peculiar result was seen in transgenic rice plants overexpressing OsNAC6 that showed increased tolerance to water scarcity to high-salinity situations, but it also led to growth retardation and low yield also, while this low yield was reverted in transgenic rice plants piloted by a stress-inducible promoter. These results designate that *OsNAC6* acts as a transcriptional activator in abiotic stress situations (Chung et al. 2009). With other experiments of induction of *OsNAC5* expression when induced with dehydration, cold, ABA, and MeJA treatments were also promising. This was done by a comparative study of growth of rice plants overexpressing *OsNAC5* to that of control plants under non-stress conditions. In another study it was demonstrated that if we restrict *OsNAC5* expression, as done in RNA interference (RNAi), transgenic rice plants lead to reduced tolerance to abiotic stresses in comparison to control plants, whereas over-expression of *OsNAC5* greatly enhances abiotic stress tolerance. *OsNAC5* not only does work in protection by only overexpression, but it also relishes accumulation of compatible solutes such as proline and soluble sugars which play a role in secondary level of plant defense line (Song et al. 2011). A recent report suggested that *GmNAC11* functions as a transcriptional activator, while *GmNAC20* functions as a temperate repressor even though its C-terminal end has transcriptional activation activity. In continuation to this, overexpression of *GmNAC20* improved salinity tolerance and cold tolerance in transgenic *Arabidopsis* plants. Overexpression of *GmNAC20* also magnifies lateral root formation which helps plant combat water-scarce condition (Hao et al. 2011). Thus the above research suggests that *GmNAC20* may regulate stress tolerance by activating the *DREB/CBF–COR* pathway and may regulate lateral root formation by reorienting auxin signaling and related hormonal genes. Recently, through *Agrobacterium*-mediated transformation, it was attempted to introduce NAC1 gene from *E. coli* into tobacco (Ramegowda et al. 2012) to model the stress interaction on physiology. In another experiment, overexpression of *TaNAC2* resulted in increased tolerance to abiotic stress in *Arabidopsis*, which was manifested by enhanced root system, more hormone secretion, and changes in physiological parameters (Mao et al. 2012). Thus the conclusion is that NAC gene clearly plays an important role in plant responses to various stresses; not only this NAC plays significant role in senescence, but also overexpression of *AtNAP*, *ORE1* (Kim and Kim 2009; Balazadeh et al. 2010), *ORS1* (Balazadeh et al. 2011), *ANAC016* (Kim et al. 2013), and *ATAF1* (Garapati et al. 2015) resulted in precocious senescence, while hindering the function of these genes leads to lack of senescence indicating that they function as nonredundant positive regulators of senescence in *Arabidopsis*; thus global gene expression pattern of these NAC transcription factors exclusively regulates the senescence process which reveals the versatility of functions this TF performs (Skriver et al. 2015). So comprehending the molecular mechanisms of NAC TF networks consolidating multiple stress responses will be essential for the development of broad-range stress-tolerant crop plants that can cope with the recent global climatic aberrations our world is currently going through (Shao et al. 2015). The overexpression of candidate genes governing *NAC* transcription factors in *Arabidopsis thaliana* for tolerance against salinity, drought, and freezing/cold stress (Mao et al. 2016; Yu et al. 2016; Zhang et al. 2016b; Zhao et al. 2016; Gunapati et al. 2016; Wang et al. 2017b); *Artemisia annua* for salinity and osmotic stress (Lv et al. 2016); *Cicer arietinum* for drought and salinity stress (Yu et al. 2016); *Dendronthema grandiform* for salinity stress (Wang et al. 2017a, b, c); *Musa paradisiaca* for salinity and drought stress (Negi et al. 2016; Tak et al. 2017);

Nicotiana tabacum for drought stress (Tang et al. 2017a, b); rice for drought, salinity, and freezing stress (Hong et al. 2016; Huang et al. 2016; Rahman et al. 2016; Lee et al. 2017; Shen et al. 2017); *Setaria italica* for drought and salinity stress (Lina et al. 2016); *Solanum lycopersicum* for drought stress (Thirumalaikumar et al. 2017; Wang et al. 2017a); and *Tamarix hispida* (Wang et al. 2017b) has been established successfully, and the remaining work is on the way. For more information in regard to overexpressing *NAC* genes, kindly refer to Table 2.

Table 2 Transgenic plants developed for abiotic stress tolerance by the overexpression of *NAC* genes

NAC genes	Targeted plant species	Targeted trait	References
RD26	*Arabidopsis thaliana*	Drought and salinity stress tolerance	Fujita et al. (2004)
ANAC019, ANAC055, AtNAC072	*Arabidopsis thaliana*	Drought and salt stress tolerance	Tran et al. (2004)
AtNAC2, AtNAC055	*Arabidopsis thaliana*	Drought and salinity stress tolerance	He et al. (2005)
AtNAC019, ATAF1	*Arabidopsis thaliana*	Drought and salt stress tolerance	Lu et al. (2007)
ATAF1	*Arabidopsis thaliana*	Drought stress tolerance	Wu et al. (2009)
ONAC063	*Arabidopsis thaliana*	High-salinity and osmotic stress tolerance	Yokotani et al. (2009)
ANAC019	*Arabidopsis thaliana*	Cold stress tolerance	Jensen et al. (2010)
GmNAC20	*Arabidopsis thaliana*	Salinity and freezing stress tolerance	Hao et al. (2011)
AhNAC2	*Arabidopsis thaliana*	Drought and salt tress tolerance	Liu et al. (2011a)
NTL6, SnRK2.8	*Arabidopsis thaliana*	Drought stress tolerance	Kim et al. (2012)
ZmSNAC1	*Arabidopsis thaliana*	Cold, high-salinity, and drought stress tolerance	Lu et al. (2012)
TaNAC2	*Arabidopsis thaliana*	Drought, salt, and freezing stress tolerance	Mao et al. (2012)
BnNAC5	*Arabidopsis thaliana*	Salinity stress tolerance	Zhong et al. (2012)
ANAC042	*Arabidopsis thaliana*	Heat stress tolerance	Shahnejat-Bushehri et al. (2012)
ANAC013	*Arabidopsis thaliana*	Drought and salt stress tolerance	De Clercq et al. (2013)
RhNAC3	*Arabidopsis thaliana*	Drought tress tolerance	Jiang et al. (2014a, b)

(continued)

Table 2 (continued)

NAC genes	Targeted plant species	Targeted trait	References
TaNAC67	*Arabidopsis thaliana*	Cold, salinity, and drought stress tolerance	Mao et al. (2014)
TaNAC29	*Arabidopsis thaliana*	Salt and drought tolerance	Huang et al. (2015)
TaNAC29	*Arabidopsis thaliana*	Salt stress tolerance	Xu et al. (2015)
MLNAC5	*Arabidopsis thaliana*	Drought and cold stress tolerance	Yang et al. (2015)
ZmNAC55	*Arabidopsis thaliana*	Drought stress tolerance	Mao et al. (2016)
CarNAC5	*Arabidopsis thaliana*	Drought stress tolerance	Yu et al. (2016)
TaNAC47	*Arabidopsis thaliana*	Drought, freezing, and salt stress tolerance	Zhang et al. (2016b)
MlNAC9	*Arabidopsis thaliana*	Drought and cold stress tolerance	Zhao et al. (2016)
GhNAC2	*Arabidopsis thaliana*	Drought stress tolerance	Gunapati et al. (2016)
ThNAC13	*Arabidopsis thaliana, Tamarix hispida*	Salt and osmotic stress tolerance	Wang et al. (2017c)
MuNAC4	*Arachis hypogea*	Drought stress tolerance	Pandurangaiah et al. (2017)
AhNAC2	*Arachis pintoi*	Salinity stress tolerance	Liu et al. (2011a)
AaNAC1	*Artemisia annua*	Drought stress tolerance	Lv et al. (2016)
OsNAC6	*Allium cepa*	Salinity stress tolerance	Nakashima et al. (2007)
OsSNAC1	*Boehmeria nivea*	Drought and salt stress tolerance	An et al. (2015)
CsNAM	*Camellia sinensis*	Drought, salt, and heat stress tolerance	Paul et al. (2012)
CarNAC5	*Cicer arietinum*	Drought and heat stress tolerance	Peng et al. (2009)
CarNAC5	*Cicer arietinum*	Drought and salt stress tolerance	Yu et al. (2016)
DgNAC1	*Dendronthema grandiform*	Salinity stress tolerance	Wang et al. (2017a)
GhNAC3 GhNAC4 GhNAC5 GhNAC6	*Gossypium hirsutum*	Drought cold and salt stress tolerance	Meng et al. (2009)
OsSNAC1	*Gossypium hirsutum*	Drought and salinity stress tolerance	Liu et al. (2014)
GmNAC11, GmNAC20	*Glycine max*	Drought and salinity stress tolerance	Hao et al. (2011)
HvSNAC1	*Hordeum vulgare*	Drought stress tolerance	Abdallat et al. (2014)

(continued)

Table 2 (continued)

NAC genes	Targeted plant species	Targeted trait	References
MusaNAC68	*Musa paradisiaca*	Salinity and drought stress tolerance	Negi et al. (2016)
MusaNAC042	*Musa paradisiaca*	Salinity and drought stress tolerance	Tak et al. (2017)
DgNAC1	*Nicotiana tabacum*	Salt, drought, and cold stress tolerance	Liu et al. (2011b, c)
EcNAC1	*Nicotiana tabacum*	Water-deficit, salt stress tolerance	Ramegowda et al. (2012)
TaNAC2a	*Nicotiana tabacum*	Drought stress tolerance	Tang et al. (2012)
AhNAC3	*Nicotiana tabacum*	Drought stress tolerance	Liu et al. (2013)
AhNAC4	*Nicotiana tabacum*	Drought stress tolerance	Tang et al. (2017a)
SNAC1	*Oryza sativa*	Cold, drought, and salt stress tolerance	Hu et al. (2006)
OsNAC6	*Oryza sativa*	Drought and salt stress tolerance	Nakahsima et al. (2007)
SNAC2	*Oryza sativa*	Drought, salinity, and cold stress tolerance	Sindhu et al. (2008)
OsNAC5	*Oryza sativa*	Salt and cold stress tolerance	Sperotto et al. (2009)
OsNAC45	*Oryza sativa*	Drought, salt, and cold stress tolerance	Zheng et al. (2009)
OsNAC10	*Oryza sativa*	Drought, salinity, and cold stress tolerance	Jeong et al., 2010
OsNAC5	*Oryza sativa*	Salinity stress tolerance	Takasaki et al., 2010
ONAC045	*Oryza sativa*	Drought and salt stress tolerance	Song et al. 2011
OsNAC9	*Oryza sativa*	Drought stress tolerance	Kim et al. (2012)
OsNAC9	*Oryza sativa*	Drought stress tolerance	Redillas et al. (2012)
OsOAT, SNAC2	*Oryza sativa*	Drought and oxidative stress tolerance	You et al. (2013)
OsNAP	*Oryza sativa*	Cold, salinity, and drought stress tolerance	Chen et al. (2014)
SNAC3	*Oryza sativa*	Heat and drought stress tolerance	Fang et al. (2015a)
ONAC022	*Oryza sativa*	Drought and salt tress tolerance	Hong et al. (2016)
ONAC095	*Oryza sativa*	Drought and cold stress tolerance	Huang et al. (2016)

(continued)

Table 2 (continued)

NAC genes	Targeted plant species	Targeted trait	References
EcNAC67	*Oryza sativa*	Salinity and drought stress tolerance	Rahman et al. (2016)
OsNAC6	*Oryza sativa*	Drought stress tolerance	Lee et al. (2017)
OsNAC2	*Oryza sativa*	Drought and salt stress tolerance	Shen et al. (2017)
SiNAC	*Setaria italica*	Drought and salt stress tolerance	Puranik et al. (2011)
SiNAC110	*Setaria italica*	Drought and high-salt stress tolerance	Lina et al. (2016)
SlNAC4	*Solanum lycopersicum*	Drought and salt stress tolerance	Zhu et al. (2014)
ATNAC3	*Solanum lycopersicum*	Drought and salt stress tolerance	Abdallat et al. (2015)
AtJUB1	*Solanum lycopersicum*	Drought stress tolerance	Thirumalaikumar et al. (2017)
SlNAC11	*Solanum lycopersicum*	Drought and salinity stress tolerance	Wang et al. (2017b)
TaNAC4	*Triticum aestivum*	Cold and salt stress tolerance	Xia et al. (2010)
TaNAC69	*Triticum aestivum*	Osmotic, salt stress tolerance	Xue et al. (2011)
TaNAC4a, TaNAC6	*Triticum aestivum*	Dehydration, salinity, and cold stress tolerance	Tang et al. (2012)
SNAC1	*Triticum aestivum*	Drought and salinity stress tolerance	Saad et al. (2013)
SNAC1	*Triticum aestivum*	Drought and salinity stress tolerance	Yu et al. (2013)
ZmSNAC1	*Zea mays*	Cold, drought, and salinity stress tolerance	Lu et al. (2012)

6 Conclusion and Future Prospects

As we know the primitive breeding technologies have produced a very few refined stress-tolerant cultivars; thus the implementation of marker-assisted breeding provides a better opportunity to introduce the gene of interest which is not only fast but easy. For abiotic stress engineering, we have three prospects to choose from, be it altering biochemical pathway-associated genes to change the morphology to avoid stress, altering enzymes producing protein factors to act as osmolytes/protectants, or directly regulating gene expression as regulatory proteins to alter at genetic level. As we know abiotic stress is a polygenic trait which requires a symphony of many genes working in collaboration, so for an effective result, a strong promoter will handle this stress tolerance, but it has its own side effects, as it leads to loss of energy and hampers plant's overall well-being, but if we know how to control gene expression and can fine-tune expression of stress-induced transcription factors, this in turn would control the expression of a wide array of genes downstream. Considerable information through transcriptome profiling strategies has been gained about DREBs and NAC TFs since the discovery of them, but the research in this area is still in its inceptive phase. Genome-wide identification and expression profiling have undeniably open new avenues for deciphering the functioning of these transcription factors. In particular, the stress-responsive TFs can be used as promising candidates for generation of stress-tolerant transgenic plants possessing high productivity under adverse conditions. But there are still some problems to be solved as some studies have shown. Firstly, the collective overexpression of TF genes seldomly may lead to negative effects in transgenic plants such as lower yields, late flowering, and dwarfing. Secondly, the transgenic plants overexpressing genes may occasionally have antagonistic responses to different stresses. Thirdly, there are still less evidence of DREB and NAC gene expression model plants on field trails so far. Thus as everyone knows, the prime aim for plant stress research is to provide targets for the improvement of stress tolerance in crop plants. With the unforeseen changes in climatic conditions that will lead to more complex stress environment in the fields, we will face new hurdles in creating multiple stress-tolerant crops. Breeding such plants will depend on understanding the crucial stress-regulatory networks and the potential effects of different combinations of adverse conditions. It's a new arena still open for positive changes to achieve our prime aim to develop stress-tolerant plant fit to survive the environment without having any antagonist effect. And targeting transcription factors is safe as it is just like epigenetic switching on or off of the gene-expressing machinery. Future research should focus on additional approaches including (1) better understanding of the intricate network of genes and the effect it produces on metabolic and physiological level during stress condition, (2) a better understanding of the molecular machinery as well as its response toward the overall health of plants and its development, (3) consolidating the numerous approaches for engineering these genes to focus on the desired gene affecting the concerned stress, and (4) successfully using the related techniques such as transcriptomics, proteomics, and metabolomics targeting TF to reduce the anomalies a plant withstands through the induced gene to withstand stress. Finally, the myriad of efforts of plant molecular biologists, physiologists, and breeders are required to generate stress-tolerant grasses through genetic engineering. It is hoped that, in the future, all these collective efforts will definitely put forward its contribution to sustainable food production in the world, thus putting a bandage on global-scale environmental damage caused by abiotic stresses.

References

Abdallat AMA, Ayad JY, Elenein JMA, Ajlouni Z, Al WA (2014) Harwood: overexpression of the transcription factor *HvSNAC1* improves drought tolerance in barley (*Hordeum vulgare* L.) Mol Breed 33(2):401–414

Abdallat AMA, Ali Sheikh OMA, Alnemer LM (2015) Overexpression of two *ATNAC3* related genes improves drought and salt tolerance in tomato (*Solanum lycopersicum* L.) Plant Cell Tissue Organ Cult 120(3):989–1001

Abogadallah GM, Nada RM, Malinowski R, Quick P (2011) Overexpression of *HARDY*, an *AP2/ERF* gene from *Arabidopsis*, improves drought and salt tolerance by reducing transpiration and sodium uptake in transgenic *Trifolium alexandrinum* L. Planta 233:1265–1276

Agarwal PK, Agarwal P, Reddy MK, Sopory SK (2006) Role of DREB transcription factors in abiotic and biotic stress tolerance in plants. Plant Cell Rep 25:1263–1274

Agarwal P, Agarwal PK, Joshi AJ, Reddy MK, Sopory SK (2010) Overexpression of *PgDREB2A* transcription factor enhances abiotic stress tolerance and activates downstream stress-responsive genes. Mol Biol Rep 37:1125–1135

Aida M, Ishida T, Fukaki H, Fujisawa H, Tasaka M (1997) Genes involved in organ separation in Arabidopsis: an analysis of the cup-shaped cotyledon mutant. Plant Cell 9:841–857

Allen MD, Yamasaki K, Ohme-Takagi M, Tateno M, Suzuki M (1998) A novel mode of DNA recognition by a β-sheet revealed by the solution structure of the GCC-box binding domain in complex with DNA. EMBO J 17:5484–5496

Amudha J, Chhajed S, Balasubramani G (2014) Cotton transgenic plants with Dre-binding transcription factor gene (*DREB 1A*) confers enhanced tolerance to drought. Int J Adv Biotechnol Res 5(4):635–648

An X, Liao Y, Zhang J, Dai L, Zhang N, Wang B, Liu L, Peng D (2015) Overexpression of rice NAC gene *SNAC1* in ramie improves drought and salt tolerance. Plant Growth Regul 76(2):211–223

An D, Ma Q, Yan W, Zhou W, Liu G, Zhang P (2016) Divergent regulation of *CBF* regulon on cold tolerance and plant phenotype in cassava overexpressing *Arabidopsis CBF3* Gene. Front Plant Sci 7:1866

Anbazhagan K, Bhatnagar-Mathur P, Vadez V, Dumbala SR, Kavi Kishor PB, Sharma KK (2015) *DREB1A* overexpression in transgenic chickpea alters key traits influencing plant water budget across water regimes. Plant Cell Rep 34(2):199–210

Arroyo-Herrera A, Figueroa-Yanez L, Castano E, Santamarıa J, Pereira-Santana A, Espadas-Alcocer J, Sanchez-Teyer F, Espadas-Gil F, Alcaraz LD, Lopez-Gomez R, Sanchez-Calderon L, Rodrıguez-Zapata LC (2016) A novel *Dreb2*-type gene from *Carica papaya* confers tolerance under abiotic stress. Plant Cell Tissue Organ Cult 125(1):119–133

Augustine SM, Narayan JA, Syamaladevi DP, Appunu C, Chakravarthi M, Ravichandran V, Tuteja N, Subramonian N (2015) Overexpression of *EaDREB2* and pyramiding of *EaDREB2* with the pea DNA helicase gene (*PDH45*) enhance drought and salinity tolerance in sugarcane (*Saccharum* spp. hybrid). Plant Cell Rep 34(2):247–263

Balazadeh S, Siddiqui H, Allu AD, Matallana-Ramirez LP, Caldana C, Mehrnia M, Zanor MI, Kahler B, Mueller-Roeber B (2010) A gene regulatory network controlled by the NAC transcription factor ANAC092/AtNAC2/ORE1 during salt promoted senescence. Plant J 62:250–264

Balazadeh S, Kwasniewski M, Caldana C, Mehrnia M, Zanor MI, Xue GP, Roeber MB (2011) ORS1, an H_2O_2-responsive NAC transcription factor, controls senescence in *Arabidopsis thaliana*. Mol Plant 4(2):346–360

Ban Q, Liu G, Wang Y (2011) A *DREB* gene from *Limonium bicolor* mediates molecular and physiological responses to copper stress in transgenic tobacco. J Plant Physiol 168(5):449–458

Basu U (2012) Identification of molecular processes underlying abiotic stress plants adaptation using "omics" technologies. In: Benkeblia N (ed) Sustainable Agricul-ture and new Technologie. CRC Press, Boca Raton, pp 149–172

Behnam B, Kikuchi A, Celebi-Toprak F, Yamanaka S, Kasuga M, Kazuko Y, Watanabe KN (2006) The *Arabidopsis DREB1A* gene driven by the stress-inducible rd29A promoter increases salt-stress tolerance in proportion to its copy number in tetrasomic tetraploid potato (*Solanum tuberosum* L.) Plant Biotechnol 23:169–177

Bhatnagar-Mathur P, Devi MJ, Reddy DS, Lavanya M, Vadez V, Serraj R, Yamaguchi-Shinozaki K, Sharma KK (2007) Stress inducible expression of *At DREB1A* in transgenic peanut *(Arachis hypogaea)* increases transpiration efficiency. Plant Cell Rep 26(12):2071–2082

Bhatnagar-Mathur P, Rao JS, Vadez V, Dumbala SR, Rathore A, Yamaguchi-Shinozaki K, Sharma KK (2014) Transgenic peanut overexpressing the *DREB1A* transcription factor has higher yields under drought stress. Mol Breed 33:327–340

Bihani P, Char B, Bhargava S (2011) Transgenic expression of sorghum *DREB2* in rice improves tolerance and yield under water limitation. J Agric Sci 149:95–101

Bohnert HJ, Ayoubi P, Borchert C, Bressan RA, Burnap RL, Cushman JC, Cushman MA, Deyholos M, Fisher R, Galbraith DW, Hasegawa PM, Jenks M, Kawasaki S, Koiwa H, Kore-eda S, Lee BH, Michalowski CB, Misawa E, Nomura M, Ozturk N, Postier B, Prade R, Song CP, Tanaka Y, Wang H, Zhu JK (2001) A genomic approach towards salt stress tolerance. Plant Physiol Biochem 39:295–311

Bouaziz D, Pirrello J, Amor BH, Hammami A, Charfeddine M, Dhieb A, Bouzayen M, GargouriBouzid R (2012) Ectopic expression of dehydration responsive element binding proteins (*StDREB2*) confers higher tolerance to salt stress in potato. Plant Physiol Biochem 60:98–108

Bouaziz D, Pirrello J, Charfeddine M, Hammami A, Jbir R, Dhieb A, Bouzayen M, Gargouri-Bouzid R (2013) Overexpression of *StDREB1* transcription factor increases tolerance to salt in transgenic potato plants. Mol Biotechnol 54(3):803–817

Bouaziz D, Charfeddine M, Jbir R, Saidi MN, Pirrello J, Charfeddine S, Bouzayen M, Gargouri-Bouzid R (2015a) Identification and functional characterization of ten *AP2/ERF* genes in potato. Plant Cell Tissue Organ Cult 123(1):155–172

Bouaziz D, Jbir R, Charfeddine S, Saidi MN, Gargouri R (2015b) The *StDREB1* transcription factor is involved in oxidative stress response and enhances tolerance to salt stress. Plant Cell Tissue Organ Cult 121(1):237–248

Bouaziz D, Charfeddine S, Jbir R, Hammami A, Kamoun L, Gargouri-Bouzid R (2017) Effects of phosphogypsum on the growth of potato plants overexpressing the *StDREB1* transcription factor. Plant Cell Tiss Organ Cult 130:1–11

Bowman JL (2000) The YABBY gene family and abaxial cell fate. Curr Opin Plant Biol 3:17–22

Bowman JL, Eshed Y, Baum SF (2002) Establishment of polarity in angiosperm lateral organs. Trends Genet 18:134–141

Byun MY, Lee J, Cui LH, Kang Y, Oha TK, Park H, Lee H, Kim WT (2015) Constitutive expression of *DaCBF7*, an Antarctic vascular plant *Deschampsia antarctica CBF* homolog, resulted in improved cold tolerance in transgenic rice plants. Plant Sci 236:61–74

Celebi-Toprak F, Behnam B, Serrano G, Kasuga M, Yamaguchi-Shinozaki K, Naka H, Watanabe JA, Yamanaka S, Watanabe KN (2005) Tolerance to salt stress of the transgenic tetrasomic tetraploid potato, *Solanum tuberosum* cv. Desiree appears to be induced by the *DREB1A* gene and *rd29A* promoter of *Arabidopsis thaliana*. Breed Sci 55:310–319

Chao DY, Luo YH, Shi M, Luo D, Lin HX (2005) Salt-responsive genes in rice revealed by cDNA microarray analysis. Cell Res 15:796–810

Charfeddine M, Charfeddine S, Bouaziz D, Messaoud RB, Bouzid RG (2017) The effect of cadmium on transgenic potato (*Solanum tuberosum*) plants overexpressing the *StDREB* transcription factors. Plant Cell Tissue Organ Cult 128(3):521–541

Chaves MM, Costa JM, Saibo NJM (2011) Recent advances in photosynthesis under drought and salinity. Adv Bot Res 57:50–83

Chen M, Wang QY, Cheng XG, Xu ZS, Li LC, Ye XG (2007) *GmDREB2*, a soybean *DRE* binding transcription factor, conferred drought and high salt tolerance in transgenic plants. Biochem Biophys Res Commun 353:299–305

Chen JQ, Meng XP, Zhang Y, Xia M, Wang XP (2008) Overexpression of *OsDREB* genes lead to enhanced drought tolerance in rice. Biotechnol Lett 30(12):2191–2198

Chen M, Xu Z, Xia L, Li L, Cheng X, Dong J, Wang Q, Ma Y (2009a) Cold-induced modulation and functional analyses of the *DRE*-binding transcription factor gene, *GmDREB3*, in soybean (*Glycine max* L.) J Exp Bot 60(1):121–135

Chen J, Xia X, Yin W (2009b) Expression profiling and functional characterization of a *DREB2* type gene from *Populus euphratica*. Biochem Biophys Res Commun 378:483–487

Chen J, Xia X, Yin W (2009c) Expression profiling and functional characterization of a *DREB2*-type gene from *Populus euphratica*. Biochem Biophys Res Commun 378:483–487

Chen JR, Lu JJ, Liu R, Xiong XY, Wang TX, Chen SY (2010) *DREB1C* from *Medicago truncatula* enhances freezing tolerance in transgenic *M. truncatula* and China rose (*Rosa chinensis* Jacq.) Plant Growth Regul 60:199–211

Chen J, Xia X, Yin W (2011) A poplar *DRE* binding protein gene, *PeDREB2L*, is involved in regulation of defense response against abiotic stress. Gene 483(12):36–42

Chen X, Wang Y, Lv B, Li J, Luo L, Lu S (2014) The *NAC* family transcription factor *OsNAP* confers abiotic stress response through the ABA pathway. Plant Cell Physiol 55:604–619

Chen H, Liu L, Wang L, Wang S, Cheng X (2016) *VrDREB2A*, a *DREB*-binding transcription factor from *Vigna radiata*, increased drought and high salt tolerance in transgenic *Arabidopsis thaliana*. J Plant Res 129:263–273

Cheng L, Hui L, Li L, Chen SX, Li L (2015) Overexpression of *NnDREB2*, isolated from lotus improves salt tolerance in transgenic *Arabidopsis thaliana*. Acta Physiol Plant 37:261–272

Chung PJ, Kim YS, Jeong JS, Park SH, Nahm BH, Kim JK (2009) The histone deacetylase OsHDAC1 epigenetically regulates the OsNAC6 gene that controls seedling root growth in rice. Plant J 59(5):764–776

Cong L, Chai T, Zhang Y (2008) Characterization of the novel gene *BjDREB1B* encoding a *DRE* binding transcription factor from *Brassica juncea* L. Biochem Biophys Res Commun 371(4):702–706

Cong L, Jing Y, Wu X, Cong X, Jingjuan Y (2014) An ABA-responsive *DRE*-binding protein gene from *Setaria italica*, *SiARDP*, the target gene of *SiAREB*, plays a critical role under drought stress. J Exp Bot 65(18):5415–5427

Cui M, Zhang W, Zhang Q, Xu Z, Zhu Z, Duan F, Wu R (2011) Induced overexpression of the transcription factor *OsDREB2A* improves drought tolerance in rice. Plant Physiol Biochem 49(12):1384–1391

Cutler SR, Rodriguez PL, Finkelstein RR, Abrams SR (2010) Abscisic acid: emergence of a core signaling network. Ann Rev Plant Biol 61:651–679

Datta K, Niranjan B, Moumita G, Sellapna K, Yamaguchi-Shinozaki K, Datta SK (2012) Overexpression of *Arabidopsis* and rice stress genes inducible transcription factor confers drought and salinity tolerance to rice. Plant Biotechnol J 10:579–586

De Clercq I, Vermeirssen V, Aken OV, Vandepoele K, Murcha MW, Law SR, Inze A, Ng S, Ivanova A, Rombaut D, Cotte BV, Jaspers P, Peer YV, Kangasjarvi J, Whelan J, Breusegema FV (2013) The membrane-bound *NAC* transcription factor *ANAC013* functions in mitochondrial retrograde regulation of the oxidative stress response in *Arabidopsis*. Plant Cell 25:3472–3490

De Paiva Rolla AA, de Fatima Correa Carvalho J, FugantiPagliarini R, Engels C, do Rio A, Marin SR, de Oliveira MC, Beneventi MA, Marcelino-Guimaraes FC, Farias JR, Neumaier N, Nakashima K, Yamaguchi-Shinozaki K, Nepomuceno AL (2014) Phenotyping soybean plants transformed with *rd29A:AtDREB1A* for drought tolerance in the greenhouse and field. Transgenic Res 23(1):75–87

Dubouzet JG, Sakuma Y, Ito Y, Kasuga M, Dubouzet EG, Miura S (2003) *OsDREB* genes in rice (*Oryza sativa* L.) encode transcription activators that function in drought, high-salt- and cold-responsive gene expression. Plant J 33:751–763

Duval M, Hsieh TF, Kim SY, Thomas TL (2002) Molecular characterization of AtNAM: a member of the Arabidopsis NAC domain super family. Plant Mol Biol 50:237–248

Engels C, Fuganti-Pagliarini R, Marin SR, Marcelino-Guimaraes FC, Oliveira MC, Kanamori N, Mizoi J, Nakashima K, Yamaguchi-Shinozaki K, Nepomuceno AL (2013) Introduction of the *rd29A:AtDREB2A CA* gene into soybean (*Glycine max* L. Merril) and its molecular characterization in leaves and roots during dehydration. Genet Mol Res 36(4):556–565

Ernst HA, Olsen AN, Larsen S, Lo-Leggio L (2004) Structure of the conserved domain of ANAC, a member of the NAC family of transcription factors. EMBO Rep 5:297–303

Fang Y, You J, Xie K, Xie W, Xiong L (2008) Systematic sequence analysis and identification of tissue-specific or stress-responsive genes of NAC transcription factor family in rice. Mol Gen Genomics 280:547–563

Fang Y, Liao K, Du H, Xu Y, Song H, Li X, Xiong L (2015a) A stress responsive *NAC* transcription factor *SNAC3* confers heat and drought tolerance through modulation of reactive oxygen species in rice. J Exp Bot 66:6803–6817

Fang ZW, Xu XY, Gao JF, Wang PK, Liu ZX, Feng BL (2015b) Characterization of *FeDREB1* promoter involved in cold and drought-inducible expression from common buckwheat (*Fagopyrum esculentum*). Genet Mol Res 14(3):7990–8000

Finkelstein RR, Gampala SS, Rock CD (2002) Abscisic acid signaling in seeds and seedlings. Plant Cell 14:S15–S45

Flowers TJ, Yeo AR (1995) Breeding for salinity resistance in crop plants-where next? Aust J Plant Physiol 22:876–884

Fowler S, Thomashow MF (2002) *Arabidopsis* transcriptome profiling indicates that multiple regulatory pathways are activated during cold acclimation in addition to the CBF cold response pathway. Plant Cell 14:1675–1690

Fujita M, Fujita Y, Maruyama K, Seki M, Hiratsu K, Ohme-Takagi M (2004) A dehydration-induced NAC protein, *RD26*, is involved in a novel ABA-dependent stress-signaling pathway. Plant J 39:863–876

Gao SQ, Chen M, Xia LQ, Xiu HJ, Xu ZS, Li LC, Zhao CP, Cheng XG, Ma YZ (2009) A cotton (*Gossypium hirsutum*) *DRE* binding transcription factor gene, *GhDREB*, confers enhanced tolerance to drought, high salt, and freezing stresses in transgenic wheat. Plant Cell Rep 28:301–311

Garapati P, Xue GP, Munne-Bosch S, Balazadeh S (2015) Transcription factor ATAF1 in Arabidopsis promotes senescence by direct regulation of key chloroplast maintenance and senescence transcriptional cascades. Plant Physiol 168(3):1122–1139

Gilmour SJ, Hajela RK, Thomashow MF (1998) Cold acclimation in *Arabidopsis thaliana.* Plant Physiol 87:745–750

Gilmour SJ, Sebolt AM, Salazar MP, Everard JD, Thomashow MF (2000) Overexpression of the *Arabidopsis CBF3* transcriptional activator mimics multiple biochemical changes associated with cold acclimation. Plant Physiol 124:1854–1865

Goldshmidt A, Alvarez JP, Bowman JL, Eshed Y (2008) Signals derived from YABBY gene activities in organ primordia regulate growth and partitioning of *Arabidopsis* shoot apical meristems. Plant Cell 20:1217–1230

Grover A, Kapoor A, Laksmi OS, Agarwal S, Sahi C, Katiyar-Agarwal S, Agarwal M, Dubey H (2001) Understanding molecular alphabets of the plant abiotic stress responses. Curr Sci 80(2):206–216

Gu YQ, Yang C, Thara VK, Zhou J, Martin GB (2000) *Pti4* is induced by ethylene and salicylic acid, and its product is phosphorylated by the *Pto* kinase. Plant Cell 12:771–785

Gunapati S, Naresh R, Ranjan S, Nigam D, Hans A, Verma PC, Gadre R, Pathre UV, Sane AP, Sane VA (2016) Expression of *GhNAC2* from *G. herbaceum*, improves root growth and imparts tolerance to drought in transgenic cotton and *Arabidopsis*. Sci Rep 6:24978

Guo Z, Chen X, Wu X, Ling J, Xu P (2004) Overexpression of the *AP2/EREBP* transcription factor *OPBP1* enhances disease resistance and salt tolerance in tobacco. Plant Mol Biol 55:607–618

Gupta K, Jha B, Agarwal PK (2014) A dehydration-responsive element binding (*DREB*) transcription factor from the succulent halophyte *Salicornia brachiata* enhances abiotic stress tolerance in transgenic tobacco. Mar Biotechnol 16(6):657–673

Gutha LR, Reddy AR (2008) Rice *DREB1B* promoter shows distinct stress specific responses, and the overexpression of cDNA in tobacco confers improved abiotic and biotic stress tolerance. Plant Mol Biol 68(6):533–555

Haake V, Cook D, Riechmann JL, Pineda O, Thomashow MF, Zhang JZ (2002) Transcription factor *CBF4* is a regulator of drought adaptation in *Arabidopsis*. Plant Physiol 130:639–648

Hao YJ, Wei W, Song QX, Chen HW, Zhang YQ, Wang F (2011) Soybean NAC transcription factors promote abiotic stress tolerance and lateral root formation in transgenic plants. Plant J 68(2):302–313

He XJ, Mu RL, Cao WH, Zhang ZG, Zhang JS, Chen SY (2005) *AtNAC2*, a transcription factor downstream of ethylene and auxin signaling pathways, is involved in salt stress response and lateral root development. Plant J 44(6):903–916

Hichri I, Muhovski Y, Clippe A, Zizkova E, Dobrev PI, Motyka V, Lutts S (2016) *SlDREB2*, a tomato dehydration-responsive element-binding2 transcription factor, mediates salt stress tolerance in *tomato and Arabidopsis*. Plant Cell Environ 39:62–79

Hong B, Tong Z, Ma N, Li J, Kasuga M, Yamaguchi-Shinozaki K, Gao J (2006) Heterologous expression of the *AtDREB1A* gene in *Chrysanthemum* increases drought and salt stress tolerance. Sci China C Life Sci 49:436–445

Hong B, Ma C, Yang Y, Wang T, Yamaguchi-Shinozaki K, Gao J (2009) Over-expression of *AtDREB1A* in *Chrysanthemum* enhances tolerance to heat stress. Plant Mol Biol 70:231–240

Hong Y, Zhang H, Huang L, Li D, Song F (2016) Overexpression of a stress-responsive NAC transcription factor gene *ONAC022* improves drought and salt tolerance in rice. Front Plant Sci 7:4

Hsieh TH, Lee JT, Yang PT, Chiu LH, Charng YY, Wang YC, Chan MT (2002) Heterology expression of the *Arabidopsis* C-repeat/dehydration response element binding factor 1 gene confers elevated tolerance to chilling and oxidative stresses in transgenic tomato. Plant Physiol 129:1086–1094

Hu H, Dai M, Yao J, Li BX, Zhang XQ (2006) Overexpressing a NAM, ATAF, and CUC (NAC) transcription factor enhances drought resistance and salt tolerance in rice. Proc Natl Acad Sci U S A 103:12987–12992

Hu H, You J, Fang Y, Zhu X, Qi Z, Xiong L (2008) Characterization of transcription factor gene SNAC2 conferring cold and salt tolerance in rice. Plant Mol Biol 67:169–181

Hu R, Qi G, Kong Y, Kong D, Gao Q, Zhou G (2010) Comprehensive analysis of NAC domain transcription factor gene family in *Populus trichocarpa*. BMC Plant Biol 10:145

Huang Q, Wang Y, Li B, Chang J, Chen M, Li K, Yang G, He G (2015) *TaNAC29*, a *NAC* transcription factor from wheat, enhances salt and drought tolerance in transgenic *Arabidopsis*. BMC Plant Biol 15:268

Huang L, Hong Y, Zhang H, Li D, Song F (2016) Rice NAC transcription factor *ONAC095* plays opposite roles in drought and cold stress tolerance. BMC Plant Biol 16:203

Hussain Z, Ali S, Hayat Z, Zia MA, Iqbal A, Ali GM (2014) *Agrobacterium* mediated transformation of *DREB1A* gene for improved drought tolerance in rice cultivars (*Oryza sativa* L.) Aust J Crop Sci 8(7):1114–1123

Ito Y, Katsura K, Maruyama K, Taji T, Kobayashi M, Seki M (2006) Functional analysis of rice *DREB1/CBF* type transcription factors involved in cold responsive gene expression in transgenic rice. Plant Cell Physiol 47:141–153

Jaglo KR, Kleff S, Amundsen KL, Zhang X, Haake V, Zhang JZ, Deits T, Thomashow MF (2001) Components of the *Arabidopsis* C repeat/dehydration-responsive element binding factor cold-response pathway are conserved in *Brassica napus* and other plant species. Plant Physiol 127:910–917

Jaglo-Ottosen KR, Gilmour SJ, Zarka DG, Schabenberger O, Thomashow MF (1998) *Arabidopsis CBF1* overexpression induces cor genes and enhances freezing tolerance. Science 280:104–106

James VA, Neibaur I, Altpeter F (2008) Stress inducible expression of the *DREB1A* transcription factor from xeric, *Hordeum spontaneum* L. in turf and forage grass (*Paspalum notatum* Flugge) enhances abiotic stress tolerance. Transgenic Res 17(1):93–104

Jensen MK, Kjaersgaard T, Nielsen MM, Galberg P, Petersen K, O'Shea C (2010) The *Arabidopsis thaliana* NAC transcription factor family: structure-function relationships and determinants of *ANAC019* stress signalling. Biochem J 426:183–196

Jeong JS, Kim YS, Baek KH, Jung H, Ha SH, Choi YD, Kim M, Reuzeau C, Kim JK (2010) Root-specific expression of *OsNAC10* improves drought tolerance and grain yield in rice under field drought condition. Plant Physiol 153:185–197

Jiang Y, Deyholos MK (2006) Comprehensive transcriptional profiling of NaCl-stressed Arabidopsis roots reveals novel classes of responsive genes. BMC Plant Biol 6:25

Jiang X, Zhang C, Lu P, Jiang G, Liu X, Dai F, Gao J (2014a) *RhNAC3*, a stress-associated NAC transcription factor, has a role in dehydration tolerance through regulating osmotic stress-related genes in rose petals. Plant Biotechnol J 12:38–48

Jiang QY, Hu Z, Zhang H, Ma YZ (2014b) Overexpression of *GmDREB1* improves salt tolerance in transgenic wheat and leaf protein response to high salinity. Crop J 2:120–131

Jiang L, Wang Y, Zhang S, He R, Li W, Han J, Cheng X (2016) Tomato *SlDREB1* gene conferred the transcriptional activation of drought induced gene and an enhanced tolerance of the transgenic *Arabidopsis* to drought stress. Plant Growth Regul 81(1):131–145

Jin T, Chang Q, Li W, Yin D, Li Z, Wang D, Liu B, Liu L (2010) Stress inducible expression of *GmDREB1* conferred salt tolerance in transgenic alfalfa. Plant Cell Tissue Organ Cult 100(2):219–227

Jun-Wei W, Feng-Ping Y, Xu-Qing C, Rong-Qi L, Li-Quan Z, Dong-Mei G, Xiao-Dong Z, Ya-Zhen S, Gai-Sheng Z (2006) Induced expression of *DREB* transcriptional factor and study on its physiological effects of drought tolerance in transgenic wheat. Acta Genet Sin 33(5):468–476

Kasuga M, Liu Q, Miura S, Yamaguchi-Shinozaki K, Shinozaki K (1999) Improving plant drought, salt, and freezing tolerance by gene transfer of a single stress-inducible transcription factor. Nat Biotechnol 17:287–291

Kasuga M, Miura S, Shinozaki K, Kazuko Y (2004) A combination of the *Arabidopsis DREB1A* gene and stress-inducible *rd29A* promoter improved drought and low temperature stress tolerance in tobacco by gene transfer. Plant Cell Physiol 45:346–350

Katerji N, Van Horn JW, Hamdy A, Mastrorilli M (2003) Salinity effect on crop development and yield, analysis of salt tolerance according to several classification methods. Agric Water Manag 62:37–66

Katerji N, Van Horn JW, Hamdy A, Mastrorilli M (2004) Comparison of corn yield response to plant water stress caused by salinity and drought. Agric Water Manag 65:95–101

Kavar T, Maras M, Kidric M, Sustar-Vozlic J, Meglic V (2007) Identification of genes involved in the response of leaves of *Phaseolus vulgaris* to drought stress. Mol Breed 21:159–172

Kidokoro S, Watanabe K, Ohori T, Moriwaki T, Maruyama K, Mizoi J, Nang MPSH, Yasunari F, Sachiko S, Kazuo S, Kazuko Y (2015) Soybean *DREB1/CBF* type transcription factors function in heat and drought as well as cold stress-responsive gene expression. Plant J 81(3):505–518

Kilian J, Peschke F, Berendzen KW, Harter K, Wanke D (2012) Prerequisites, performance and profits of transcriptional profiling the abiotic stress response. Biochim Biophys Acta 1819:166–175

Kim JK, Jung YSH, Bang SW, Choi YD, Ha SH, Reuzeau C, Mark CFR, Jin S, Jeong R (2012) The overexpression of *OsNAC9* alters the root architecture of rice plants enhancing drought resistance and grain yield under field conditions. Plant Biotechnol J 10:792–805

Kim YS, Kim JK (2009) Rice transcription factor transcription factor *AP37* involved in grain yield increase under drought stress. Plant Signal Behav 4:735–736

Kim YS, Sakuraba Y, Han SH, Yoo SC, Paek NC (2013) Mutation of the Arabidopsis NAC016 transcription factor delays leaf senescence. Plant Cell Physiol 54:1660–1672

Kjaersgaard T, Jensen MK, Christiansen MW, Gregersen P, Kragelund BB, Skriver K (2011) Senescenceassociated barley NAC (NAM, ATAF1,2, CUC) transcription factor interacts with radical-induced cell death 1 through a disordered regulatory domain. J Biol Chem 286:35418–35429

Kitano H (2002) Systems biology: a brief overview. Science 295(5560):1662–1664

Kobayashi F, Ishibashi M, Takumi S (2008) Transcriptional activation of Cor/Lea genes and increase in abiotic stress tolerance through expression of a wheat *DREB2* homolog in transgenic tobacco. Transgenic Res 17(5):755–767

Kovalchuk N, Jia W, Eini O, Morran S, Pyvovarenko T, Fletcher S, Bazanova N, Harris J, BeckOldach K, Shavrukov Y, Langridge P, Lopato S (2013) Optimization of *TaDREB3* gene expression in transgenic barley using cold inducible promoters. Plant Biotechnol J 11:659–670

Kudo M, Kidokoro S, Yoshida T, Mizoi J, Todaka D, Fernie AR, Shinozaki K, Yamaguchi-Shinozaki K (2017) Double overexpression of *DREB* and *PIF* transcription factors improves drought stress tolerance and cell elongation in transgenic plants. Plant Biotechnol J 15:458–471

Lata C, Prasad M (2011) Role of *DREBs* in regulation of abiotic stress responses in plants. J Exp Bot 62:4731–4748

Le DT, Nishiyama R, Watanabe Y, Mochida K, Yamaguchi-Shinozaki K, Shinozaki K (2011) Genome-wide survey and expression analysis of the plant-specific NAC transcription factor family in soybean during development and dehydration stress. DNA Res 18:263–276

Lee SC, Huh KW, An K, An G, Kim SR (2004) Ectopic expression of a cold-inducible transcription factor, *CBF1/DREB1b*, in transgenic rice (*Oryza sativa* L.) Mol Cell 18:107–114

Lee HE, Shin D, Park SR, Han SE, Jeong MJ, Kwon TR, Lee SK, Park SC, Yi BY, Kwon HB, Byun MO (2007) Ethylene responsive element binding protein 1 (*StEREBP1*) from *Solanum tuberosum* increases tolerance to abiotic stress in transgenic potato plants. Biochem Biophys Res Commun 353:863–868

Lee DK, Chung PJ, Jeong JS, Jang G, Bang SW, Jung H, Kim YS, Ha SH, Choi YD, Kim JK (2017) The rice *OsNAC6* transcription factor orchestrates multiple molecular mechanisms involving root structural adaptions and nicotianamine biosynthesis for drought tolerance. Plant Biotechnol J 15:754–764

Li X, Cheng X, Liu J, Zeng H, Han L, Tang W (2011) Heterologous expression of the *Arabidopsis DREB1A/CBF3* gene enhances drought and freezing tolerance in transgenic *Lolium perenne* plants. Plant Biotechnol Rep 5(1):61–69

Li JT, Wang N, Xin HP, Li SH (2013) Overexpression of *VaCBF4,* a transcription factor from *Vitis amurensis*, improves cold tolerance accompanying increased resistance to drought and salinity in *Arabidopsis*. Plant Mol Biol Report 31:1518–1528

Li X, Zhang D, Li H, Wang Y, Zhang Y, Wood AJ (2014) *EsDREB2B*, a novel truncated *DREB2* type transcription factor in the desert legume *Eremosparton songoricum*, enhances tolerance to multiple abiotic stresses in yeast and transgenic tobacco. BMC Plant Biol 14:44–59

Li H, Zhang Y, Guo Q, Yao W (2017) Molecular characterization of a *DREB* gene from *Sophora moorcroftiana*, an endemic species of plateau. Protoplasma 254(4):1735–1741

Liao X, Guo X, Wang Q, Wang Y, Zhao D, Yao L, Wang S, Liu G, Li T (2017) Overexpression of *MsDREB6.2* results in cytokinin-deficient developmental phenotypes and enhances drought tolerance in transgenic apple plants. Plant J 89(3):510–526

Lim CJ, Hwang JE, Chen H, Hong JK, Yang KA, Choi MS, Lee KO, Chung WS, Lee SY, Lim CO (2007) Over-expression of the *Arabidopsis DRE/CRT*-binding transcription factor *DREB2C* enhances thermo-tolerance. Biochem Biophys Res Commun 362:431–436

Lina X, Ming C, Dong-hong M, Lu F, Zhao-shi X, Yongbin Z, Dong-bei X, Lian-cheng L, You-zhi M, Xiao-hong Z (2016) The NAC-like transcription factor *SiNAC110* in foxtail millet (*Setaria italica* L.) confers tolerance to drought and high salt stress through an ABA independent signaling pathway. J Integr Agric 15:60345–60347

Liu Q, Kasuga M, Sakuma Y, Abe H, Miura S, Goda H, Shimada Y, Yoshida S, Shinozaki K, Kazuko Y (1998) Two transcription factors, *DREB1* and *DREB2*, with an *EREBP/AP2* DNA binding domain separate two cellular signal transduction pathways in drought- and low-temperature-responsive gene expression, respectively, in *Arabidopsis*. Plant Cell 10:391–406

Liu N, Zhong N, Wang G, Li L, Liu X, He Y, Xia G (2007) Cloning and functional characterization of *PpDBF1* gene encoding a *DRE* binding transcription factor from *Physcomitrella patens*. Planta 226:827–838

Liu X, Hong L, Li XY, Yao Y, Hu B, Li L (2011a) Improved drought and salt tolerance in transgenic *Arabidopsis* overexpressing a NAC transcriptional factor from *Arachis hypogaea*. Biosci Biotechnol Biochem 75:443–450

Liu L, Wang Y, Wang N, Dong YY, Fan XD, Liu XM, Yang J, Li HY (2011b) Cloning of a vacuolar H^+ pyrophosphatase gene from the halophyte *Suaeda corniculata* whose heterologous overexpression improves salt, saline-alkali and drought tolerance in *Arabidopsis*. J Integr Plant Biol 53:731–742

Liu QL, Xu KD, Zhao LJ, Pan YZ, Jiang BB, Zhang HQ, Liu GL (2011c) Overexpression of a novel chrysanthemum NAC transcription factor gene enhances salt tolerance in tobacco. Biotechnol Lett 33:2073–2082

Liu D, Chen X, Liu J, Ye J, Guo Z (2012) The rice *ERF* transcription factor *OsERF922* negatively regulates resistance to *Magnaporthe oryzae* and salt tolerance. J Exp Bot 63:3899–3911

Liu X, Liu S, Wu J, Zhang B, Li X, Yan Y (2013) Overexpression of *Arachis hypogea NAC3* in tobacco enhances dehydration and drought tolerance by increasing superoxide scavenging. Plant Physiol Biochem 70:354–359

Liu G, Li X, Jin S, Liu X, Zhu L, Nie Y, Zhang X (2014) Overexpression of rice NAC gene *SNAC1* improves drought and salt tolerance by enhancing root development and reducing transpiration rate in transgenic cotton. PLoS One 9(1):e86895

Lu PL, Chen NZ, An R, Su Z, Qi BS, Ren F, Chen J, Wg XC (2007) A novel drought inducible gene, ATAF1, encodes a NAC family protein that negatively regulates the expression of stress-responsive genes in *Arabidopsis*. Plant Mol Biol 63:289–305

Lu M, Ying S, Zhang DF, Shi YS, Song YC, Wang TY (2012) A maize stress-responsive NAC transcription factor, *ZmSNAC1*, confers enhanced tolerance to dehydration in transgenic *Arabidopsis*. Plant Cell Rep 31:1701–1711

Lv Z, Wang S, Zhang F, Chen L, Hao X, Pan Q, Fu X, Li L, Sun X, Tang K (2016) Overexpression of a novel NAC domain-containing transcription factor gene (*AaNAC1*) enhances the content of artemisinin and increases tolerance to drought and *Botrytis cinerea* in *Artemisia annua*. Plant Cell Physiol 57(9):1961–1971

Magnani E, Sjolander K, Hake S (2004) From endonuclease to transcription factors: Evolution of the *AP2* DNA binding domain in plants. Plant Cell 16:2265–2277

Mallikarjuna G, Mallikarjuna K, Reddy MK, Kaul T (2011) Expression of *OsDREB2A* transcription factor confers enhanced dehydration and salt stress tolerance in rice (*Oryza sativa* L.) Biotechnol Lett 33:1689–1697

Manavalan LP, Guttikonda SK, Tran L-SP, Nguyen HT (2009) Physiological and molecular approaches to improve drought resistance in soybean. Plant Cell Physiol 50:1260–1276

Mao X, Zhang H, Qian X, Zhao AG, Jing R (2012) *TaNAC2*, a NAC type wheat transcription factor conferring enhanced multiple abiotic stress tolerances in *Arabidopsis*. J Exp Bot 63(8):2933–2946

Mao X, Chen S, Li A, Zhai C, Jing R (2014) Novel NAC transcription factor *TaNAC67* confers enhanced multi-abiotic stress tolerances in *Arabidopsis*. PLoS One 9(1):e84359

Mao H, Yu L, Han R, Li Z, Liu H (2016) *ZmNAC55*, a maize stress responsive NAC transcription factor, confers drought resistance in transgenic *Arabidopsis*. Plant Physiol Biochem 105:55–66

Maruyama K, Sakuma Y, Kasuga M, Ito Y, Seki M, Goda H, Shimada Y, Yoshida S, Shinozaki K, Kazuko Y (2004) Identification of cold-inducible downstream genes of the *Arabidopsis DREB1A/CBF3* transcriptional factor using two microarray systems. *The*. Plant J 38:982–993

Maruyama K, Takeda M, Kidokoro S (2009) Metabolic pathways involved in cold acclimation identified by integrated analysis of metabolites and transcripts regulated by *DREB1A* and *DREB2A*. Plant Physiol 150:1972–1980

Matsui A, Ishida J, Morosawa T, Mochizuki Y, Kaminuma E, Endo TA, Okamoto M, Maiko EN, Kawashima MN, Nakajima SMM, Kim MJ, Kobayashi N, Toyoda T, Shinozaki S, Seki M (2008) Arabidopsis transcriptome analysis under drought, cold, high-salinity and ABA treatment conditions using a tiling array. Plant Cell Physiol 49(8):1135–1149

Matsukura S, Mizoi J, Yoshida T, Todaka D, Ito Y, Maruyama K (2010) Comprehensive analysis of rice *DREB2*-type genes that encode transcription factors involved in the expression of abiotic stress-responsive genes. Mol Gen Genomics 283:185–196

Meiru L, Yan L, Hongqing L, Guojiang W (2012) Improvement of paper mulberry tolerance to abiotic stresses by ectopic expression of tall fescue *FaDREB1*. Tree Physiol 32(1):104–113

Meng C, Cai C, Zhang T, Guo W (2009) Characterization of six novel NAC genes and their responses to abiotic stresses in *Gossypium hirsutum* L. Plant Sci 176(3):352–359

Mishra S, Phukan UJ, Tripathi V, Singh DK, Luqman S, Shukla RK (2015) *PsAP2* an *AP2/ERF* family transcription factor from *Papaver somniferum* enhances abiotic and biotic stress tolerance in transgenic tobacco. Plant Mol Biol 89:173–186

Mizoi J, Shinozaki K, Yamaguchi-Shinozaki K (2011) *AP2/ERF* family transcription factors in plant abiotic stress responses. Biochim Biophys Acta 1819:86–96

Mizoi J, Ohori T, Moriwaki T, Kidokoro S, Todaka D, Maruyama K, Kusakabe K, Osakabe Y, Shinozaki K, Kazuko Y (2013) *GmDREB2A*: a canonical *DREB2*-type transcription factor in soybean is post-translationally regulated and mediates *DRE*-dependent gene expression. Plant Physiol 161(1):346–361

Morran S, Eini O, Pyvovarenko T, Parent B, Singh R, Ismagul A, Eliby S, Shirley N, Langridge P, Lopato S (2011) Improvement of stress tolerance of wheat and barley by modulation of expression of *DREB/CBF* factors. Plant Biotechnol J 9(2):230–249

Movahedi S, Tabatabaei BES, Alizade H, Ghobadi C, Yamchi A, Khaksar G (2012) Constitutive expression of *Arabidopsis DREB1B* in transgenic potato enhances drought and freezing tolerance. Biol Plant 56:37–42

Nakashima K, Tran LS, Van Nguyen D, Fujita M, Maruyama K, Todaka D, Ito Y, Hayashi N, Shinozaki K, Yamaguchi-Shinozaki K (2007) Functional analysis of a NAC-type transcription factor *OsNAC6* involved in abiotic and biotic stress-responsive gene expression in rice. Plant J 51:617–630

Negi S, Tak H, Ganapathi TR (2016) Expression analysis of *MusaNAC68* transcription factor and its functional analysis by overexpression in transgenic banana plants. Plant Cell Tissue Organ Cult 125(1):59–70

Nuruzzaman M, Manimekalai R, Sharoni AM, Satoh K, Kondoh H, Ooka H (2010) Genome-wide analysis of NAC transcription factor family in rice. Gene 465:30–44

Nuruzzaman M, Sharoni AM, Satoh K, Kondoh H, Hosaka A, Kikuchi S (2012) A genome-wide survey of the NAC transcription factor family in monocots and eudicots. In: Introduction to genetics–DNA methylation, histone modification and gene regulation. iConcept Press, Hong Kong. isbn:ISBN, 978- 14775549-4-4

Nuruzzaman M, Sharoni AM, Kikuchi S (2013) Roles of NAC transcription factors in the regulation of biotic and abiotic stress responses in plants. Front Microbiol 4:1–15

Oakenfull RJ, Baxter R, Knight MR (2013) A Crepeat binding factor transcriptional activator (*CBF/DREB1*) from European bilberry (*Vaccinium myrtillus*) induces freezing tolerance when expressed in *Arabidopsis thaliana*. PLoS One 8(1):e54119

Oh SJ, Song SI, Kim YS, Jang HJ, Kim SY, Kim M, Kim YK, Nahm BH, Kim JK (2005) *Arabidopsis CBF3/DREB1A* and *ABF3* in transgenic rice increased tolerance to abiotic stress without stunting growth. Plant Physiol 138:341–335

Oh SJ, Kwon CW, Choi DW, Song SI, Kim JK (2007) Expression of barley *HvCBF4* enhances tolerance to abiotic stress in transgenic rice. Plant Biotechnol J 5:646–656

Ohnishi T, Sugahara S, Yamada T, Kikuchi K, Yoshiba Y, Hirano HY (2005) OsNAC6, a member of the NAC gene family, is induced by various stresses in rice. Genes Genet Syst 80:135–139

Olsen AN, Ernst HA, Lo Leggio L, Skriver K (2005) NAC transcription factors: structurally distinct, functionally diverse. Trends Plant Sci 10:1360–1385

Ooka H, Satoh K, Doi K, Nagata T, Otomo Y, Murakami K (2003) Comprehensive analysis of NAC family genes in *Oryza sativa* and *Arabidopsis thaliana*. DNA Res 10:239–247

Pandurangaiah M, Lokanadha Rao G, Sudhakarbabu O, Nareshkumar A, Kiranmai K, Lokesh U, Thapa G, Sudhakar C (2017) Overexpression of horsegram (*Macrotyloma uniflorum* Lam. Verdc.) NAC transcriptional factor (*MuNAC4*) in groundnut confers enhanced drought tolerance. Mol Biotechnol 56(8):758–769

Paul A, Muoki RC, Singh K, Kumar S (2012) CsNAM-like protein encodes a nuclear localized protein and responds to varied cues in tea (*Camellia sinensis* L. O. Kuntze). Gene 502:69–74

Pellegrineschi A, Reynolds M, Pacheco M, Brito RM, Almeraya R, Yamaguchi-Shinozaki K, Hoisington D (2004) Stress-induced expression in wheat of the *Arabidopsis* thaliana *DREB1A* gene delays water stress symptoms under greenhouse conditions. Genome 47:493–500

Peng H, Cheng HY, Yu XW, Shi QH, Zhang H, Li JG, Ma H (2009) Characterization of a chickpea (*Cicer arietinum* L.) *NAC* family gene, *CarNAC5*, which is both developmentally and stress-regulated. Plant Physiol Biochem 47:1037–1045

Peng T, Guo C, Yang J, Xu M, Zuo J, Bao M, Zhang J (2016) Overexpression of a Mei (*Prunus mume*) *CBF* gene confers tolerance to freezing and oxidative stress in *Arabidopsis*. Plant Cell Tissue Organ Cult 126(3):373–385

Polizel AM, Medri ME, Nakashima K, Yamanaka N, Farias JR, de Oliveira MC, Marin SR, Abdelnoor RV, Marcelino-Guimaraes FC, Fuganti R, Rodrigues FA, StolfMoreira R, Beneventi MA, Rolla AA, Neumaier N, YamaguchiShinozaki K, Carvalho JF, Nepomuceno AL (2011) Molecular, anatomical and physiological properties of a genetically modified soybean line transformed with *rd29A:AtDREB1A* for the improvement of drought tolerance. Genet Mol Res 10(4):3641–3656

Puranik S, Bahadur RP, Srivastava PS, Prasad M (2011) Molecular cloning and characterization of a membrane associated *NAC* family gene, *SiNAC* from foxtail millet [*Setaria italica* (L.) P. Beauv.] Mol Biotechnol 49:138–150

Qin F, Sakuma Y, Li J, Liu Q, Li YQ, Shinozaki K, Kazuko Y (2004) Cloning and functional analysis of a novel *DREB1/CBF* transcription factor involved in cold responsive gene expression in maize (*Zea mays* L.) Plant Cell Physiol 45:1042–1052

Qin F, Kakimoto M, Sakuma Y, Maruyama K, Osakabe Y, Tran LSP (2007) Regulation and functional analysis of *ZmDREB2A* in response to drought and heat stress in *Zea mays* L. Plant J 50:54–59

Qin F, Shinozaki K, Yamaguchi-Shinozaki K (2011) Achievements and challenges in understanding plant abiotic stress responses and tolerance. Plant Cell Physiol 5:1569–1582

Rahman H, Ramanathan V, Nallathambi J, Duraialagaraja S, Muthurajan R (2016) Over-expression of a *NAC67* transcription factor from finger millet (*Eleusine coracana* L.) confers tolerance against salinity and drought stress in rice. BMC Biotechnol 16(1):35

Ramegowda V, Senthil-Kumar M, Nataraja KN, Reddy MK, Mysore KS, Udayakumar M (2012) Expression of a finger millet transcription factor, *EcNAC1*, in tobacco confers abiotic stress-tolerance. PLoS One 7:e40397

Ravikumar G, Manimaran P, Voleti SR, Subrahmanyam D, Sundaram RM, Bansal KC, Viraktamath BC, Balachandran SM (2014) Stress inducible expression of *AtDREB1A* transcription factor greatly improves drought stress tolerance in transgenic *indica* rice. Transgenic Res 23(3):421–439

Redillas MCFR, Jeong JS, Kim YS, Jung H, Bang SW, Choi YD, Ha SH, Reuzeau C, Kim JK (2012) The overexpression of *OsNAC9* alters the root architecture of rice plants enhancing drought resistance and grain yield under field conditions. Plant Biotechnol J 10:792–805

Reis RR, Cunha BADB, Martins PK, Martins MTB, Alekcevetch JC Júnior AC, Andradec AC Ribeiroa AP, Qin F, Mizoie J, Kazuko YS, Nakashima K, Carvalho JFC, Sousaa CAF, Nepomuceno AL, Kobayashia AK, Molinaria HBC (2014) Induced overexpression of *AtDREB2A CA* improves drought tolerance in sugarcane. Plant Sci, 221 & 222:59–68

Riechmann JL, Meyerowitz EM (1998) The AP2/EREBP family of plant transcription factors. Biol Chem 379:633–646

Riechmann JL, Heard J, Martin G, Reuber L, Jiang C, Keddie J, Adam L, Pineda O, Ratcliffe OJ, Samaha RR (2000) *Arabidopsis* transcription factors: genome-wide comparative analysis among eukaryotes. Science 290:2105–2110

Rong W, Qi L, Wang A, Ye X, Du L, Liang H, Xin Z, Zhang Z (2014) The *ERF* transcription factor *TaERF3* promotes tolerance to salt and drought stresses in wheat. Plant Biotechnol J 12:468–479

Rushton PJ, Bokowiec MT, Han S, Zhang H, Brannock JF, Chen X (2008) Tobacco transcription factors: novel insights into transcriptional regulation in the solanaceae. Plant Physiol 147:280–295

Saad AS, Li X, Li HP, Huang T, Gao CS. Guo MW (2013) A rice stress-responsive NAC gene enhances tolerance of transgenic wheat to drought and salt stresses. Plant Sci 203 & 204:33–40

Sadhukhan A, Kobayashi Y, Kobayashi Y, Tokizawa M, Yamamoto YY, Iuchi S, Koyama H, Panda SK, Sahoo L (2014) *VuDREB2A*, a novel *DREB2* type transcription factor in the drought tolerant legume cowpea, mediates *DRE* dependent expression of stress responsive genes and confers enhanced drought resistance in transgenic *Arabidopsis*. Planta 240(3):645–664

Sakuma Y, Liu Q, Dubouzet JG, Abe H, Shinozaki K, Yamaguchi-Shinozaki K (2002) DNA-binding specificity of the *ERF/AP2* domain of *Arabidopsis DREBs*, transcription factors involved in dehydration- and cold-inducible gene expression. Biochem Biophys Res Commun 290:998–1009

Sakuma Y, Maruyama K, Osakabe Y, Qin F, Seki M, Shinozaki K, Kazuko Y (2006a) Functional analysis of an *Arabidopsis* transcription factor, *DREB2A*, involved in drought-responsive gene expression. Plant Cell 18:1292–1309

Sakuma Y, Maruyama K, Qin F, Osakabe Y, Shinozaki K, Kazuko Y (2006b) Dual function of an *Arabidopsis* transcription factor *DREB2A* in water-stress responsive and heat-stress-responsive gene expression. PNAS USA 103:18822–18827

Samara Reddy S, Singh B, Peter AJ, Rao TV (2016) Production of transgenic local rice cultivars (*Oryza sativa* L.) for improved drought tolerance using agrobacterium mediated transformation. Saudi J Biol Sci. https://doi.org//10.1016/j.sjbs.2016.01.035

Savitch LV, Allard G, Seki M, Robert LS, Tinker NA, Huner NPA, Shinozaki K, Singh J (2005) The effect of overexpression of two *Brassica CBF/DREB1*-like transcription factors on photosynthetic capacity and freezing tolerance in Brassica napus. Plant Cell Physiol 46:1525–1539

Seki M, Narusaka M, Ishida J, Nanjo T, Fujita M, Oono Y, Kamiya A, Nakajima M, Enju A, Sakurai T, Satou M, Akiyama K, Taji T, Yamaguchi-Shinozaki K, Carninci P, Kawai J, Hayashizaki Y, Shinozaki K (2002) Monitoring the expression profiles of 7000 *Arabidopsis* genes under drought, cold and high-salinity stresses using a full-length cDNA microarray. The Plant J 31(3):279–292

Seki M, Kamei A, Satou M, Sakurai T, Fujita M, Oono Y, Yamaguch-Shinozaki K, Shinozaki K (2003) Transcriptome analysis in abiotic stress conditions in higher plants. *Topics*. Curr Genet 4:271–295

Shah SH, Ali S, Hussain Z, Jan SA, Din JU, Ali GM (2016) Genetic improvement of tomato (*Solanum lycopersicum*) with *AtDREB1A* gene for cold stress tolerance using optimized *Agrobacterium*-mediated transformation system. Int J Agric Biol 18:471–482

Shahnejat-Bushehri S, Mueller-Roeber B, Balazadeh S (2012) *Arabidopsis* NAC transcription factor *JUNGBRUNNEN1* affects thermo-memory associated genes and enhances heat stress tolerance in primed and unprimed conditions. Plant Signal Behav 7:1518–1521

Shan DP, Huang JG, Yang YT, Guo YH, Wu CA, Yang GD, Gao Z, Zheng CC (2007) *Cotton GhDREB1* increases plant tolerance to low temperature and is negatively regulated by gibberellic acid. New Phytol 176:70–81

Shao H, Wang H, Tang X (2015) NAC transcription factors in plants multiple abiotic stress responses: progress and prospects. Front Plant Sci 6:902

Shen YG, Zhang WK, He SJ, Zhang JS, Liu Q, Chen SY (2003) An *EREBP/AP2*-type protein in *Triticum aestivum* was a *DRE*-binding transcription factor induced by cold, dehydration and ABA stress. Theor Appl Genet 106:923–930

Shen J, Shen J, Lv B, Luo L, He J, Mao C, Xi D, Ming F (2017) The NAC-type transcription factor *OsNAC2* regulates ABA-dependent genes and abiotic stress tolerance in rice. Sci Rep 7:40641

Shinozaki K, Yamaguchi-Shinozaki K (2000) Molecular responses to dehydration and low temperature: differences and cross-talk between two stress signaling pathways. Curr Opin Plant Biol 3:217–223

Shinozaki K, Yamaguchi-Shinozaki K (2007) Gene networks involved in drought stress tolerance and response. J Exp Bot 58:221–227

Shiqing G, Huijan X, Xianguo C, Ming C, Zhaosi X, Liancheng L, Xingguo Y, Lipu D, Xiaoyan H, Youzhi M (2005) Improvement of wheat drought and salt tolerance by expression of a stress inducible transcription factor *GmDREB* of soybean (*Glycine max*). Chin Sci Bull 50:2714–2723

Shukla RK, Raha S, Tripathi V, Chattopadhyay D (2006) Expression of *CAP2*, an *APETALA2*-family transcription factor from chickpea, enhances growth and tolerance to dehydration and salt stress in transgenic tobacco. Plant Physiol 142:113–123

Shukla RK, Tripathi V, Jain D, Yadav RK, Chattopadhyay D (2009) *CAP2* enhances germination of transgenic tobacco seeds at high temperature and promotes heat stress tolerance in yeast. FEBS J 276:5252–5262

Sindhu A, Chintamanani S, Brandt AS, Zanis M, Scofield SR, Johal GS (2008) A guardian of grasses: specific origin and conservation of a unique disease-resistance gene in the grass lineage. Proc Natl Acad Sci U S A 105:1762–1767

Skriver K, Podzimska-Sroka D, O'Shea C, Gregersen PL (2015) NAC transcription factors in senescence: from molecular structure to function in crops. Plants 4:412–448

Song SY, Chen Y, Chen J, Dai XY, Zhang WH (2011) Physiological mechanisms underlying *OsNAC5*-dependent tolerance of rice plants to abiotic stress. Planta 234:331–345

Souer E, Van Houwelingen A, Kloos D, Mol J, Koes R (1996) The no apical meristem gene of Petunia is required for pattern formation in embryos and flowers and is expressed at meristem and primordia boundaries. Cell 85:159–170

Sperotto RA, Ricachenevsky FK, Duarte GL, Boff T, Lopes KL, Sperb ER (2009) Identification of up-regulated genes in flag leaves during rice grain filling and characterization of *OsNAC5*, a new ABA-dependent transcription factor. Planta 230:985–1002

Stockinger EJ, Gilmour SJ, Thomashow MF (1997) *Arabidopsis thaliana CBF1* encodes an AP2 domain-containing transcriptional activator that binds to the C-repeat/DRE, a cis-acting DNA regulatory element that stimulates transcription in response to low temperature and water deficit. Proc Natl Acad Sci U S A 94:1035–1040

Sun J, Peng X, Fan W, Tanga M, Liua J, Shena S (2014a) Functional analysis of *BpDREB2* gene involved in salt and drought response from a woody plant *Broussonetia papyrifera*. Gene 535(2):140–149

Sun ZM, Zhou ML, Xiao XG, Tang YX, Wu YM (2014b) Overexpression of a *Lotus corniculatus* AP2/ERF transcription factor gene, *LcERF080*, enhances tolerance to salt stress in transgenic *Arabidopsis*. Plant Biotechnol Report 8(4):315–324

Tak H, Negi S, Ganapathi TR (2017) Banana *NAC* transcription factor *MusaNAC042* is positively associated with drought and salinity tolerance. Protoplasma 254(2):803–816

Takasaki H, Maruyama K, Kidokoro S, Ito Y, Fujita Y, Shinozaki K, Yamaguchi-Shinozaki K, Nakashima K (2010) The abiotic stress-responsive NAC-type transcription factor *OsNAC5* regulates stress-inducible genes and stress tolerance in rice. Mol Gen Genomics 284:173–183

Tan DX, Tuong HM, Thuy VTT, Son LV, Mau CH (2015) Cloning and overexpression of *gmDREB2* gene from a Vietnamese drought resistant soybean variety. Braz Arch Biol Technol 58(5):651–657

Tang M, Liu X, Deng H, Shena S (2011) Over-expression of *JcDREB*, a putative *AP2/EREBP* domain-containing transcription factor gene in woody biodiesel plant *Jatropha curcas*, enhances salt and freezing tolerance in transgenic *Arabidopsis thaliana*. Plant Sci 181:623–631

Tang Y, Liu M, Gao S, Zhang Z, Zhao X, Zhao C (2012) Molecular characterization of novel *TaNAC* genes in wheat and overexpression of *TaNAC2a* confers drought tolerance in tobacco. Physiol Plant 144:210–224

Tang GY, Shao FX, Xu PL, Shan L, Liu ZJ (2017a) Overexpression of a peanut NAC gene, *AhNAC4*, confers enhanced drought tolerance in tobacco. Russ J Plant Physiol 64(4):525–535

Tang Y, Liu K, Zhang J, Li X, Xu K, Zhang Y, Qi J, Yu D, Wang J, Li C (2017b) *JcDREB2*, a physic nut *AP2/ERF* gene, alters plant growth and salinity stress responses in transgenic rice. Front Plant Sci 8:306

Tran LS, Nakashima K, Sakuma Y, Simpson SD, Fujita Y, Maruyama K (2004) Isolation and functional analysis of Arabidopsis stress-inducible NAC transcription factors that bind to a drought-responsive cis-element in the early responsive to dehydration stress 1 promoter. Plant Cell 16:2481–2498

Thirumalaikumar VP, Devkar V, Mehterov N, Ali S, Ozgur R, Turkan I, Mueller-Roeber B, Balazadeh S (2017) NAC transcription factor *JUNGBRUNNEN1* enhances drought tolerance in tomato. Plant Biotechnol J:1–13

Thomashow MF (1999) Plant cold acclimation: freezing tolerance genes and regulatory mechanisms. Annu Rev Plant Physiol Plant Mol Biol 50:571–599

Tian Z, He Q, Wang H, Liu Y, Zhang Y, Shao F, Xie C (2015) The potato *ERF* transcription factor *StERF3* negatively regulates resistance to *Phytophthora infestans* and salt tolerance in potato. Plant Cell Physiol 56:992–1005

Tian Q, Chen J, Wang D, Wang H, Liu C, Wang S, Xia X, Yin W (2017) Over expression of a *Populus euphratica CBF4* gene in poplar confers tolerance to multiple stresses. Plant Cell Tissue Organ Cult 128(2):391–407

Tong Z, Hong B, Yang Y, Li Q, Ma N, Ma C (2009) Overexpression of two *Chrysanthemum DgDREB1* group genes causing delayed flowering or dwarfism in *Arabidopsis*. Plant Mol Biol 71:115–129

Udvardi MK, Kakar K, Wandrey M, Montanari O, Murray J, Andriankaja A, Zhang JY, Benedito V, Hofer JMI, Chueng F (2007) Legume transcription factors: global regulators of plant development and response to the environment. Plant Physiol 144:538–549

Umezawa T, Fujita M, Fujita Y, Yamaguchi-Shinozaki K, Shinozaki K (2006) Engineering drought tolerance in plants: discovering and tailoring genes to unlock the future. Curr Opin Plant Biotechnol 17(2):113–122

Wang CT, Dong YM (2009) Overexpression of maize *ZmDBP3* enhances tolerance to drought and cold stress in transgenic *Arabidopsis* plants. Biologia 64(6):1108–1114

Wang Q, Guan Y, Wu Y, Chen H, Chen F, Chu C (2008) Overexpression of a rice *OsDREB1F* gene increases salt, drought, and low temperature tolerance in both *Arabidopsis* and rice. Plant Mol Biol 67:589–602

Wang X, Dong J, Liu Y, Gao H (2010) A novel dehydration responsive element binding protein from *Caragana korshinskii* is involved in the response to multiple abiotic stresses and enhances stress tolerance in transgenic tobacco. Plant Mol Biol Report 28(4):664–675

Wang X, Chen X, Liu Y, Gao H, Wang Z, Sun G (2011a) *CkDREB* gene in *Caragana korshinskii* is involved in the regulation of stress response to multiple abiotic stresses as an *AP2/EREBP* transcription factor. Mol Biol Rep 38:2801–2811

Wang CT, Yang Q, Wang CT (2011b) Isolation and functional characterization of *ZmDBP2* encoding a dehydration responsive element-binding protein in *Zea mays*. Plant Mol Biol Report 29(1):60–68

Wang CT, Yang Q, Yang YM (2011c) Characterization of the *ZmDBP4* gene encoding a *CRT/DRE* binding protein responsive to drought and cold stress in maize. Acta Physiol Plant 33(2):575–583

Wang K, Zhong M, Wu Y, Bai Z, Liang Q, Liu Q, Pan Y, Zhang L, Jiang B, Jia Y, Liu G (2017a) Overexpression of a *Chrysanthemum* transcription factor gene *DgNAC1* improves the salinity tolerance in *Chrysanthemum*. Plant Cell Rep 36(4):571–581

Wang L, Hu Z, Zhu M, Zhu Z, Hu J, Qanmber G, Chen G (2017b) The abiotic stress responsive NAC transcription factor *SlNAC11* is involved in drought and salt response in tomato (*Solanum lycopersicum* L.) Plant Cell Tissue Organ Cult 129(1):161–174

Wang L, Li Z, Lu M, Wang Y (2017c) *ThNAC13*, a NAC transcription factor from *Tamarix hispida*, confers salt and osmotic stress tolerance to transgenic *Tamarix* and *Arabidopsis*. Front Plant Sci 8:635

Waterer D, Benning NT, Wu G, Luo X, Liu X, Gusta M, McHughen A, Gusta LV (2010) Evaluation of abiotic stress tolerance of genetically modified potatoes (*Solanum tuberosum* cv. Desiree). Mol Breed 25(3):527–540

Wei T, Deng K, Gao Y, Liu Y, Yang M, Zhang L, Zheng X, Wang C, Song W, Chen C, Zhang Y (2016a) *Arabidopsis DREB1B* in transgenic *Salvia miltiorrhiza* increased tolerance to drought stress without stunting growth. Plant Physiol Biochem 104:1728

Wei T, Deng K, Liu D, Gao Y, Liu Y, Yang M, Zhang L, Zheng X, Wang C, Song W, Chen C, Zhang Y (2016b) Ectopic expression of *DREB* transcription factor, *AtDREB1A*, confers tolerance to drought in transgenic *Salvia miltiorrhiza*. Plant Cell Physiol 57(8):1593–1609

Wei T, Deng K, Zhang Q, Gao Y, Liu Y, Yang M, Zhang L, Zheng X, Wang C, Liu Z, Chen C, Zhang Y (2017) Modulating *AtDREB1C* expression improves drought tolerance in *Salvia miltiorrhiza*. Front Plant Sci 8:52

Wu Y, Deng Z, Lai J, Zhang Y, Yang C, Yin B (2009) Dual function of *Arabidopsis ATAF1* in abiotic and biotic stress responses. Cell Res 19:1279–1290

Wu J, Folta KM, Xie Y, Jiang W, Lu J, Zhang Y (2017) Overexpression of *Muscadinia rotundifolia CBF2* gene enhances biotic and abiotic stress tolerance in *Arabidopsis*. Protoplasma 254(1):239–251

Xia N, Zhang G, Liu XY, Deng L, Cai GL, Zhang Y, Wang XJ, Zhao J, Huang LL, Kang ZS (2010) Characterization of a novel wheat NAC transcription factor gene involved in defense response against stripe rust pathogen infection and abiotic stresses. Mol Biol Rep 37:3703–3712

Xianjun P, Xingyong M, Weihong F, Man S, Liqin C, Alam I (2011) Improved drought and salt tolerance of *Arabidopsis thaliana* by transgenic expression of a novel *DREB* gene from *Leymus chinensis*. Plant Cell Rep 30:1493–1502

Xie Q, Frugis G, Colgan D, Chua N (2000) Arabidopsis NAC1 transduces auxin signal downstream of TIR1 to promote lateral root development. Genes Dev 14:3024–3036

Xiong L, Fang Y, Liao K, Du H, Xu Y, Song H, Li X (2015) A stress-responsive NAC transcription factor SNAC3 confers heat and drought tolerance through modulation of reactive oxygen species in rice. J Exp Bot 66(21):6803–6817

Xiu Y, Iqbal A, Zhu C, Wu G, Chang Y, Li N, Cao Y, Zhang W, Zeng H, Chen S, Wang H (2016) Improvement and transcriptome analysis of root architecture by overexpression of *Fraxinus pennsylvanica DREB2A* transcription factor in *Robinia pseudoacacia* L. 'Idaho'. Plant Biotechnol J 14(6):1456–1469

Xu ZS, Xia LQ, Chen M, Cheng XG, Zhang RY, Li LC (2007) Isolation and molecular characterization of the *Triticum aestivum* L. ethylene-responsive factor 1 (*TaERF1*) that increases multiple stress tolerance. Plant Mol Biol 65:719–732

Xu ZS, Ni ZY, Li ZY, Li LC, Chen M, Gao DY (2009) Isolation and functional characterization of *HvDREB1*, a gene encoding a dehydration-responsive element binding protein in *Hordeum vulgare*. J Plant Res 122:121–130

Xu M, Li L, Fan Y, Wan J, Wang L (2011) *ZmCBF3* overexpression improves tolerance to abiotic stress in transgenic rice (*Oryza sativa* L.) without yield penalty. Plant Cell Rep 30:1949–1957

Xu Z, Gongbuzhaxi CW, Xue F, Zhang H, Ji W (2015) Wheat NAC transcription factor *TaNAC29* is involved in response to salt stress. Plant Physiol Biochem 96:356–363

Xue GP, Way HM, Richardson T, Drenth J, Joyce PA, McIntyre CL (2011) Overexpression of *TaNAC69* leads to enhanced transcript levels of stress up-regulated genes and dehydration tolerance in bread wheat. Mol Plant 4(4):697–712

Xue Y, Wang YY, Peng RH, Zhen JL, Gao B, Zhu JJ, Zhao W, Han HJ, Yao QH (2014) Transcription factor *MdCBF1* gene increases freezing stress tolerance in transgenic *Arabidopsis thaliana*. Biol Plant 58:499–506

Yamaguchi-Shinozaki K, Shinozaki K (1994) A novel cis-acting element in an Arabidopsis gene is involved in responsiveness to drought, low-temperature, or high-salt stress. Plant Cell 6:251–264

Yang W, Liu XD, Chi XJ, Wu CA, Li YZ, Song LL, Liu XM, Wang YF, Wang FW, Zhang C (2011) Dwarf apple *MbDREB1* enhances plant tolerance to low temperature, drought, and salt stress via both ABA dependent and ABA independent pathways. Planta 233(2):219–229

Yang X, Wang X, Lu J, Yi Z, Fu C, Ran J (2015) Overexpression of a *Miscanthus lutarioriparius* NAC gene *MlNAC5* confers enhanced drought and cold tolerance in *Arabidopsis*. Plant Cell Rep 34:943–958

Yin Y, Ma Q, Zhu Z, Cui Q, Chen C, Chen X, Fang W, Li X (2016) Functional analysis of *CsCBF3* transcription factor in tea plant (*Camellia sinensis*) under cold stress. Plant Growth Regul 80(3):335–343

Yokotani N, Ichikawa T, Kondou Y, Matsui M, Hirochika H, Iwabuchi M (2009) Tolerance to various environmental stresses conferred by the salt-responsive rice gene *ONAC063* in transgenic *Arabidopsis*. Planta 229:1065–1075

You J, Zong W, Li X, Ning J, Hu H, Li X (2013) The SNAC1-targeted gene OsSRO1c modulates stomatal closure and oxidative stress tolerance by regulating hydrogen peroxide in rice. J Exp Bot 64:569–583

Yu CL, Abu S, Xu L, He-Ping L, Tao H, Chun-Sheng G, Mao-Wei G, Wei C, Guang-Yao Z (2013) A rice stress-responsive *NAC* gene enhances tolerance of transgenic wheat to drought and salt stresses. Plant Sci 203 & 204:33–40

Yu X, Liu Y, Wang S, Tao Y, Wang Z, Mijiti A, Wang Z, Zhang H, Ma H (2016) A chickpea stress responsive NAC transcription factor, *CarNAC5*, confers enhanced tolerance to drought stress in transgenic *Arabidopsis*. Plant Growth Regul 79(2):187–197

Zhai Y, Wang Y, Li Y, Lei T, Yan F, Su L, Li X, Zhao Y, Sun X, Li J, Wang Q (2013) Isolation and molecular characterization of *GmERF7*, a soybean ethylene-response factor that increases salt stress tolerance in tobacco. Gene 513:174–183

Zhang JZ, Creelman RA, Zhu JK (2004) From laboratory to field: Using information from *Arabidopsis* to engineer salt, cold, and drought tolerance in crops. Plant Physiol 135:615–621

Zhang Y, Chen C, Jin XF, Xiong AS, Peng RH, Hong YH (2009) Expression of a rice *DREB1* gene, *OsDREB1D*, enhances cold and high salt tolerance in transgenic *Arabidopsis*. Biochem Mol Biol Rep 42:486–492

Zhang X, Tang Y, Ma Q, Yang C, Mu Y, Suo H (2013) *OsDREB2A*, a rice transcription factor, significantly affects salt tolerance in transgenic soybean. PLoS One 8(12):83011–83020

Zhang X, Liu X, Wu L, Yu G, Wang X, Ma H (2015) The *SsDREB* transcription factor from the succulent halophyte *Suaeda salsa* enhances abiotic stress tolerance in transgenic tobacco. Int J Genomics (15):1–13

Zhang W, Yang G, Mu D, Li H, Zang D, Xu H, Zou X, Wang Y (2016a) An ethylene-responsive factor *BpERF11* negatively modulates salt and osmotic tolerance in *Betula platyphylla*. Sci Rep 6:23085

Zhang L, Zhang L, Xia C, Zhao G, Jia J, Kong X (2016b) the novel wheat transcription factor *TaNAC47* enhances multiple abiotic stress tolerances in transgenic plants. Front Plant Sci 6:1174

Zhao JS, Ren W, Zhi DY, Wang L, Xia GM (2007) *Arabidopsis DREB1A/CBF3* bestowed transgenic tall fescue increased tolerance to drought stress. Plant Cell Rep 26(9):1521–1528

Zhao XJ, Lei HJ, Zhao K, Yuan HZ, Li TH (2012) Isolation and characterization of a dehydration responsive element binding factor *MsDREBA5* in *Malus sieversii* Roem. Sci Hortic 142:212–220

Zhao K, Shen X, Yuan H, Liu Y, Liao X, Wang Q, Liu L, Li F, Li T (2013) Isolation and characterization of dehydration-responsive element-binding factor 2C (*MsDREB2C*) from *Malus sieversii* Roem. Plant Cell Physiol 54(9):1415–1430

Zhao X, Yang X, Pei S, He G, Wang X, Tang Q, Jia C, Lud Y, Hu R, Zhou G (2016) The *Miscanthus* NAC transcription factor *MlNAC9* enhances abiotic stress tolerance in transgenic *Arabidopsis*. Gene 586(1):158–169

Zheng X, Chen B, Lu G, Han B (2009) Overexpression of a NAC transcription factor enhances rice drought and salt tolerance. Biochem Biophys Res Commun 379(4):985–989

Zhong H, Guo QQ, Chen L, Ren F, Wang QQ, Zheng Y (2012) Two *Brassica napus* genes encoding NAC transcription factors are involved in response to high-salinity stress. Plant Cell Rep 31:1991–2003

Zhou M, Maa J, Zhao Y, Wei Y, Tanga Y, Wua Y (2012) Improvement of drought and salt tolerance in *Arabidopsis thaliana* and *Lotus corniculatus* by overexpression of a novel *DREB* transcription factor from *Populus euphratica*. Gene 506(1):10–17

Zhu M, Chen G, Zhang J, Zhang Y, Xie Q, Zhao Z, Pan Y, Hu Z (2014) The Abiotic stress responsive *NAC* type transcription factor *SlNAC4* regulates salt and drought tolerance and stress related genes in tomato *S. Lycopersicum*. Plant Cell Rep 33(11):1851–1863

Zhuang J, Wang F, Xu ZS, Xiong AS (2015) Microarray analysis of different expression profiles between wild type and transgenic rice seedlings over expressing *OsDREB1BI* gene. Biologia 70(6):760–770

Zong JM, Li XW, Zhou YH, Wang FW, Wang N, Dong YY, Yuan YX, Chen H, Liu XM, Yao N, Li HY (2016) The *AaDREB1* transcription factor from the cold-tolerant plant *Adonis amurensis* enhances abiotic stress tolerance in transgenic plant. Int J Mol Sci 17:611–625

Perspectives in High-Throughput Phenotyping of Qualitative Traits at the Whole-Plant Level

Oksana Sytar, Marek Zivcak, Katarina Olsovska, and Marian Brestic

Abstract

Recent advances in technology have enabled the rapid development of high-throughput automated and semi-automated field and laboratory phenotyping platforms worldwide. In this review, we discuss possible ways of matching the qualitative traits of the above-ground parts of crop plants, also defining the target traits and possible approaches that would be useful in automated phenotyping systems. Optical tools based on light reflectance are presented as a high-throughput and low-cost alternative to some destructive analytical methods. Special attention is paid to hyperspectral imaging and its integration in high-throughput phenotyping systems, as well as its special applications for the assessment of specific plant material traits associated with food quality.

Keywords

Phenotyping · Phenomics · Hyperspectral imaging · Qualitative traits

O. Sytar (✉)
Department of Plant Physiology, Slovak University of Agriculture in Nitra, Nitra, Slovak Republic

Plant Physiology and Ecology Department, Institute of Biology, Taras Shevchenko National University of Kyiv, Kyiv, Ukraine
e-mail: oksana.sytar@gmail.com

M. Zivcak · K. Olsovska · M. Brestic
Slovak University of Agriculture in Nitra, Nitra, Slovak Republic

R. S. Sengar, A. Singh (eds.), *Eco-friendly Agro-biological Techniques for Enhancing Crop Productivity*, https://doi.org/10.1007/978-981-10-6934-5_10

Abbreviations

DESI-MSI	Desorption electrospray ionization mass spectrometry imaging
DNA	Deoxyribonucleic acid
ESI	Electrospray ionization
FAO	Food and Agriculture Organization of the United Nations
HPLC-MS	High-performance liquid chromatography–mass spectrometry
HSI	Hyperspectral imaging
IR light	Infrared light
LEDI	Lettuce decay indices
MALDI	Matrix-assisted laser desorption ionization
MS	Mass spectrometry
MSI	Mass spectrometry imaging
NIR light	Near-infrared light
NMR	Nuclear magnetic resonance
PCA	Principal component analysis
PMT	Photomultiplier tube
QTL	Quantitative trait locus
RGB camera	Red–green–blue camera
SVDD	Support vector data description
SWIR light	Short-wave infrared light
TIR light	Thermal infrared light
TNC	Total nitrogen content
TOF	Time-of-flight (mass spectrometer)

1 Introduction

At present, there are intense discussions regarding the many problematic challenges that agriculture may face in the coming decades. The United Nations Food and Agriculture Organization (FAO) considers that the world's population in 2050 will be 34% greater than it is today. Currently, 49% of the world's population lives in urban areas, while in 2050 this proportion will be closer to 70% (Alexandratos and Bruinsma 2012). During this time period, climate change and the development of biofuel production will present major risks to long-term food security. Population growth, urban civilization, and climate change can stimulate high plant crop concurrence as possible resources of food, fiber, bioenergic products, and other industrial needs in future (http://www.fao.org). Such resource factors will require new approaches to the genetic and agronomic components of crop production systems (Fiorani and Schurr 2013). Continual management of sustainable water and soil resources and more adequate utilization of genetic diversity are basic steps that are necessary to increase productivity (McCouch et al. 2012; Bakker et al. 2012; Shao et al. 2005, 2009).

Currently, society has the technology and resources to provide long-term food security for all, despite some risks and challenges. In the past few years, many approaches have been discussed and incorporated into the constantly improving new process of highly automated, non-destructive phenotyping of plants. Plant phenotyping is the extensive evaluation of multiple plant features, such as architecture, cultivation, development, resistance, ecophysiology, and harvest, as well as the analysis of particular significant specifications for quantity analysis of specific biological active compounds. Detailed and specific phenotyping strategies are required to research genome-wide associations and permit high-resolution linkage mapping investigations, and also to create models of genomic selection for plant development. To simplify assessments and understanding of the phenotype based on the metabolome, metabolomics experiments are required to answer questions that underlie genomics, transcriptomics, and proteomics. Three types of metabolomic tools are available. First is a direct tool with which it is possible to identify the spectrum of secondary metabolites. Second is a tool that performs epigenetic gene amplification at the level of the whole phenome, enabling the description of the entire genotype by a metabolome marker characteristic. Third is a tool that identifies a marker for the optimal adaptation of a specific genotype to the environment (Khakimov et al. 2013; Sytar et al. 2016).

Improvements in the specificity and throughput of phenotypic assessment at all biological levels—phenomics, metabolomics, and genetics—are the main objectives of modern phenotyping. Because of their automation, improved data integration, remote sensing, and experimental design, phenotyping systems minimize labor and costs (Cobb et al. 2013; Chen et al. 2002). These systems developed for phenotyping are important for representing the full set of genetic factors that play roles in the phenotypic variation of quantitative parameters for cells, tissues, and organs; these plant phenotyping systems also play roles in identifying stages of development, plant species, and environments (Cobb et al. 2013). The newly developed plant phenotyping platforms produce significantly more data than the initial platforms did, and they need special systems for data management, access, and storage. New statistical tools are needed for enhancing the experimental design and for making greater use of ideas to promote the integration of data, as well as for deriving biologically significant signals from experimental and environmental noise (Cobb et al. 2013).The phenomics field faces two major challenges. First is the features analysis of a large quantity of genetic lines, and second is the replication of measurements of dynamic traits (i.e., traits whose phenotype changes during the vegetation period) (Araus and Cairns 2014). To characterize a genotype, it may be more valuable to determine levels of a few key compounds in several organs and at several time-points, rather than obtaining the full metabolic profile for just one sample. For phenotyping projects that engage several hundred samples, it can be useful to use more oriented, less costly methods for all samples with full-scale metabolomics results. Preliminary metabolomics work might indicate which substances are the main contributors to the phenotype of interest and lead to a focused analysis that is more productive and cost effective for phenotyping than previous methods.

The following high-parameterized platforms for phenotyping have been installed in the field or laboratory (Table 1): (semi)-automatic evaluation of morphometric parameters using red-green-blue (RGB) image analysis; chlorophyll and

Table 1 Automated and semi-automated high-throughput plant phenotyping platforms

Location/Producer	Platform	Features	URL
PSI (Drasov, Czech Republic)	PlantScreen™	Conveyor phenotyping system in controlled environmental conditions with analysis of chlorophyll fluorescence, kinetic and thermal imaging, morphometric and RGB analysis, and hyperspectral and NIR imaging; uses an automated weighing and watering system.	http://www.psi.cz
LemnaTec (Aachen, Germany)	Scanalyzer^{3D}	Comprehensive non-destructive 2D-3D assessment of plant physiological traits in controlled environmental conditions.	http://www.lemnatec.com
INRA (Montpellier, France)	Phenopsis	Specific platform for phenotyping *Arabidopsis* plant growth under controlled environmental conditions.	http://bioweb.supagro.inra.fr/phenopsis
INRA (Versailles, France)	Phenoscope	Automated phenotyping device to handle and monitor hundreds of individual pots.	https://phenoscope.versailles.inra.fr
INRA (Montpellier, France)	Phenodyn	Temporal analyses of growth rate and transpiration of hundreds of monocot crop species.	https://www.phenome-fppn.fr/phenome_eng/Facilities/Montpellier-Controlled/PHENODYN
INRA (Montpellier, France)	Phenoarch	Automated platform based on a LemnaTec system to analyze the genetic determinants of plant responses to environmental conditions.	https://www.phenome-fppn.fr/phenome_eng/Facilities/Montpellier-Controlled/PHENOARCH
Phenospex (Heerlen, Netherlands)	FieldScan	Phenotyping under field- or semi-field conditions that is designed to screen large populations.	http://phenospex.com
WPS (De Lier, Netherlands)	WPS	Fully automated digital phenotyping system using high-throughput RGB sensors.	http://www.wps.eu/en/plant-phenotyping
Keygene (Wageningen, Netherlands)	PhenoFab^{R}	Greenhouse service operation that combines phenotyping technology with trait interpretation to exploit phenotypic variation.	http://www.keygene.com/products-tech/digital-phenotyping/
Jülich Plant Phenotyping Centre (Julich, Germany)	Growscreen	Non-invasive methods designed to quantify shoot morphometric and functional parameters and root architecture.	http://www.fz-juelich.de/ibg/ibg-2/EN/methods_jppc/GROWSCREEN/_node.html

(continued)

Table 1 (continued)

Location/Producer	Platform	Features	URL
Wageningen UR (Wageningen, Netherlands)	PhenoBot	Autonomous mobile robot with camera promises to output direct registered depth and color image for morphometric analysis.	http://www.wageningenur.nl
Wiwam (Gent, Belgium)	Wiwam Conveyor	Integrated robotic system for phenotyping of larger plants with automated irrigation and measurement of a variety of plant growth parameters at regular time intervals.	http://wiwam.be
Australian Plant Phenomics Facility (Adelaide, Canberra, Australia)	PlantScan	Provides non-destructive analyses of plant morphology, structure, and function by using high-resolution cameras with cutting-edge information technology.	http://www.plantphenomics.org.au/services/plantscan/

2D-3D Two-dimensional-three dimensional, *NIR* near infrared, *RGB* red-green-blue

fluorescence kinetic imaging; and hyperspectral, thermal, and near-infrared light (NIR) imagings. These systems are composed of controlled watering and nutrient regimes regulated by automatic weight systems and environmental controls in the imaging cabinet (in laboratory platforms). The control and programming of platform systems, as well as data analysis, is performed with sophisticated and user-friendly software packages. Such newly developed phenotyping systems have tools for assessing many photosynthetic parameters; RGB systems can assess plant morphometry, IR-thermal cameras can evaluate stomatal conductance, and hyperspectral imaging (HSI) systems can evaluate the metabolomics of experimental plants at different stages of growth.

Currently, as part of the phenotyping system, HSI is a promising novel technology that requires the development of unique methodologies in order to make use of all spectrum data (Kuska et al. 2015). In this review, we will present information about phenotyping research using HSI and other imaging techniques. For assessing qualitative traits using the phenotyping system, we recommend choosing a plant that has a wide range of metabolites and one that is resistant to some stress factors. Among known crop plants that have been proposed as models for phenotyping research, it is important to choose one that has undergone genetic analysis, because genetic variations, particularly those characterized by fast growth, can be useful for model plants. Improving the breeding process of cultivated plants requires a unified mix of classical breeding approaches, together with one or more categories of phenotyping, in order to identify genotypes with the desired features (Tester and Langridge 2010). Only by clearly specifying and differentiating the physiological phenotyping approach at every breeding step will it be possible to effectively address the challenge of integrating phenotyping into a breeding program. Accordingly, the aim of this review is to identify perspectives in the high-throughput phenotyping of qualitative traits at the plant level.

2 Phenotyping of Qualitative Traits

2.1 Definitions of Target Traits for Phenomics, with Emphasis on Qualitative Traits and the Schema of Leaf Traits

A known limitation of the spectroradiometric approaches used for identifying the developmental stages of plants and for identifying differences in plant architecture is related to the following characteristics of the leaf surface: epicuticular waxes, roughness, and trichomes (Fig. 1). Spectroradiometric approaches can negatively influence the measurement of some features (e.g., biochemical parameters) (Cobb et al. 2013). The effect of leaf anisotropy is more visible in phenotyping systems with remote sensing-based approaches that use reflected rather than transmitted radiation. Furthermore, it is useful to use plants at the same developmental stage in order to standardize measurements. Variability in plant architecture should be minimized because the sun's elevation must be taken into account, as many spectroradiometric evaluations in the field use passive sensors (Araus and Cairns 2014).

Phenotyping platforms can be specific to plant parts and their growth periods (Chen et al. 2014). Phenotyping platforms use model-assisted methods (phenomics, metabolomics, proteomics, and genomics) that are specific to the developmental level and architecture of different plants (Fig. 1).

Genetic diversity in some plants has been described with plant morphology methods, randomly amplified polymorphic deoxyribonucleic acid (DNA), chloroplast and DNA sequence analysis, microsatellite markers, seed protein allozymes, and restriction fragment length polymorphisms (Mensack et al. 2010). Currently, accessible bioinformatics approaches and databases have been used in plants for metabolic quantitative trait locus (QTL) and comparative pathway analyses (Nadella et al. 2012). Genetic loci that affect quantitative features such as biomass or metabolite content can be connected to molecular markers (Fumio et al. 2012). This method

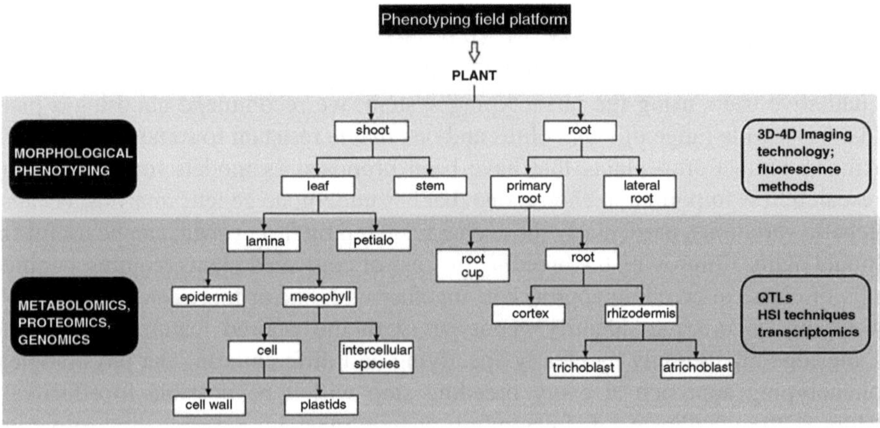

Fig. 1 Definitions of target traits for phenomics, with emphasis on qualitative traits and leaf traits

is called QTL mapping, and it can simplify the identification of new genes in biochemical metabolic pathways. For the QTL mapping of metabolites and for phenotyping experiments, *Arabidopsis thaliana* is a good choice as a model seed plant, because of its highly advanced genetic and molecular characteristics. In mQTL investigations, the total number of genes that have been molecularly estimated remains limited. To improve the network of known metabolic reactions, multi-parallel analytic approaches for the assessment of gene expression, rapid gene identification, and determination of metabolite and protein content would be useful (Brotman et al. 2011).

For example, an analysis of the complete metabolite and phenotype profile that was undertaken with the aim of improving the quality of tomatoes found that nearly 50% of the metabolic loci were combined with QTLs that were associated with whole-plant-yield pecularities. Correlation analysis established a cartographic network that showed associations with the phenotype of the whole plant and independent relationships with metabolites. This cartographic network also contained information about connections with important organoleptic and nutritional metabolites (Schauer and Fernie 2006). The utility of combining agronomic and metabolomic features of wheat plants was evaluated by a QTL study that aimed to determine potential targets for the improvement of breeding selection. The usefulness of including plant phenotyping in the assessment of the practical features of the wheat mapping population has been confirmed (Hill et al. 2015). In this case, the genetic-metabolic synergism and its connection to crop phenotypes can be used as a target with which to improve breeding selection strategies and efficiency.

2.2 Possible Approaches to Matching Qualitative Traits

Three types of parameters can be evaluated in the phenotype. These are threshold, quantitative, and qualitative parameters. A qualitative parameter can be expressed qualitatively, which means that the phenotype varies by categories that do not need to have an exact form. The pattern is typically influenced by a single gene (monogenetic). Plant metabolomic methods have facilitated large-scale analyses that have determined quantitative and qualitative variations of metabolic content in different types of plants (Tang et al. 2014). The genetic factors identified through genetic mapping approaches have been connected to plant variation through the linkage of genetic architecture to the plant metabolome (Brestic et al. 2015). However, in an experiment with rice grains, the content of many metabolites was found to be highly sensitive to environmental factors, but only weakly sensitive to genetic background (Matsuda et al. 2012). A one-line analysis of physiological performance and morphological phenotype features obtained further results that increased our understanding of the complex molecular mechanisms that regulate quantitative parameters (Carreno-Quintero et al. 2013). To improve such investigations, methods allowing the non-destructive, fast, and simultaneous identification of primary and secondary metabolites have been developed with phenotyping systems using HSI.

It is known that many experiments are managed using plants grown under more-or-less controlled environmental conditions (Chen et al. 2013; Yan et al. 2011, 2012, 2013). The goal of many such studies is to make phenotypes of various plant species comparable, in order to study plant changes under a range of suboptimal conditions, or to study genotypes in a specific environment (Poorter et al. 2012.). New approaches are emerging that focus on the traits and phenotyping needed. In novel plant phenotyping, the aim of selecting optimal features rather than directly studying the crop plant yield is related to precise phenotyping in the field or in controlled conditions, or in networks and models of field experiments (Tardieu and Tuberosa 2010). The discovery of complete genomes and the rapid improvement of tools with which to characterize gene expression and proteins complements the already developed objective methods for primary and secondary metabolite detection (Roessner et al. 2002).

It is important to remember that plant metabolism during growth is highly dependent on light parameters and the time of day. Diurnal fluctuation occurs when the levels of some metabolites differ during the day and night (Urbanczyk-Wochniak et al. 2005). Carbon availability can affect the number of metabolic pathways and photosynthesis and respiration rates, which change with light availability. Therefore, special treatment must be used when harvesting samples of plant tissue. In ideal conditions, all sampling would be performed at the same time of day or within a very small time window in order to minimize biological variation. When samples are collected for a metabolomic experiment, it is necessary to quickly suppress metabolism during the extraction procedures and biochemical analysis.

It is necessary to examine the cleanliness of samples harvested for analyses of metabolites (Verhoeven et al. 2006). Contamination of plant tissue with pests (e.g., aphids), bacteria, or fungi can be prevented with the prelimiary data analysis during plant growth, especially if the samples are collected from plants under natural growth conditions and not under controlled laboratory conditions. The application of chemicals to control pests such as sciarid flies and aphids in laboratory conditions should also be considered when analyzing metabolites obtained from plants that have received chemical treatment.

In a field experiment performed by Ossipov et al. (2008), frozen samples of leaves were placed directly in liquid nitrogen for transport to the laboratory. Although metabolite analysis can be performed on site, the cost can be problematic. For example, concentrations of carotenoids, anthocyanins, and chlorophylls can be estimated on site from cut leaf discs by spectrophotometry, via basic solvent extraction (Sims and Gamon 2002). Bruneti et al. (2013) provided detailed descriptions of sampling and processing assays and emphasized the importance of *in situ* techniques in field sites. However, in laboratory-based settings, such techniques are more applicable for secondary metabolite analysis.

Many plant metabolomics techniques have used whole plants, organs such as leaves or roots, and tissue parts, which are homogenized until the samples can be extracted. The methodology for tissue analysis should be decided at the start of the experiment, as the methodology can affect the analysis protocol. For example, typically an amount of 10–100 mg of fresh tissue is acceptable for most analytical methods, including gas chromatography, mass spectrometry (MS), and nuclear magnetic resonance. The heterogeneity of cell types in plant organs is represented by more than 40 different cell types. Normally, any metabolomic analysis of a whole organ shows more than one metabolite and reflects the metabolome of a particular cell type.

Current improvements in plant metabolomics techniques focus on using either laser micro-dissection technology or *in situ* MS. Laser capture micro-dissection technology is used to investigate specific kinds of cells, which are analyzed by common metabolomics approaches. For metabolite imaging, *in situ* MS can be used for a particular tissue. The most recently developed approaches include mounting tissue sections on a matrix-assisted laser desorption ionization (MALDI) plate and exposing them to laser pulses. Molecules are vaporized by the laser and ionized by the matrix prior to being transferred into the mass spectrometer for analysis. The MS analysis characterizes the molecules present in the area that is subjected to laser pulses, which is approximately 50 μm in size. Specific software then allows the deconvolution of individual mass spectra across the tissue and the generation of tissue images (Roessner and Bacic 2009).

2.3 Optical Tools Based on Light Reflectance as a High-Throughput and Low-Cost Alternative to Destructive Analytical Methods

Optical detection approaches adapted to microfluidic biosensors, based on methods such as chemiluminescence, fluorescence, and surface plasmon resonance are being developed (Pires et al. 2014). For actual-time scanning, small-size detection tools have been developedfor point-of-use in certain samples. The size of these devices can be reduced using microfluidic chip-based technologies, which combine a few laboratory targets on a separate chip. This methodology addresses the distribution of small volumes of fluids in microchannels (Pires et al. 2014).

Schwarz and Hauser (2001) have used optical and electrochemical constructions for investigating high-sensitivity microfluidic analytical systems. These detailed constructions were later described by Mogensen et al. (2004). Further, for many common biochemical analyses, there have been innovations in micro- and nanomechanical sensors (Waggoner and Craighead 2007).

Optical methods have been used to estimate the sensitivity of the refractive index, light intensity variations, and interference patterns (Fiorani and Schurr 2013). Positive features of optical methods include real-time detection (in the range of hundredths of seconds), a low quantity of required sample, and the ubiquity of these methods in the laboratory. However, limitations include the complexity of the set-up of the optical tools and the expense of conventiona l optical instrumentation.

Chemiluminescence uses sensor technologies such as microplate readers, photomultiplier tubes (PMTs), charge-coupled device (CCD) cameras, inorganic photodiodes or photoconductors, and organic photodiodes. Fluorescence uses the same CCD cameras, organic photodiodes, and inorganic photodiodes, as well as complementary metal-oxide semiconductor image sensors. Absorbance is measured using a visual sensor, a CMOS image sensor, and a CCD camera. The surface plasmon resonance sensor opto-microfluidic detection method mostly uses an infrared camera and a CCD camera (Pires et al. 2014).

All conventional optical detection methods have been used in microfluidic biosensors (Jiseok et al. 2013). As long as spectrophotometers, microscopes, lasers, PMTs, and CCDs are precisely linked to Lens Opacity Classification System, such systems are hard to convert to portable detection devices, which are low in price. Rather, with the goal of decreasing the cost of diagnostic platforms, the waveguides and photodiodes used in optoelectronic technology have been profitably joined to microfluidic systems (Baker 2008; Chen et al. 2014). Ultra-sensitive methods for the detection of bio-analytes at nano- and microscales are being developed, and integrated optical microfluidic platforms can be used with these methods (Fahlgren et al. 2015).

Another optical tool is Raman imaging, which has been used to develop an understanding of various real-time pretreatment strategies, while providing integral chemical information (Lupoi et al. 2015). Raman spectroscopy is a spectroscopic technique with vibrational features that are measured as diffused photons, which are produced during the interaction between matter and light. During the development of this methodology, C. V. Raman used sunlight as the light source. Modern Raman spectroscopy techniques use lasers in visible (VIS), ultraviolet (UV), or near-infrared (NIR) spectra. A significant parameter of the Raman phenomenon compared with IR spectroscopy is that molecular compounds are advanced to virtual vibrational, short-lived levels. A molecule is recognized as "active" for Raman spectroscopy when the interchange of light with the molecule creates changes in the polarizability of the electron. The archetypal Raman spectrum would be expected to be C–C, C=C, C–H, C–O, H–C–C, C–O–H, H–C–H, and other vibrational modes (Larsen and Barsberg 2011; Agarwal and Atalla 2010). The choice of the excitement wavelength for obtaining high-quality Raman spectra is difficult, as the Raman intensity is precisely equivalent to the fourth power of the incident frequency.

In recent years, different approaches have been developed to test the difficulty of obtaining Raman spectra for molecules with highly fluorescent capacities, such as lignin. The 1064-nm neodymium-doped yttrium orthovanadate (Nd: YVO4) or neodymium-doped yttrium aluminum garnet (Nd: Y3Al5O12) lasers, which are sources of NIR technologies, can be used in some improved Raman spectra methodologies (Meyer et al. 2011; Lupoi and Smith 2012). Phenolic components define lignin solubilization and reactivity (Zakis 1994). The use of alkaline media results in vibrational lignin modes in pulps, with a concomitant shift expressed in the ionization of the corresponding phenolic lignin model compounds (Warsta et al. 2012). In wood pulp analysis, a low noticeable shift developed because of the lower content of functional phenolic compounds in the pulps compared with that in the model compounds. It was found that the shift appeared to be due to ionization of the phenolic group. Improved band intensities also appeared after the pH increased. However, no shift was detected during the measurement of 3,4-dimethoxytoluene of non-phenolic background.

Mass spectrometry imaging (MSI) underwent an important trial in the mapping of targeted compound distributions s in plant tissue (Todd et al. 2001). Recently MALDI-MSI has been used in the direct profiling and imaging of biomolecules in thin tissue sections (Jackson et al. 2005). Currently, MSI applications can be used to estimate and characterize the proteomes expressed by tissues and organs (Chaurand et al. 2002). The preparation of samples for metabolite analysis by MALDI-MSI is more complicated than that required for proteins or peptides, because metabolites in the sample can be diffuse, can easily circulate via metabolic changes, or can be eliminated during sample preparation (Lee et al. 2012). Organic solvents can be used to help in the estimation of lipids. This has specific implications, as organic solvents are often employed for localizing proteins and peptides in the preparation of biological samples. It is generally recommended to avoid the washing or rinsing steps during sample preparation in order to avoid removing small metabolic compounds in the sample. After harvesting, to avoid metabolite degradation, plant tissue samples should be freeze-dried or rapidly dried in a vacuum desiccator on the imaging plate. For plant IMS applications, it is recommended to use fresh tissue because the effects of histological tissue fixation are not well studied.

Another promising application of high-throughput screening is the droplet-based microfluidic system (Lim et al. 2013). The fluorescence intensity measurements are non-destructive, with photomultipliers or photodiode signal detection, and images are created by high-speed or standard cameras (Millera et al. 2012). Any type of droplet-based microfluidics which uses simple alignment procedures can be combined with microoptics. This combination, by focusing light with a single element, results in high sensitivity and resolution. Higher throughput is attained by lenses combined in an array. This combination can be used for cells and fluorescent particle analysis too, as it may support the creation of high-quality fluorescence signals and a significant reduction of analytical costs.

Despite the development of modern high-throughput methods and sample preparation, the analyses remain relatively costly and laborious. However, non-invasive fluorescence-based methods, which are typically used in plant phenomics and may serve as early steps in metabolomic screening targeted at nutritionally and pharmaceutically important phenolic compounds, are now part of modern research activities. Sytar et al. (2015) presented the results of *in situ* measurements from many plant species, and they indicated high interspecific variability, which seems to be promising for further studies.

In general, the term 'metabolomics' covers two basic approaches. First, the non-targeted approach, which aims to determine as many compounds in the sample as possible. This approach may lead to discoveries of new active molecules, but it is slower and more expensive than the second approach. The second, targeted, approach aims at research for practical applications, such as food or pharmaceutical needs, targeting a single, or relatively narrow, well-defined group of compounds (e.g., amino acids, phenolics) (Verpoorte et al. 2005). Depending on the target metabolites, several possible non-invasive techniques may be used in the early stages of metabolomics research. These techniques are based on the simultaneous assessment of multispectrally induced chlorophyll fluorescence and HSI findings that can be part of the phenotyping system. These techniques, though not yet widely used, have become more popular in the past decade, owing to the introduction of commercially available devices.

3 Hyperspectral Imaging as a Component of Non-destructive Plant Metabolomics

3.1 Principles of the Hyperspectral Method and Basic Information about the Hyperspectral Method

Hyperspectral imaging has been widely accepted as a non-destructive, rapid, and safe method of qualitative analysis, and the method has been used for evaluation of the qualitative traits of a wide spectrum of food products (Elmasry et al. 2012). Hyperspectral imaging has been specifically improved for remote sensing tools (Goetz et al. 1985). Its capacity to consolidate imaging techniques and spectroscopy is its main benefit. Hyperspectral imaging systems have been adapted to make direct synchronous measurements of different components and to identify the spatial distribution of measured components in tested products or plant tissues (Gamalero et al. 2004). The characteristics of molecular structure are related to the absorption, reflection, and transmission of electromagnetic radiation under certain wavelengths (Sun 2010). It is known that UV, VIS, NIR, mid-IR, and far-IR lights are electromagnetic waves. Each region of the electromagnetic waves is linked to a precise type of transition on atomic or molecular levels corresponding to various degrees of repetitiveness. Plant tissues, like any other biological material, may be bound together by various molecular forces and connections. The bonds C–H and O–H are mostly present in water, carbohydrates, and fats. The bonds N–H and C–H are typical for derivatives of petroleum products and organic compounds. Electromagnetic waves are transmitted through a plant sample when it is exposed to light. The energy of the incident electromagnetic wave in the plant samples is modified by the extension and rotational fluctuation of some bonds, such as N–H, O–H, and C–H. Therefore, for the investigation of some parameters and the precise identification of plant examples, transformations in levels of molecular energy can be determined by spectroscopy.

At the macro level, electromagnetic waves are seen as light. Electromagnetic wave transitions are exposed as light reflection or scattering, and also as light transmission. The emission and absorption strength and wavelengths depend on the chemical and physical parameters of the plant material, because the consumed light partly diffuses into the sample of plant tissue. The obtained incident light is transformed to a spectrum and to images converted by hyperspectrometers. Images received from the hyperspectral camera can characterize the plant compounds and the physical characteristics of the plant samples. However, for such characterizations, HSI technology, which produces large datasets, requires improvements in appropriate image processing and multivariate data analysis tools.

3.2 Target Bands and Indices Based on HSI

The efficiency of agriculture crop classification obtained by using hyperspectral narrowband data is substantially higher than that of broadband data analysis (Thenkabail et al. 2004a, b; Schull et al. 2011). In experimental research with different plants, the optimal hyperspectral narrow bands were identified, based on a comprehensive analysis of the literature that combined classifications of redundant bands (Table 2).

Table 2 Recommended optimal hyperspectral narrow bands for different plants

Waveband interval	Waveband center (nm)	Importance in vegetation and agricultural cropland studies	References
Blue bands			
	466	Chlorophylls: Chlorophyll *a* and *b* in vegetation	[1]
Green bands			
	515	Leaf nitrogen	[3]
	510/530	Anthocyanin	[3]
	520/442	β-Carotene	[4]
	550	Chlorophyll and biomass: Total chlorophyll, chlorophyll/carotenoid ratio, vegetation nutritional and fertility level	[4], [5]
	550	Reflectance in the green region of the spectrum proved to be sensitive to anthocyanin content	[6]
	584–609	Determination of total flavonoid content of leaf samples in two colors (green-yellowish and yellow)	[7]
	662–688	Determination of total flavonoid content of leaf samples in two colors (green-yellowish and yellow)	[7]
Red bands			
	675	Chloropyll absorption maxima: Greatest crop-soil contrast is around this band for most crops in most growing conditions. Strong correlation with chlorophyll *a* and *b*	[2]
	682	Biophysical quantities and yield and chlorophyll absorbtion: Leaf area index, wet and dry biomass, plant height, crop type, crop discrimination. Strong correlation with chlorophyll *a* and *b*	[1], [2]
Red-edge bands			
	700, 720	Stress and chlorophyll: Nitrogen stress, crop stress, crop growth stage studies	[1], [8],[9]
	759	Nitrogen	[10]
	778/658	Phytochrome P660	[3]; [11]
Near-infrared (NIR) bands			
	845	Biophysical quantities and yield: Leaf area index, wet and dry biomass, plant height, grain yield, crop type and crop discrimination. Often used in combination with an index from 682 nm	[8]
	845/730	Phytochrome P730	[3]
Far NIR (FNIR) bands			
	1028/2101	Lignin	[3]
	1100	Biophysics quantities: Sensitive to biomass and leaf area index. The point of the most rapid rise in spectra with unit change in wavelength in FNIR bands	[12]

(continued)

Table 2 (continued)

Waveband interval	Waveband center (nm)	Importance in vegetation and agricultural cropland studies	References
	1215	Moisture and biomass: The point of the most rapid fall in spectra with unit change in wavelength in FNIR bands	[13]
Short-wave infrared (SWIR) bands			
	1316	Nitrogen: Leaf nitrogen content of crops	[14]
	1731/1691	Nitrogen	[3]
	1725	Lignin, biomass, starch, moisture	[8], [13]
	1954	Crude fat	[10]
	2211/2400	Cellulose	[3]
	2359	Cellulose, protein, nitrogen: Sensitive to crop stress, lignin, and starch	[1], [9]
	2370	Crude fiber	[10]

References: [1] Thenkabail et al. (2004a); [2] Thenkabail et al. (2004b); [3] Almeida and De Souza Filho (2004); [4] Chan and Paelinckx (2008); [5] Yang et al. (2010); [6] Merzlyak et al. (2003); [7] Shi et al. (2012); [8] Le Maire et al. (2008); [9] Penuelas et al. (1995); [10] Yi et al. (2008); [11]; [12] Abdel-Rahman et al. (2010); [13] Yao et al. (2010a, b)

Spectral data can present much information about the plant parameters. Usually, the dataset is very large and needs to be analyzed by appropriate multivariate and machine-learning methods. Such methods can aid in the extraction of useful information from spectral data and can also avoid multicollinearity (Witten and Frank 2005). For example, both a stay-green mutation and the effect of megadose methylprednisolone treatment on the compositional consequences of *Lolium temulentum* senescence have been studied. The investigated spectral parameters of the leaf tissue were estimated non-destructively and interpreted by various methods. Principal component analysis (PCA) was used to assess differences in leaf reflectance spectra between genotypes, types of treatment, and time points to demonstrate the disruption of pigment metabolism during senescence caused by the stay-green mutation. The stay-green mutation is caused by the inhibition of protein synthesis and by the combination of types of treatment and time points (Gay et al. 2008).

3.3 Integration of HSI in High-Throughput Phenotyping Systems

With increasing options for image capture and open-source analysis tools, the field of high-throughput plant phenomics is poised to enter a phase of fast development. High-throughput image-based phenotyping is described as a technology that can create hundreds of images of plants per day (Messina et al. 2011). Such volumes allow the analysis of mutant populations, the detection of QTLs, and the discovery of genes by the investigation of various environmental factors (Noah et al. 2015).

The phenotyping system PlantScreen™ has components that can capture broad spectral information. A type of imaging methodology in such phenotyping platforms is used for data collection in quantitative studies. This methodology is a complex feature related to growth parameters, increased yield, and adaptation to abiotic or biotic stresses. The imaging techniques include VIS imaging, imaging spectroscopy, fluorescence and thermal infrared (TIR) imagings, plus tomographic (positron emission tomography [PET], magnetic resonance tomography [MRT] and computerized tomography [CT]) and three-dimensional (3D) imaging (Li et al. 2014). Many image-based phenotyping approaches have originated from the use of non-imaging sensors that have been applied in the field, such as thermography point sensors. Other methods have been introduced from the field of remote sensing, such as the satellite-based calculation of spectral indices (Walter et al. 2015).

Greenhouse-based and growth-chamber phenotyping platforms have the advantage of increased experimental cycling and environmental control, but they are often defined by pot growth and the spectrum of environmental conditions that can be measured. Controlled environmental conditions are also well suited for the investigation of root phenotypes (Bucksch et al. 2014; Moore et al. 2013; Topp et al. 2013). In addition to the development of design strategies that support experimentation, camera technologies are needed to obtain important data related to metabolomics and plant physiology.

The target of plant imaging and analysis is to measure the physiological and qualitative development, growth, and other phenotypic properties of plants through automated processes. Hyperspectral imaging camera technology has become relatively inexpensive and universal, and it stands out among a recent surge of high-throughput phenotyping systems that utilize plant imaging to describe data. Standard consumer cameras use silicon-based sensors that are responsive to light wavelengths in the 400- to 1000-nm range. However, color cameras can be used in the 400- to 700-nm range, which is visible to humans with the inclusion of an IR-blocking filter (VIS camera). VIS camera sensor arrays have three color sensors (red, blue, and green) that are used to assess the true color of each pixel. VIS cameras are used to measure the morphological, geometric, and color properties of plants in the phenotyping system (Tessmer et al. 2013; Yang et al. 2014; Neilson et al. 2015). Infrared (IR) cameras detect NIR light and are used for night imaging (Matos et al. 2014). NIR cameras detect NIR and short-wave infrared (SWIR) light in a region that is useful for detecting leaf water content (Seelig et al. 2008). Thermal infrared (TIR) cameras detect long-wave infrared light, which is emitted by leaves at a temperature-dependent intensity (Sirault et al. 2009). Hyperspectral cameras that detect hundreds of spectral bands with nm-level resolution between 350 and 2500 nm are currently being developed to detect plant stress (Mahlein et al. 2013). Additionally, specialized imaging systems can measure chlorophyll fluorescence after excitation (Chen et al. 2014).

For phenotyping platforms, VIS cameras and additional camera systems can accompany color imaging. A disadvantage of VIS and NIR cameras is that the detection sensors are sensitive to relatively broad ranges of the electromagnetic spectrum, and specific wavelength information is lost in the output image. In contrast, state-of-the-art hyperspectral cameras can measure hundreds of spectral bands between 350 and 2500 nm at nm-level resolution for each image pixel (Mahlein et al. 2012a, b). Hyperspectral imaging is an encouraging approach for the discovery of both abiotic (Römer et al. 2012) and biotic stresses (Mahlein et al. 2012a, b). The images produced by various high-throughput phenotyping platforms and camera types can present challenges for common image processing and trait extraction.

3.4 Recent Advances in the Use of HSI at the Plant Level

Hyperspectral and fluorescent imaging may provide a means to directly and non-invasively detect and quantify secondary metabolites such as flavonoids and terpenoids (Table 3). In 2015, Deepak et al. compared different imaging approaches and conventional chemical analysis by high performance liquid chromatography (HPLC)-MS for the detection of secondary compounds of birch leaf. Clonally regenerated silver birch from three natural Finnish populations (latitudes 60°, 62°, and 66°) and four genotypes from each population growing in the Botanic Garden, Joensuu, Finland (62°35' N, 29°46' E) were used as the plant material. Chemical analysis by HPLC-MS revealed genetic differences within populations from different provenances based on the secondary chemistry. Visible, SWIR, and mid-infrared imagings were used to study spectral reflectance differences in leaf material from individual trees. Fluorescence approaches and UV reflectance imaging at the macroscopic scale were used to localize within-leaf spatial differences for UV-absorbing and fluorescent flavonoids (Deepak et al. 2015).

The following crop plants have been used for phenotyping projects related to breeding strategy and the influence of stress factors (especially drought): rice (Fischer et al. 2003), wheat (Monneveux et al. 2012; Rajendran et al. 2009), barley (Hartmann et al. 2011), maize (Dignat et al. 2013; Winterhalter et al. 2011), sorghum (Schaffert et al. 2011), and pearl millet (Vadez et al. 2012). Some are minor crops that are considered to be high in value but are not widely grown in some countries. It should be noted that a minor use in one country may be regarded as a major use in another country. Thus, many leguminous plants, such as buckwheat, can be minor crops and also *Pseudocereals.* Among leguminous plants, common beans (Beebe et al. 2013), chickpeas (Upadhyaya et al. 2012), cowpeas (Hall 2012), and groundnuts (Sheshshayee et al. 2011) have been used for phenotyping. Clonal plants known for phenotyping include cassava (Okogbenin et al. 2013), sweet potatoes (Vilaró 2011), and bananas (Ravi et al. 2013).

Arabidopsis plants are used as model plants for phenotyping because they have known genotype characteristics and can be easily manipulated with known gene parameters. Additionally, it is easy to identify connections between phenotyping technology, genetic background, and plant metabolomics. The Plant Image Analysis phenotyping system (http://www.plant-image-analysis.org/) is a convenient database of commercial and open-source image analysis tools (Lobet et al. 2013). The

Table 3 Application of hyperspectral imaging (HIS) for different plant parts and needs

Plant species	Part of plant	Use of his	References
Birch (*Betula* spp.)	Leaves	Detect and quantify secondary metabolites such as flavonoids	[1], [2]
Portobello mushroom (*Agaricus bisporus*)	Whole plant	Food quality evaluation	[3], [4]
Wheat (*Triticum* spp.)	Kernels	Food quality evaluation	[5]
Arabidopsis thaliana	Whole plant	Detection of ultraviolet (UV)-absorbing secondary metabolites	[6]
Hypericum (*Hypericum* spp.)	Whole plant	Detection of UV-absorbing secondary metabolites	[6]
Barley (*Hordeum vulgare* L.)	Leaves	Determining nitrogen levels in plants	[7]
Pepper (*Capsicum* spp.)	Leaves, whole plant	Determining nitrogen levels in plants	[8]
Barley (*Hordeum vulgare* L.)	Grains	Food quality evaluation	[9]
Cotton (*Gossypium hirsutum* L.)	Leaves	Determining nitrogen levels in plants	[10]
Cotton (*Gossypium hirsutum* L.)	Leaves	Determining chlorophyll concentrations	[10]
Eucalyptus (*Eucalyptus* spp.)	Canopy	Determining chlorophyll concentrations	[11]
Soybean (*Glycine max.* L. Merr.)	Canopy	Determining chlorophyll concentrations	[12]
Apple (*Malus domestica*)	Fruits	Defect detection	[13]
Apple (*Malus domestica*)	Fruits	Detection of fecal contamination	[14]
Lettuce (*Lactuca sativa*)	Leaves	Defect detection	[15]
Citrus (*Citrus* spp.)	Fruits	Detection of fungal infections	[16]
Sugar beet (*Beta vulgaris* L.)	Leaves	Detection of fungal infections	[17], [18], [19]
Maize (*Zea mays* L.)	Kernels	Detection of *Fusarium* damage	[20], [21]
Wheat (*Triticum* spp.)	Kernels	Detection of *Fusarium* damage	[22], [23]
Maize (*Zea mays* L.)	Kernels	Determination of mycotoxin contamination	[24], [25]

[1] Deepak et al. (2015); [2] Matros and Mock (2013); [3] Gaston et al. (2010); [4]Taghizadeh et al. (2011); [5] Jaya et al. (2010) [6] Hölscher et al. (2009); [7] Pytela et al. (2015); [8] Yu et al. (2014); [9] Arngren et al. (2011); [10] Zhao et al. (2005); [11] Coops et al. (2003); [12] Ferri et al. (2004); [13] Baranowski et al. (2012); [14] Yang et al. (2011); [15] Simko et al. (2015); [16] Lorente et al. (2013); [17] Mahlein et al. (2010); [18] Mahlein et al. (2012a, b); [19] Hillnhütter et al. (2011); [20] Williams et al. (2010); [21] Williams et al. (2012); [22] Shahin and Symons (2011); [23] Bauriegel et al. (2014); [24] Firrao et al. (2010); [25] Yao et al. (2010a, b)

majority of open-source tools found in the Plant Image Analysis database (Lobet et al. 2013) focus on the phenotyping of plants with rosette architecture (*Arabidopsis thaliana*) and specific plant organs such as excised leaves, although some experiments were conducted with HSI analysis (Table 3).

Spectral reflectance is also known as a fast method for determining nitrogen levels in plants. The general principle of spectral analysis involves reflectance values measured at different wavelengths. Nitrogen content is predicted from the linear dependence of reflectance and reference values of leaf nitrogen content. Reflectance spectra were determined with the PolyPen RP400 (Photon Systems, Drasov, Czech repubic) on the second and third youngest leaves of barley after 10 days of nitrogen treatment, as an average value for each nitrogen nutrition group. An average reflectance spectrum in the interval of 500–650 nm was negatively linked to the dose of nitrogen added to the medium. A similar degree of correlation was found when reflectance spectra were related to the weight of shoot biomass or the nitrogen content of the leaves. The reflectance parameter green normalized difference vegetation index, determined from the second leaves of individual plants, was dependent on the nitrogen content in the respective plants, with linear regression showing $r^2 = 0.612$ (Pytela et al. 2015).

Mass spectrometry-based imaging techniques can also be used as part of HSI to provide knowledge of proteins and other metabolites in plant tissues. Graphs of different types of compounds can be made from primary and secondary metabolism findings. Differential dissemination diagrams have been created for some compounds, particularly lipids, sugars, and amino acids, and for polyphenols, anthocyanins, glucosinolates, and alkaloids as secondary metabolites from different plant varieties (Matros and Mock 2013). Matrix-free laser desorption/ionization-MSI was used at the single-cell level to analyze the allocation of secondary metabolites with UV-absorbing capacities in *Hypericum* species (phloroglucinols and naphthodianthrones) and in *Arabidopsis thaliana* (kaempferol derivatives) (Hölscher et al. 2009). The preparation of samples and choosing of matrix substances and function is important for imaging analysis. The parameters are critical for obtaining images of sufficient quality using MALDI-MSI; also, desorption electrospray ionization (DESI)-MSI, combined with the above-mentioned imaging advances, provides strategies for integral applications. Molecular ions caused by MALDI time-of-flight (TOF), ESI-TOF mass spectrometry, and objective metabolite analysis were used to detect fructans, hexoses, and sucrose accumulation during the development of barley grain (Peukert et al. 2014).

The early detection of decay, and its gradual development, have been assessed in lettuce plants using chlorophyll fluorescence and hyperspectral imaging techniques. Two lettuce decay indices (LEDI) for the detection of leaf tissue decay have been proposed. The background of one of the indices ($LEDI_4$) is based on three identified hyperspectral imaging wavelengths. Chlorophyll fluorescence imagingwas used for establishing the other index ($LEDI_{CF}$). Both indexes, $LEDI_4$ and $LEDI_{CF,}$ when analyzed on yellow, light green, dark green, green, and red leaves, demonstrated near 97% validity in classifying tissues as being withered or fresh (Simko et al. 2015).

The typical green leaf reflectance spectrum is in the range of the VIS (400- to 700 nm) and NIR (700- to1000 nm) wavebands for hyperspectral analytical approaches. Few classes of spectral variables have been found to be valuable for describing the way in which reflectance spectra shift owing to changes in pigment concentrations. However, the concentrations of pigments and spectral variables have shown empirical relationships. An alternative technique for HSI analysis aimed at estimating the quantity of pigments has been established in canopy radiative transfer models and in various model versions of the physical leaf (Blackburn 2007). The high absorption spectral ranges have a high sensitivity to low concentrations of pigments, whereas reflectance can be more sensitive to high concentrations (Yamada and Fujimura 1991).

The presence of chlorophyll is the major feature that encompasses the relation between anthocyanin content and reflectance (Sytar et al. 2017a). Two narrow wavebands with reflectance ratios from the VIS, red edge, and NIR regions were selected for the assessment of chlorophyll in leaves (Zhao et al. 2005) and also at the canopy range (Coops et al. 2003; Ferri et al. 2004). The establishment of universal chlorophyll indices in plant reflectance data is necessary to cover more sources of spectral variability by using several narrow wavebands (Thenkabail et al. 2000). The calibration of reflectance data in a more detailed predictive model for a specific plant type is possible using just two wavebands. To resolve the problem of chlorophyll reflectance and anthocyanin identification, an index that uses the green narrow band (550 nm) and the reflectance of the early red edge band (700 nm) has been suggested (Gitelson et al. 2001).

Hyperspectral imaging has shown high effectiveness for assessing the quality of fruit and vegetables and their safety regarding surface defects, contamination, starch index, bruises, sugar content, freeze damage, firmness, and bitter pit (Nicolai et al. 2007; Lorente et al. 2012). Defect detection with HSI analysis is based on identifying the spectral trait wavelengths for the defect and then using these spectral parameters to create discrimination models. Thermal imaging (3500–5000 nm) was integrated with HSI (400–2500 nm) to detect bruises created 1 h before the assessment of bruising (Baranowski et al. 2012). Hyperspectral imaging has been adopted for the automatic detection of the early stages of fungal infections in post-harvest citrus fruit. To choose the optimal parameters for the determination of decay in citrus fruit generated by *Penicillium digitatum* and *Penicillium italicum* fungi, a receiver operating characteristic curve has been proposed that considers 74 parameters, covering 57 purely spectral variables and 17 spectral indexes (Lorente et al. 2013).

The detection of fecal contamination of vegetables and fruits is necessary because feces can contain different pathogens. An HSI system developed in 2011 by Yang et al. (2011) used a camera with electron multiplying CCD technology, which is able to achieve high-resolution images at high speed under low illumination. Fluorescence methods that could identify almost 99% of the fecal spots were used at wavebands of 680, 684, 720, and 780 nm (Yang et al. 2012).

To detect insect damage, classification models have been established using various methods of analysis—quadratic discriminant, linear discriminant, a support vector data description (SVDD) classifier for stepwise partial least squares (PLS) discriminant analysis, and a distribution-based Bayesian discriminant model. HIS has revealed better detection capacity for fruit quality and safety analysis compared with the standard investigation (Wu and Sun 2013).

Determination of moisture content and surface color is also needed for the assessment of fruit and vegetable quality. Current color assessments have been regularly achieved by the collection of images from a hyperspectral camera in reflectance mode in the spectrum area between 400 and 1000 nm, which also covers the visible spectrum (Taghizadeh et al. 2011). Color was predicted in two ways; the first method used the calculation of tristimulus values from the hyperspectral images, and the second used a multivariate PLS regression algorithm (Taghizadeh et al. 2011). The capacity of HSI to determine color was similar to that of a colorimeter (Ariana and Lu 2010), and HIS seemed to obtain more information about the surface than that obtained from RGB camera images (Taghizadeh et al. 2011).

Since 1998 many HIS studies of grain analysis have been conducted, consisting of the identification of sound or stained grains, analysis of grain color, determination of grain kernel vitreousness, and wheat class bias (Jayas et al. 2010).

HSI has been used to classify crop seeds, including maize, barley, rice, groats, oats, soybeans, and wheat seed. An NIR spectrum range (900–1700 nm) HSI system was used for mathematical modeling to analyze early pregerminated barley in order to separate low- and high-quality barley kernels (Arngren et al. 2011). The model used a direct classification structure established on the basis of traits of insensitivity to kernel orientation.

Toxigenic fungi in seeds are toxic for both animals and humans. HSI systems in the VIS and short-wave NIR (400–1000 nm) zones or long-wave NIR zones (960–1662 nm and 1000–2498 nm) have been used for the indirect detection of *Fusarium* damage in maize (Williams et al. 2010) and wheat (Shahin and Symons 2011), as well as for the investigation of fungal expansion (Williams et al. 2012). The symptoms of *Fusarium* disease are easily identified by the analysis of hyperspectral images. Therefore, this methodology can help to map the presence and assess the level of *Fusarium* infection. From this point of view, independent field harvests can be considered. *Fusarium* detection in wheat is established by HSI and chlorophyll fluorescence methods, according to the requirements, parameters, and limitations of the specifications of these methods. The adjustment of spectral changes caused by fungal infection is detected by HSI. The low physiological activity of tissues which occurs under the influence of *Fusarium* is the basis for confirmatory factor analyses (Bauriegel and Herppich 2014).

Measurement of the quantitative parameters of mycotoxin infection is a specific topic in the qualitative analysis of seed and grain production. Of note, HSI has been used to detect fumonisin contamination produced by *Fusarium* spp. in milled maize (Firrao et al. 2010). HIS has also been used to determine the concentration of aflatoxin in corn kernels after their inoculation with *Aspergillus flavus* spores (Yao et al. 2010a, b). Multiple linear regression was adopted to calculate correlations ($r^2 = 0.44$ for fumonisin analysis and $r^2 = 0.72$ for aflatoxin analysis).

In 2010, Gaston et al. used HSI to predict polyphenol oxidase activity in mushrooms; PCA was used for data analysis. The results led to the probability of establishing a sensor that could rapidly detect mushrooms that have a high possibility of developing enzymatic browning. Certainly, the discussed research covers the utility of HIS in quality management and food industry safety (Gaston et al. 2010).

Based on the full spectra of HSI and the important wavelengths when using the Dumas combustion method, quantitative relationships between spectral data and the total nitrogen content (TNC) in plant organs (root, stem, and leaf), and also in whole plants, were separately determined with the help of PLS regression. The PLS regression model, constructed using the relevant wavebands for the estimation of TNC in whole plants, showed a promising correlation coefficient for prediction and the root mean square error for prognosis. Each TNC pixel within the region of interest of the pepper plant sample was evaluated to make a structural distribution map of the TNC. These accomplishments indicate that HSI is encouraging and presents a powerful potential tool with which to estiamte the structural distribution of nitrogen content in plants (Yu et al. 2014).

HSI at the leaf level has proven to be specifically relevant for the estimation and quantification of fungal invasion. Each fungal infection affects sugar beet tissue in a specific form, and the resulting spectral reflectance shows fungal infection-specific spectral markers. Mahlein et al. (2010) and Hillnhütter et al. (2011) have characterized similar effects for foliar and soil-borne infections of sugar beet, using non-imaging hyperspectrometry.

HSI has high potential as a non-destructive diagnostic tool for disease detection. In 2012, Mahlein et al. investigated spectral reflectance and the peculiarities of leaves diseased with *Cercospora* at different stages of development. Light microscopy was used to characterize changes in the host tissues, at the morphological level, caused by colonization with the pathogen. An HSI line scan spectrometer (ImSpector V10E, Specim Spectral Imaging Ltd., Oulu, Finland), which has a spectral resolution of 2.8 nm from 400 to 1000 nm, was used under controlled conditions for the extended monitoring of infection syndromes at the time of the pathogenetic process. Pixel-wise mapping in the VIS and NIR areas for spectral reflectance allowed the detailed characterization and detection of disease at the leaf tissue level. The leaf architecture was related to the leaf patterns of spectral reflectance. The separation and identification of symptoms in regions showing all stages of ontogenesis, from youth to maturity, can assist in mapping classification. Better knowledge of leaf reflectance changes induced by plant diseases was achieved via pure spectral markers pixel-wise extraction using HSI. The process of pixel-wise extraction noticeably raises the specificity and sensitivity of hyperspectrometry in the immediate sensing of plant infection and developing plant disease symptoms (Mahlein et al. 2012a, b).

3.5 Future Perspectives of HSI at the Plant Level

Quantitative studies of plant systems and functions support the better adaptation of plants to conditions of low-input agriculture and resource-limited qualitative breeding processes (Fabio and Ulrich 2013). Hyperspectral imaging is a component of multidisciplinary research on plant phenotyping, and it focuses on features that can help in genotype selection with expanded efficiency. Hyperspectral imaging technology is evolving rapidly; however, parallel achievements in large-scale phenotyping applications, detailed coverage of minimum points of information regarding protocols for different experiments, and integration with modeling and data management schemes are required.

Hyperspectral imaging was basically advanced for remote sensing and is currently available as an efficient analytical approach for the non-destructive investigation of plants and food. Methods of using HSI techniques, with different sensors for hyperspectral reflectance imaging and fluorescence imaging, and hyperspectral transmission imaging techniques for plant process monitoring are being developed (Gowen et al. 2007). Currently, these HSI techniques are mostly used for food quality control and less for studying plant metabolomics and phenomics.

The HSI system is a valuable piece of emerging technology for the evaluation of food quality, and it can also be used for some commercial needs. It is foreseen that future methodological and technical improvements in HSI systems that suit the pharmaceutical industry will lead to the development of low-cost systems that are also suitable for food industry utilization. Comparatively lengthy times are needed for hypercube image addition, processing, and classification, as well as for other defining factors (Chen et al. 2002). The acquisition time can range from 2 to 4 min, built upon on the resolution of image and target size, but processing and classification times are closely related to computer hardware and software capacities. However, further improvements are needed in phenotyping system constituents, such as faster hardware in cameras and more detailed and efficient algorithms, which can shorten processing time and enable real-time HSI quality monitoring. It is important to develop HSI methodology for the theoretical needs of plant physiology, with the main goal being the creation of a standard spectrum database for the identification of as many biochemical compounds as possible. Using an HSI system as part of the phenotyping system will help to support the complex analysis of plant reactions to different stressors, which could be useful for building promising models of plant behavior under changing environmental conditions (Sytar et al. 2017b). Thus, it is time to develop HSI techniques and methods for the food and pharmaceutical industries, while at the same time building background HSI methodology for plant sustainability and biodiversity as the main goal in plant-level phenotyping.

4 Conclusions

In this review, we have presented information regarding the use of HSI in plant phenotyping. Hyperspectral imaging techniques have been used in experiments on food quality, evaluations of fruit defects, and the detection of fungal infections, as well as the identification of fecal contamination and *Fusarium* damage. The determination of chlorophyll concentrations, mycotoxin contamination, plant nitrogen levels, and UV-absorbing secondary metabolites is part of the metabolic analysis that is performed with non-destructive HSI methods. Hyperspectral imaging cameras can be used to assess qualitative traits in the phenotyping system. The methodology for analyzing biochemical compounds, other than those mentioned here, with HSI cameras requires further development.

Acknowledgments This work was supported by the research project of the Scientific Grant Agency of the Slovak Republic VEGA- 1-0923-16 and APVV-15-0721.

Competing Interests The authors declare no financial conflict of interests.

Conflict of Interests There are no conflicts of interest for this article.

Ethical Approval The presented research does not require ethical approval.

References

Abdel-Rahman EM, Fethi BA, van den Berg M (2010) Estimation of sugarcane leaf nitrogen concentration using in situ spectroscopy. Int J Appl Earth Obs Geoinf 12(1):S52–S57. https://doi.org/10.1016/j.jag.2009.11.003

Agarwal UP, Atalla RH (2010) Vibrational spectroscopy. In: Heitner C, Dimmel DR, Schmidt JA (eds) Lignin and Lignans: advances in chemistry. CRC Press, Boca Raton, pp 103–136

Alexandratos N, Bruinsma J (2012) Report. Title: World agriculture towards 2030/2050: the 2012 revision. Authors: Publisher: Food and Agriculture Organization of the United Nations. Institution: FAO. Report number: ESA Working Paper No. 12–03

Almeida TIR, De Souza Filho CR (2004) Principal component analysis applied to feature-oriented band ratios of hyperspectral data: a tool for vegetation studies. Int J Remote Sens 25:5005–50023

Araus JL, Cairns JE (2014) Field high-throughput phenotyping: the new crop breeding frontier. Trends Plant Sci 19(1):52–61. https://doi.org/10.1016/j.tplants.2013.09.008

Ariana DP, Lu RF (2010) Evaluation of internal defect and surface color of whole pickles using hyperspectral imaging. J Food Eng 96(4):583–590. https://doi.org/10.1016/j.jfoodeng.2009.09.005

Arngren M, Hansen PW, Eriksen B, Larsen J, Larsen R (2011) Analysis of pregerminated barley using hyperspectral image analysis. J Agric Food Chem 59:11385–11394. https://doi.org/10.1021/jf202122y

Baker NR (2008) Chlorophyll fluorescence: a probe of photosynthesis in vivo. Annu Rev Plant Biol 59:89–113. https://doi.org/10.1146/annurev.arplant.59.032607.092759

Bakker MG, Manter DK, Sheflin AM, Weir TL, Vivanco JM (2012) Harnessing the rhizosphere microbiome through plant breeding and agricultural management. Plant Soil 360:1. https://doi.org/10.1007/s11104-012-1361-x

Baranowski P, Mazurek W, Wozniak J, Majewska U (2012) Detection of early bruises in apples using hyperspectral data and thermal imaging. J Food Eng 110:345–355. https://doi.org/10.1016/j.jfoodeng.2011.12.038

Bauriegel E, Herppich WB (2014) Hyperspectral and chlorophyll fluorescence imaging for early detection of plant diseases, with special reference to *Fusarium* spec. Infections on wheat. Agriculture 4:32–57. https://doi.org/10.3390/agriculture4010032

Bauriegel E, Brabandt H, Gärber U, Herppich WB (2014) Chlorophyll fluorescence imaging to facilitate breeding of Bremia lactucae-resistant lettuce cultivars. Comput Electron Agric 105:74–82. https://doi.org/10.1016/j.compag.2014.04.010

Beebe SE, Rao IM, Blair MW, Acosta-Gallegos JA (2013) Phenotyping common beans for adaptation to drought. Front Physiol 4:35. https://doi.org/10.3389/fphys.2013.00035

Blackburn GA (2007) Hyperspectral remote sensing of plant pigments. J Exp Bot 58(4):855–867. https://doi.org/10.1093/jxb/erl123

Brestic M, Zivcak M, Datko M, Sytar O, Olsovska K, Shao H (2015) Novel resistance mechanism of barley Chlorina f104 antenna mutant against photoinhibition: possible role of new identified

chloroplastic cpNrp protein. Theoret Exp Plant Physiol 27(1):75–85. https://doi.org/10.1007/s40626-015-0033-7
Brotman Y, Riewe D, Lisec J, Meyer RC, Willmitzer L, Altmann T (2011) Identification of enzymatic and regulatory genes of plant metabolism through QTL analysis in *Arabidopsis*. J Plant Physiol 168(12):1387–1394. https://doi.org/10.1016/j.jplph.2011.03.008
Brunetti C, George RM, Tattini M, Field K, Davey MP (2013) Metabolomics in plant environmental physiology. J Exp Bot 64(13):4011–4020. https://doi.org/10.1093/jxb/ert244
Bucksch A, Burridge J, York LM, Das A, Nord E, Weitz JS, Lynch JP (2014) Image-based high-throughput field phenotyping of crop roots. Plant Physiol 166:470–486. https://doi.org/10.1104/pp.114.243519
Carreno-Quintero N, Bouwmeester HJ, Keurentjes JJ (2013) Genetic analysis of metabolome-phenotype interactions: from model to crop species. Trends Genet 29(1):41–50. https://doi.org/10.1016/j.tig.2012.09.006
Chan JC-W, Paelinckx D (2008) Evaluation of random forest and adaboost tree-based ensemble classification and spectral band selection for ecotope mapping using airborne hyperspectral imagery. Remote Sens Environ 112:2999–3011. https://doi.org/10.1016/j.rse.2008.02.011
Chaurand P, Schwartz SA, Caprioli RM (2002) Imaging mass spectrometry: a new tool to investigate the spatial organization of peptides and proteins in mammalian tissue sections. Curr Opin Chem Biol 6(5):676–681
Chen YR, Chao K, Kim MS (2002) Machine vision technology for agricultural applications. Comput Electron Agric 36(2):173–191. https://doi.org/10.1016/S0168-1699(02)00100-X
Chen P, Yan K, Shao H, Zhao S (2013) Physiological mechanisms for high salt tolerance in wild soybean (*Glycine soja*) from Yellow River delta, China: photosynthesis, osmotic regulation, ion flux and antioxidant capacity. PLoS One 8(12):e83227. https://doi.org/10.1371/journal.pone.0083227
Chen D, Neumann K, Friedel S, Kilian B, Chen M, Altmann T, Klukas C (2014) Dissecting the phenotypic components of crop plant growth and drought responses based on high-throughput image analysis. Plant Cell 26(12):4636–4655. https://doi.org/10.1105/tpc.114.129601
Cobb JN, DeClerck G, Greenberg A, Clark R, McCouch S (2013) Next-generation phenotyping: requirements and strategies for enhancing our understanding of genotype–phenotype relationships and its relevance to crop improvement. Theor Appl Genet 126:867. https://doi.org/10.1007/s00122-013-2066-0
Coops NC, Stone C, Culvenor DS, Chisholm LA, Merton RN (2003) Chlorophyll content in eucalypt vegetation at the leaf and canopy scales as derived from high spectral resolution data. Tree Physiol 23:23–31. https://doi.org/10.1093/treephys/23.1.23
Deepak M, Fauch L, Keski-Saari S, Kontunen-Soppela S, Oksanen E, Keinanen M (2015) Variation in the secondary compounds of the silver birch leaves by chemical and imaging techniques. IPAP 2015, International Plant and Algal Phenomics Meeting, 27th–30th June 2015, Prague, Czech Republic, 37
Dignat G, Welcker C, Sawkins M, Ribaut JM, Tardieu F (2013) The growths of leaves, shoots, roots and reproductive organs partly share their genetic control in maize plants. Plant, Cell Environ 36(6):1105–1119. https://doi.org/10.1111/pce.12045
Elmasry G, Kamruzzaman M, Sun D-W, Allen P (2012) Principles and applications of hyperspectral imaging in quality evaluation of agro-food products: a review. Crit Rev Food Sci Nutr 52(11):999–1023
Fabio F, Ulrich S (2013) Future scenarios for plant phenotyping. Annu Rev Plant Biol 64(1):267
Fahlgren N, Gehan MA, Baxter I (2015) Lights, camera, action: high-throughput plant phenotyping is ready for a close-up. Curr Opin Plant Biol 24:93–99. https://doi.org/10.1016/j.pbi.2015.02.006
Ferri CP, Formaggio AR, Schiavinato MA (2004) Narrow band spectral indexes for chlorophyll determination in soybean canopies [*Glycinemax* (L.) Merril]. Braz J Plant Physiol 16:131–136. https://doi.org/10.1590/S1677-04202004000300002
Fiorani F, Schurr U (2013) Future scenarios for plant phenotyping. Annu Rev Plant Biol 64:267–291. https://doi.org/10.1146/annurev-arplant-050312-120137

Firrao G, Torelli E, Gobbi E, Raranciuc S, Bianchi G, Locci R (2010) Prediction of milled maize fumonisin contamination by multispectral image analysis. J Cereal Sci 52(2):327–330. https://doi.org/10.1016/j.jcs.2010.06.017

Fischer KS, Lafitte R, Fukai S, Atlin G, Hardy B (eds) (2003) Breeding rice for drought-prone environments. IRRI, Los Baños, The Philippines, 98 pp. knowledgebank.irri. org/drought/drought.pdf

Fumio M, Okazaki Y, Oikawa A, Kusano M, Nakabayashi R, Kikuchi J, Yonemaru J-I, Ebana K, Yano M, Saito K (2012) Dissection of genotype–phenotype associations in rice grains using metabolome quantitative trait loci analysis. Plant J 70(4):624–636. https://doi.org/10.1111/j.1365-313X.2012.04903.x

Gamalero E, Trotta A, Massa N, Copetta A, Martinotti MG, Berta G (2004) Impact of two fluorescent pseudomonads and an arbuscular mycorrhizal fungus on tomato plant growth, root architecture and P acquisition. Mycorrhiza 14:233–251

Gaston E, Frias JM, Cullen PJ, O'Donnell CP, Gowen AA (2010) Prediction of polyphenol oxidase activity using visible near-infrared hyperspectral imaging on mushroom (*Agaricus bisporus*) caps. J Agric Food Chem 58:6226–6233. https://doi.org/10.1021/jf100501q

Gay A, Thomas H, Roca M, James C, Taylor J, Rowland J, Ougham H (2008) Nondestructive analysis of senescence in mesophyll cells by spectral resolution of protein synthesis-dependent pigment metabolism. New Phytol 179(3):663–674. https://doi.org/10.1111/j.1469-8137.2008.02412.x

Gitelson AA, Merzlyak MN, Chivkunova OB (2001) Optical properties and nondestructive estimation of anthocyanin content in plant leaves. Photochem Photobiol 74:38–45. https://doi.org/10.1562/0031-8655(2001)0740038OPANEO2.0.CO2

Goetz AFH, Vane G, Solomon JE, Rock BN (1985) Imaging spectrometry for earth remote-sensing. Science 228:1147–1153. https://doi.org/10.1126/science.228.4704.1147

Gowen AA, O'Donnell CP, Cullen PJ, Downey G, Frias JM (2007) Hyperspectral imaging – an emerging process analytical tool for food quality and safety control. Trends Food Sci Technol 18(12):590–598. https://doi.org/10.1016/j.tifs.2007.06.001

Hall AE (2012) Phenotyping cowpeas for adaptation to drought. Front Physiol 3:155. https://doi.org/10.3389/fphys.2012.00155

Hartmann A, Czaudern T, Hoffmann R, Stein N, Schreiber F (2011) HTPheno. An image analysis pipeline for high-throughput plant phenotyping. BMC Bioinf 12:148. https://doi.org/10.1186/1471-2105-12-148

Hill CB, Julian D, Taylor JD, James E, Mather D, Langridge P, Bacic A, Roessne U (2015) Detection of QTL for metabolic and agronomic traits in wheat with adjustments for variation at genetic loci that affect plant phenology. Plant Sci 233:143–154. https://doi.org/10.1016/j.plantsci.2015.01.008

Hillnhütter C, Mahlein A-K, Sikora RA, Oerke E-C (2011) Remote sensing to detect plant stress induced by *Heterodera schachtii* and *Rhizoctonia solani* in sugar beet fields. Field Crop Res 122:70–77. https://doi.org/10.1016/j.fcr.2011.02.007

Hölscher D, Shroff R, Knop K, Gottschaldt M, Crecelius A, Schneider B, Heckel DG, Schubert US, Svatoš A (2009) Matrix-free UV-laser desorption/ionization (LDI) mass spectrometric imaging at the single-cell level: distribution of secondary metabolites of Arabidopsis thaliana and Hypericum species. Plant J 60:907–918. https://doi.org/10.1111/j.1365-313X.2009.04012.x

Jackson SN, Wang HY, Woods AS, Ugarov M, Egan T, Schultz JA (2005) Direct tissue analysis of phospholipids in rat brain using MALDI-TOFMS and MALDI-ion mobility-TOFMS. J Am Soc Mass Spectrom 16(2):133–138

Jayas DS, Singh CB, Paliwal J (2010) Classification of wheat kernels using near-infrared reflectance hyperspectral imaging. In: Sun D-W (ed) Hyperspectral imaging for food quality analysis and control, 1st edn. Academic/Elsevier, San Diego, pp 449–470

Jiseok L, Philipp G, Manfred K, Baret J-C (2013) Micro-optical lens array for fluorescence detection in droplet-based microfluidics. Lab Chip 13:1472–1475. https://doi.org/10.1039/C3LC41329B

Khakimov B, Bak S, Engelsen SB (2013) High-throughput cereal metabolomics: current analytical technologies, challenges and perspectives. J Cereal Sci 59(3):393–418. https://doi.org/10.1016/j.jcs.2013.10.002

Kuska M, Wahabzada M, Leucker M, Dehne H-W, Kersting K, Oerke E-C, Steiner U, Mahlein A-K (2015) Hyperspectral phenotyping on the microscopic scale: towards automated characterization of plant-pathogen interactions. Plant Methods 11:28. https://doi.org/10.1186/s13007-015-0073-7

Larsen KL, Barsberg S (2011) Environmental effects on the lignin model monomer, vanillyl alcohol, studied by Raman spectroscopy. J Phys Chem B 115:11470–11480. https://doi.org/10.1021/jp203910h

Le Maire G, François C, Soudani K, Berveiller D, Pontailler JY, Bréda N, Genet H, Davi H, Dufrêne E (2008) Calibration and validation of hyperspectral indices for the estimation of broadleaved forest leaf chlorophyll content, leaf mass per area, leaf area index and leaf canopy biomass. Remote Sens Environ 112(10):3846–3864. https://doi.org/10.1016/j.rse.2008.06.005

Lee YJ, Perdian DC, Song Z, Yeung ES, Nikolau BJ (2012) Use of mass spectrometry for imaging metabolites in plants. Plant J 70:81–95. https://doi.org/10.1111/j.1365-313X.2012.04899.x

Li L, Zhang Q, Huang D (2014) A review of imaging techniques for plant phenotyping. Sensors 14:20079–20111. https://doi.org/10.3390/s141120078

Lim J, Gruner P, Konrad M, Baret JC (2013) Micro-optical lens array for fluorescence detection in dropletbased microfluidics. Lab on a Chip 13(8):1472–1475. https://doi.org/10.1039/c3lc41329b

Lobet G, Draye X, Périlleux C (2013) An online database for plant image analysis software tools. Plant Methods 9:38. https://doi.org/10.1186/1746-4811-9-38

Lorente D, Aleixos N, Gómez-Sanchis J, Cubero S, García-Navarrete OL, Blasco J (2012) Recent advances and applications of hyperspectral imaging for fruit and vegetable quality assessment. Food Bioprocess Technol 5:1121. https://doi.org/10.1007/s11947-011-0725-1

Lorente D, Aleixos N, Gómez-Sanchis J, Cubero S, Blasco J (2013) Selection of optimal wavelength features for decay detection in citrus fruit using the ROC curve and neural networks. Food Bioprocess Technol 6:530–541. https://doi.org/10.1007/s11947-011-0737-x

Lupoi JS, Smith EA (2012) Characterization of woody and herbaceous biomasses lignin composition with 1064 nm dispersive multichannel Raman spectroscopy. Appl Spectrosc 66:903–910. https://doi.org/10.1366/12-06621

Lupoi JS, Gjersing E, Davis Mark F (2015) Evaluating lignocellulosic biomass, its derivatives, and downstream products with Raman spectroscopy. Front Bioeng Biotechnol 3:50. https://doi.org/10.3389/fbioe.2015.00050

McCouch SR, McNally KL, Wang W, Sackville Hamilton R (2012) Genomics of gene banks: a case study in rice. Am J Bot 99(2):407–423. https://doi.org/10.3732/ajb.1100385

Mahlein AK, Steiner U, Dehne HW, Oerke EC (2010) Spectral signatures of sugar beet leaves for the detection and differentiation of diseases. Precis Agric 11:413–431. https://doi.org/10.1016/j.rse.2012.09.019

Mahlein A-K, Oerke E-C, Steiner U, Dehne H-W (2012a) Recent advances in sensing plant diseases for precision crop protection. Eur J Plant Pathol 133:197–120. https://doi.org/10.1007/s10658-011-9878-z

Mahlein A-K, Steiner U, Hillnhütter C, Dehne H-W, Oerke E-C (2012b) Hyperspectral imaging for small-scale analysis of symptoms caused by different sugar beet diseases. Plant Methods 8:3. https://doi.org/10.1186/1746-4811-8-3

Mahlein A-K, Rumpf T, Welke P, Dehne H-W, Plümer L, Steiner U, Oerke E-C (2013) Development of spectral indices for detecting and identifying plant diseases. Remote Sens Environ 128:21–30. https://doi.org/10.1016/j.rse.2012.09.019

Matos DA, Cole BJ, Whitney IP, MacKinnon KJ-M, Kay SA, Hazen SP (2014) Daily changes in temperature, not the circadian clock, regulate growth rate in Brachypodium distachyon. PLoS One 9:e100072. https://doi.org/10.1371/journal.pone.0100072

Matros A, Mock H-P (2013) Mass spectrometry based imaging techniques for spatially resolved analysis of molecules. Front Plant Sci 4(19):89. https://doi.org/10.3389/fpls.2013.00089

Matsuda F, Okazaki Y, Oikawa A, Kusano M, Nakabayashi R, Kikuchi J, Yonemaru J, Ebana K, Yano M, Saito K (2012) Dissection of genotype-phenotype associations in rice grains using metabolome quantitative trait loci analysis. Plant J 70(4):624–636. https://doi.org/10.1111/j.1365-313X.2012.04903.x

Mensack MM, Fitzgerald VK, Ryan EP, Lewis MR, Thompson HJ, Brick MA (2010) Evaluation of diversity among common beans (*Phaseolus vulgaris* L.) from two centers of domestication using 'omics' technologies. BMC Genomics 11:686. https://doi.org/10.1186/1471-2164-11-686

Merzlyak MN, Solovchenko AE, Gitelson AA (2003) Reflectance spectral features and non-destructive estimation of chlorophyll, carotenoid and anthocyanin content in apple fruit. Postharvest Biol Technol 27(2):197–211. https://doi.org/10.1016/S0925-5214(02)00066-2

Messina CD, Podlich D, Dong ZS, Samples M, Cooper M (2011) Yield-trait performance landscapes: from theory to application in breeding maize for drought tolerance. J Exp Bot 62:855–868. https://doi.org/10.1093/jxb/erq329

Meyer MW, Lupoi JS, Smith EA (2011) 1064 nm dispersive multichannel Raman spectroscopy for the analysis of plant lignin. Anal Chim Acta 706:164–170. https://doi.org/10.1016/j.aca.2011.08.031

Millera OJ, Harrakb AE, Mangeatb T, Bareta J-C, Frenza L, Debsa BE, Mayota E, Samuelsc ML, Rooneyd EK, Dieue P, Galvand M, Linkc DR, Griffiths AD (2012) High-resolution dose–response screening using droplet-based microfluidics. PNAS 109(2):378–383. https://doi.org/10.1073/pnas.1113324109

Mogensen KB, Klank H, Kutter JP (2004) Recent developments in detection for microfluidic systems. Electrophoresis 25:3498–3512

Monneveux P, Ruilian J, Misra SC (2012) Phenotyping wheat for adaptation to drought. Front Physiol 3:429. https://doi.org/10.3389/fphys.2012.00429

Moore CR, Johnson LS, Kwak I-Y, Livny M, Broman KW, Spalding EP (2013) High-throughput computer vision introduces the time axis to a quantitative trait map of a plant growth response. Genetics 195:1077–1086. https://doi.org/10.1534/genetics.113.153346

Nadella KD, Marla SS, Ananda Kumar P (2012) Metabolomics in agriculture. OMICS: J Integr Biol 16(4):149–159. https://doi.org/10.1089/omi.2011.0067

Neilson EH, Edwards AM, Blomstedt CK, Berger B, Møller BL, Gleadow RM (2015) Utilization of a highthroughput shoot imaging system to examine the dynamic phenotypic responses of a C4 cereal crop plant to nitrogen and water deficiency over time. J Exp Bot 66:1817–1832

Nicolai BM, Beullens K, Bobelyn E, Peirs A, Saeys W, Theron KI et al (2007) Nondestructive measurement of fruit and vegetable quality by means of NIR spectroscopy: a review. Postharvest Biol Technol 46(2):99–118. https://doi.org/10.1016/j.postharvbio.2007.06.024

Noah AD, Macho AP, Zipfel C (2015) Plant PRRs and the activation of innate immune signaling. Mol Cell 54:263–272

Okogbenin E, Setter TL, Ferguson M, Mutegi R, Ceballos H, Olasanmi B, Fregene M (2013) Phenotypic approaches to drought in cassava: review. Front Physiol 4:93. https://doi.org/10.3389/fphys.2013.00093

Ossipov V, Ossipova S, Bykov V, Oksanen E, Koricheva J, Haukioja E (2008) Application of metabolomics to genotype and phenotype discrimination of birch trees grown in a long-term openfield experiment. Metabolomics 4:39–51. https://doi.org/10.1007/s11306-007-0097-8

Penuelas J, Filella I, Lloret P, Munoz F, Vilajeliu M (1995) Reflectance assessment of mite effects on apple trees. Int J Remote Sens 16(14):2727–2733. https://doi.org/10.1080/01431169508954588. PMCID: PMC3360494

Peukert M, Thiel J, Peshev D, Weschke W, Van den Ende W, Mock HP, Matros A (2014) Spatio-temporal dynamics of fructan metabolism in developing barley grains. Plant Cell 26:3728–3744. https://doi.org/10.1105/tpc.114.130211

Pires NMM, Dong T, Hanke U, Hoivik N (2014) Recent developments in optical detection technologies in lab-ona-chip devices for biosensing applications. Sensors 14(8):15458–15479. https://doi.org/10.3390/s140815458

Poorter H, Fiorani F, Stitt M, Schurr U, Finck A, Gibon Y, Usadel B, Munns R, Atkin OK, Pons TL (2012) The art of growing plants for experimental purposes: a practical guide for the plant biologist (review). Funct Plant Biol 39(11):821–838. https://doi.org/10.1071/FP12028

Pytela J, Panzarova K, Chmelik D, Trtilek M (2015) Non-invasive spectral analysis of nitrogen content in barley leaves. IPAP 2015, International Plant and Algal Phenomics Meeting, 27th–30th June 2015, Prague, Czech Republic, P.44–45

Rajendran K, Tester M, Roy SJ (2009) Quantifying the three main components of salinity tolerance in cereals. Plant Cell Environ 32:237–249. https://doi.org/10.1111/j.1365-3040.2008.01916.x

Ravi I, Uma S, Vaganan MM, Mustaffa MM (2013) Phenotyping bananas for drought resistance. Front Physiol 4:9. https://doi.org/10.3389/fphys.2013.00009

Roessner U, Bacic A (2009) Metabolomics in plant research. Aust Biochem 40(3):9–20

Roessner U,·Willmitzer L, Fernie AR (2002) Metabolic profiling and biochemical phenotyping of plant systems. Plant Cell Rep 21:189–196. doi: https://doi.org/10.1007/s00299-002-0510-8

Römer C, Wahabzada M, Ballvora A, Pinto F, Rossini M, Panigada C, Behmann J, Léon J, Thurau C, Bauckhage C et al (2012) Early drought stress detection in cereals: simplex volume maximisation for hyperspectral image analysis. Funct Plant Biol 39:878. https://doi.org/10.1071/FP12060

Schaffert RE, Paulo EPA, Duarte JO, Garcia JC, Gomide RL, Guimarães CT, Magalhães PC, Magalhães JV, Queiroz Valéria AV (2011) Phenotyping sorghum for adaptation to drought. In: Monneveux P, Ribaut J-M (eds) Drought phenotyping in crops: from theory to practice. Frontiers Media SA, Lausanne, pp 287–299

Schauer N, Fernie AR (2006) Plant metabolomics: towards biological function and mechanism. Trends Plant Sci 11:508–516. https://doi.org/10.1016/j.tplants.2006.08.007

Schull MA, Knyazikhin Y, Xu L, Samanta A, Carmona PL, Lepine L, Jenkins JP, Ganguly S, Myneni RB (2011) Canopy spectral invariants, part 2: application to classification of forest types from hyperspectral data. J Quant Spectrosc Radiat Transf 112(4):736–775

Schwarz MA, Hauser PC (2001) Recent developments in detection methods for microfabricated analytical devices. Lab Chip 1(1):1–6

Seelig H-D, Hoehn A, Stodieck LS, Klaus DM III, Adams WW, Emery WJ (2008) The assessment of leaf water content using leaf reflectance ratios in the visible, near-, and short-wave-infrared. Int J Remote Sens 29:3701–3713. https://doi.org/10.1080/01431160701772500

Shahin MA, Symons SJ (2011) Detection of Fusarium damaged kernels in Canada Western red Spring wheat using visible/near-infrared hyperspectral imaging and principal component analysis. Comput Electron Agric 75(1):107–112. https://doi.org/10.1016/j.compag.2010.10.004

Shao HB, Liang ZS, Shao MA, Sun Q, Hu ZM (2005) Investigation on dynamic changes of photosynthetic characteristics of 10 wheat (*Triticum aestivum* L.) genotypes during two vegetative-growth stages at water deficits. Biointerfaces 43:221–227. https://doi.org/10.1016/j.colsurfb.2005.05.005

Shao HB, Jaleel CA, Shao MA (2009) Understanding water deficit stress-induced changes in basic metabolisms of higher plants for biotechnologically and sustainably improving agriculture and ecoenvironment in arid regions on the globe. Crit Rev Biotechnol 29:131–151. https://doi.org/10.1080/07388550902869792

Sheshshayee MS, Parsi Shashidhar G, Madhura JN, Beena R, Prasad TG, Udayakumar M (2011) Phenotyping groundnuts for adaptation to drought. In: Monneveux P, Ribaut J-M (eds) Drought phenotyping in crops: from theory to practice. Frontiers Media SA, Lausanne, pp 371–387

Shi J-Y, Xiao-bo Z, Jie-Wen Z, Holmes M, Kai-liang W, Xue W, ChenH (2012) Determination of total flavonoids content in fresh ginkgo biloba leaf with different colors using near infrared spectroscopy. Spectrochim Acta A Mol Biomol Spectrosc 94:271–276. https://doi.org/10.1016/j.saa.2012.03.078

Simko I, Jose AJ-B, Furbank Robert T (2015) Detection of decay in fresh-cut lettuce using hyperspectral imaging and chlorophyll fluorescence imaging. Postharvest Biol Technol 106:44–52. https://doi.org/10.1016/j.postharvbio.2015.04.007

Sims DA, Gamon JA (2002) Relationships between leaf pigment content and spectral reflectance across a wide range of species, leaf structures and developmental stages. Remote Sens Environ 81:337. https://doi.org/10.1016/S0034-4257(02)00010-X

Sirault XRR, James RA, Furbank RT (2009) A new screening method for osmotic component of salinity tolerance in cereals using infrared thermography. Funct Plant Biol 36:970. https://doi.org/10.1071/FP09182

Sun D-W (2010) Hyperspectral imaging for food quality analysis and control. Academic/Elsevier, San Diego

Sytar O, Bruckova K, Hunkova E, Zivcak M, Konate K, Brestic M (2015) The application of multiplex fluorimetric sensor for the analysis of flavonoids content in the medicinal herbs family *Asteraceae, Lamiaceae, Rosaceae*. Biol Res 48:5. https://doi.org/10.1186/0717-6287-48-5

Sytar O, Brestic M, Zivcak M, Tran L-S (2016) The contribution of buckwheat genetic resources to health and dietary diversity. Curr Genomics 17(3):193–206

Sytar O, Bruckova K, Kovar M, Hemmerich I, Zivcak M, Brestic M (2017a) Nondestructive detection and biochemical quantification of buckwheat leaves using visible (VIS) and near-infrared (NIR) hyperspectral reflectance imaging. J Cent Eur Agric 18(4):864–878. https://doi.org/10.5513/JCEA01/18.4.1978

Sytar O, Brestic M, Zivcak M, Olsovska K, Kovar M, Shao H, He X (2017b) Applying hyperspectral imaging to explore natural plant diversity towards improving salt stress tolerance. Sci Total Environ 578:90–99

Taghizadeh M, Gowen AA, O'Donnell CP (2011) Comparison of hyperspectral imaging with conventional RGB imaging for quality evaluation of Agaricus bisporus mushrooms. Biosyst Eng 108(2):191–194. https://doi.org/10.1016/j.biosystemseng.2010.10.005

Tang X, Mu X, Shao H, Wang H, Brestic M (2014) Global plant-responding mechanisms to salt stress: physiological and molecular levels and implications in biotechnology. Crit Rev Biotechnol:1–13. https://doi.org/10.3109/07388551.2014.889080

Tardieu F, Tuberosa R (2010) Dissection and modelling of abiotic stress tolerance in plants. Curr Opin Plant Biol 13:206–212. https://doi.org/10.1016/j.pbi.2009.12.012

Tessmer OL, Jiao Y, Cruz JA, Kramer DM, Chen J (2013) Functional approach to high-throughput plant growth analysis. BMC systems Biology 20137(Suppl 6):S17 https://doi.org/10.1186/1752-0509-7-S6-S17

Tester M, Langridge P (2010) Breeding technologies to increase crop production in a changing world. Science 327:818–822. https://doi.org/10.1126/science.1183700

Thenkabail PS, Smith Ronald B, Pauw Eddy D (2000) Hyperspectral vegetation indices and their relationships with agricultural crop characteristics. Remote Sens Environ 71(2):158–182. https://doi.org/10.1016/S0034-4257(99)00067-X

Thenkabail PS, Enclona EA, Ashton MS, Van Der Meer V (2004a) Accuracy assessments of hyperspecral waveband performance for vegetation analysis applications. Remote Sens Environ 91(2–3):354–376. https://doi.org/10.1016/j.rse.2004.03.013

Thenkabail PS, Enclona EA, Ashton MS, Legg C, Jean De Dieu M (2004b) Hyperion, IKONOS, ALI, and ETM +sensors in the study of African rainforests. Remote Sens Environ 90:23–43. https://doi.org/10.1016/j.rse.2003.11.018

Todd PJ, Schaaff TG, Chaurand P, Caprioli RM (2001) Organic ion imaging of biological tissue with secondary ion mass spectrometry and matrix-assisted laser desorption/ionization. J Mass Spectrom 36:355–369

Topp CN, Iyer-Pascuzzi AS, Anderson JT, Lee C-R, Zurek PR, Symonova O, Zheng Y, Bucksch A, Mileyko Y, Galkovskyi T et al (2013) 3D phenotyping and quantitative trait locus mapping identify core regions of the rice genome controlling root architecture. Proc Natl Acad Sci U S A 110:E1695–E1704. https://doi.org/10.1073/pnas.1304354110

Upadhyaya HD, Kashiwagi J, Varshney RK, Gaur PM, Saxena KB, Krishnamurthy L, Gowda CL, Pundir RP, Chaturvedi SK, Basu PS, Singh IP (2012) Phenotyping chickpeas and pigeonpeas for adaptation to drought. Front Physiol 3:179. https://doi.org/10.3389/fphys.2012.00179

Urbanczyk-Wochniak E, Baxter C, Kolbe A, Kopka J, Sweetlove LJ, Fernie AR (2005) Profiling of diurnal patterns of metabolite and transcript abundance in potato (*Solanum tuberosum*) leaves. Planta 221:891–903. https://doi.org/10.1007/s00425-005-1483-y

Vadez V, Hash T, Bidinger FR, Kholova J (2012) Phenotyping pearl millet for adaptation to drought. Front Physiol 3:386. https://doi.org/10.3389/fphys.2012.00386.

Verhoeven HA, de Vos CR, Bino RJ, Hall RD (2006) Plant metabolomics strategies based upon quadrupole time of flight mass spectrometry (QTOF-MS). In: Plant metabolomics. Springer, Berlin/Heidelberg

Verpoorte R, Choi YH, Kim HK (2005) Ethnopharmacology and systems biology: a perfect holistic match. J Ethnopharmacol 100(1–2):53–56

Vilaró F (2011) Phenotyping sweet potatoes for adaptation to drought. In: Monneveux P, Ribaut J-M (eds) Drought phenotyping in crops: from theory to practice. Frontiers Media SA, Lausanne, pp 415–427

Waggoner PS, Craighead HG (2007) Micro- and nanomechanical sensors for environmental, chemical, and biological detection. Lab Chip 7(10):1238–1255

Walter A, Liebisch F, Hund A (2015) Plant phenotyping: from bean weighing to image analysis. Plant Methods 11:14. https://doi.org/10.1186/s13007-015-0056-8

Warsta E, Lahdetie A, Jaaskelainen A-S, Vuorinen T (2012) Effect of pH on lignin analysis by Raman spectroscopy. Holzforschung 66:451–457. https://doi.org/10.1515/hf.2011.176

Williams P, Manley M, Fox G, Geladi P (2010) Indirect detection of *Fusarium verticillioides* in maize (*Zea maize* L.) kernels by NIR hyperspectral imaging. J Near Infrared Spectrosc 18(1):49–58. https://doi.org/10.1255/jnirs.858

Williams PJ, Geladi P, Britz TJ, Manley M (2012) Investigation of fungal development in maize kernels using NIR hyperspectral imaging and multivariate data analysis. J Cereal Sci 55:272–278. https://doi.org/10.1016/j.jcs.2011.12.003

Winterhalter L, Mistele B, Jampatong S, Schmidhalter U (2011) High throughput phenotyping of canopy water mass and canopy temperature in well-watered and drought stressed tropical maize hybrids in the vegetative stage. Eur J Agron 35:22–32. https://doi.org/10.1016/j.eja.2011.03.004

Witten IH, Frank E (2005) Data mining. Morgan Kauffman, San Francisco

Wu D, Sun D-W (2013) Advanced applications of hyperspectral imaging technology for food quality and safety analysis and assessment: a review — part II: applications. Innovative Food Sci Emerg Technol 19:15–28. https://doi.org/10.1016/j.ifset.2013.04.016

Yamada N, Fujimura S (1991) Nondestructive measurement of chlorophyll pigment content in plant leaves from three-color reflectance and transmittance. Appl Opt 30:3964–3973. https://doi.org/10.1364/AO.30.003964

Yan K, Chen P, Shao HB, Zhang LW, Zhang LH, Xu G, Sun JN (2011) Effects of short-term high temperature on photosynthesis and photosystem II performance in sorghum. J Agron Crop Sci 197:67–74. https://doi.org/10.1111/j.1439-037X.2011.00469.x

Yan K, Chen P, Shao HB, Zhao SJ, Zhang LH, Zhang LW, Xu G, Sun JN (2012) Photosynthetic characterization of Jerusalem artichoke during leaf expansion. Acta Physiol Plant 34:353–360. https://doi.org/10.1007/s11738-011-0834-5

Yan K, Chen P, Shao HB, Shao CY, Zhao SJ, Brestic M (2013) Dissection of photosynthetic electron transport process in sweet sorghum under heat stress. PLoS One 8(5):e62100. https://doi.org/10.1371/journal.pone.0062100

Yang F, Li J, Gan X, Qian Y, Wu X, Yang Q (2010) Assessing nutritional status of Festuca arundinacea by monitoring photosynthetic pigments from hyperspectral data. Comput Electron Agric 70(1):52–59. https://doi.org/10.1016/j.compag.2009.08.010

Yang CC, Kim MS, Kang S, Tao T, Chao K, Lefcourt AM et al (2011) The development of a simple multispectral algorithm for detection of fecal contamination on apples using a hyperspectral line-scan imaging system. Sens & Instrumen Food Qual 5(1):10–18. https://doi.org/10.1007/s11694-010-9105-1

Yang C, Kim MS, Chao K (2012) Development and application of multispectral algorithms for defect apple inspection. ASABE Annual international meeting. Dallas, Texas: The American Society of Agricultural and Biological Engineers, St. Joseph, Michigan (Paper #12133701)

Yang W, Guo Z, Huang C, Duan L, Chen G, Jiang N et al (2014) Combining high-throughput phenotyping and genome-wide association studies to reveal natural genetic variation in rice. Nat Commun 5:5087. https://doi.org/10.1038/ncomms6087

Yao X, Zhu Y, Tian Y-C, Feng W, Xing Cao W (2010a) Exploring hyperspectral bands and estimation indices for leaf nitrogen accumulation in wheat. Int J Appl Earth Obs Geoinf 12(2):89–100. https://doi.org/10.1016/j.jag.2009.11.008

Yao H, Hruska Z, Kincaid R, Brown R, Cleveland T, Bhatnagar D (2010b) Correlation and classification of single kernel fluorescence hyperspectral data with aflatoxin concentration in corn kernels inoculated with Aspergillus flavus spores. Food Addit Contam 27(5):701–709. https://doi.org/10.1080/19440040903527368

Yi Q-X, Huang J-F, Wang F-M, Wang X-Z (2008) Quantifying biochemical variables of corn by hyperspectral reflectance at leaf scale. J Zhejiang Univ Sci B 9(5):378–384. https://doi.org/10.1631/jzus.B0730019

Yu K-Q, Zhao Y-R, Li X-L, Shao Y-N, Liu F, He Y (2014) Hyperspectral imaging for mapping of total nitrogen spatial distribution in pepper plant. PLoS One 9(12):e116205. https://doi.org/10.1371/journal.pone.0116205

Zakis GF (1994) Functional analysis of lignins and their derivatives. Tappi Press, Atlanta

Zhao D, Reddy KR, Kakani VG, Read JJ, Koti S (2005) Selection of optimum reflectance ratios for estimating leaf nitrogen and chlorophyll concentrations of field-grown cotton. Agron J 97:89–98. https://doi.org/10.2134/agronj2005.0089

An Overview of QTL Identification and Marker-Assisted Selection for Grain Protein Content in Wheat

Ajay Kumar, Shalu Jain, Elias M. Elias, Mohamed Ibrahim, and Lakesh K. Sharma

Abstract

Grain protein content (GPC) is one of the most important traits in both the hexaploid and durum wheat. It plays an important role in end-use quality and thus determines the economic value of the crop. Improvement in GPC is a major objective in wheat breeding programs around the world. Therefore, in the past two decades, numerous studies on genetic dissection of this trait have been conducted in wheat. These studies have identified numerous quantitative trait loci (QTL) and markers associated with GPC in wheat. The available information about the marker trait associations for GPC offer great opportunities for marker-assisted breeding for this complex quantitative trait. In this article, we summarize the information available about the molecular genetic dissection of GPC and the progresses and prospects of application of marker-assisted breeding for improvement of this trait in wheat. We also reviewed the genetic relationship between GPC and grain yield. Strategies were also suggested to improve GPC in wheat based on available genetic and genomic resources.

Keywords

Epistasis · Grain protein content · Grain yield · Marker-assisted selection · Quantitative trait loci · Durum wheat · Hexaploid wheat

A. Kumar (✉) · E. M. Elias · M. Ibrahim
Department of Plant Sciences, North Dakota State University, Fargo, ND, USA
e-mail: ajay.kumar.2@ndsu.edu

S. Jain
Department of Plant Pathology, North Dakota State University, Fargo, ND, USA

L. K. Sharma
University of Maine Cooperative Extension, Presque Isle, ME, USA

R. S. Sengar, A. Singh (eds.), *Eco-friendly Agro-biological Techniques for Enhancing Crop Productivity*, https://doi.org/10.1007/978-981-10-6934-5_11

1 Introduction

Wheat (*Triticum spp.*) is one of the major crops grown extensively around the world and forms an important staple food in many countries. It accounts for ~20% of the calories consumed by humans (http://www.wheatworld.org/wp-content/uploads/Wheat-101.pdf) and is an important source of protein, vitamins, and minerals. Bread, ready-to-eat cereals, cake, biscuits, noodles, and pasta are some of the wheat products that are consumed daily throughout the world. For wheat to continue to sustain food security globally, it is being continuously targeted for genetic improvement of yield and resistance to both biotic and abiotic stresses. While overall production should continue to increase, there is also a growing demand to produce higher-quality and more nutritious wheat grain (Battenfield et al. 2016). Because of this changing demand in both domestic and international markets, most wheat breeding programs in Australia, Canada, India, the United States, and in other parts of the world are taking up the improvement of quality traits as part of their core breeding objectives.

The endosperm of the mature wheat grain contains 55–75% carbohydrates and 10–20% protein (Gillies et al. 2012). Grain protein content is one of the most important traits, which plays an important role in end-use quality and thus determines the economic value of both the hexaploid (*Triticum aestivum* L.) and durum (*T. turgidum* L. var. *durum* Desf.) wheat crop. Grain proteins are generally classified into four groups, albumins, globulins, prolamins, and glutelins, depending on extraction methods. According to some estimates, wheat is the most important source of worldwide protein supply (FAOSTAT 2014). Therefore, improvement in GPC and alteration in composition of storage proteins in wheat grain have been a major objective in wheat breeding programs, particularly for those working toward raising the nutritional quality (Shewry 2009). Attempts made by many breeders using conventional breeding methods have resulted in slow improvement in GPC. This is because of many reasons, including (1) high influence of environment on GPC, (2) negative correlation between GPC and grain yield, and (3) quantitative genetic control of GPC involving a large number of genes/QTL with small effects and low heritability (Simmonds 1995; Groos et al. 2003; Balyan et al. 2013). Although rare examples of wheat varieties combining high GPC and high grain yield are available (Cox et al. 1985), the simultaneous improvement of these two traits in wheat using classical methods of breeding has been a difficult exercise in the past (Balyan et al. 2013). Due to the complexity of these traits, one of the ways to move forward would be the use of modern genomic tools to identify loci which independently control GPC or yield and have no negative pleiotropic effects and use such QTL/genes in breeding programs through marker-assisted selection (MAS). The application of molecular markers is now a common practice for selection of various desirable alleles and thus is an important component of modern plant breeding programs (for details, see Gupta et al. 2010a, b). These tools may help in accurately estimating the quality characteristics and potential of individual wheat lines in early generations without conducting the direct tests.

During the last few years, several studies have investigated the genetic control of grain protein in wheat (Table 1). The results clearly showed that GPC is quantitatively inherited and controlled by a large number of QTL/genes and gene networks that are greatly influenced by environmental conditions (see Sect. 2). Cytoplasmic effects on inheritance of GPC have also been documented (El-Bok et al. 2013). Few of these QTL for GPC are also being introgressed through MAS to improve GPC in elite local genotypes (see Sect. 3). However, to the best of our knowledge, there is not a single review article available, which combines the information on molecular genetic dissection of GPC and the availability of suitable molecular markers associated with this trait. Therefore, the objective of this article is to summarize the available information on the genetics of grain protein content in wheat, resulting from QTL/association mapping studies and suggest breeding strategies for improvement of this trait based on utilization of modern genomic tools. The review considers both the major wheat types: hexaploid or bread wheat which accounts for ~95% of wheat grown in the world and mainly used for bread and durum wheat which accounts ~5% of the total wheat grown in the world and is mainly used for pasta production.

2 Genetics of Grain Protein Content in Wheat

2.1 Quantitative Trait Loci/Genes for Grain Protein Content

Extensive studies in wheat have suggested that GPC is a polygenic trait controlled by many QTL/genes with smaller effects and few QTL/genes with major effects. The genes/QTL for GPC have been located on all chromosomes of both tetraploid and hexaploid wheat (Table 1). In 49 QTL mapping publications reviewed in this article, a total of 325 main effect QTL and 42 epistatic QTL were reported for GPC. Although most of those studies were based on biparental mapping populations, few studies involving diverse association mapping panels were also conducted to dissect the genetics of GPC (Bordes et al. 2011; Reif et al. 2011). Individual studies have reported one to as many as 22 QTL for GPC (Table 1). These studies have also shown a wide range of phenotypic variation (0.6 to 66%) explained by these QTL (Table 1). Although GPC is highly influenced by environmental factors, at least 80 QTL with stable effect across environments have also been reported in different studies (Table 1). In this article, a QTL was considered stable, if it was detected in at least half of the investigated environments. Some of the studies were either conducted only in single environments or used only mean GPC data across environments to identify QTL. In such cases, it was not possible to conclude if the QTL were stable or not. Grain protein content being highly influenced by the environment, the stability of the QTL across environments is of utmost importance for their successful implication in breeding programs.

Several studies have reported few QTL that explained significant amount of phenotypic variation (PV) for GPC. Major QTL were identified on 2AS (Blanco et al. 2006, 2012; Suprayogi et al. 2009), 2BS (Terasawa et al. 2016), 2BL (Blanco et al.

Table 1 A summary of 49 QTL mapping studies for grain protein content (GPC) in wheat

S. No.	No. of QTL	PVE (%)	Chr. (arm)	No. of stable QTL	Plant material	h^2 (%)	Correlation with yield	References
1.	6	–	4BS, 5AL, 6AS (2 loci), 6BS, 7BS	–	65 recombinant inbred lines (RILs) developed from Messapia (durum wheat cv.) × MG4343 (wild tetraploid accession)	59–78	−0.02 to −0.57	Blanco et al. (1996)
2.	1	66	6BS	1 (6BS)	85 recombinant inbred chromosome lines (RICLs) for *Triticum turgidum* (L.) var. *dicoccoides* chromosome 6B	–	−0.06 to 0.07	Joppa et al. (1997)
3.	1	18.73	2DL	–	100 RILs developed from WL711 (Indian bread wheat cultivar) × PH132 (breeding line)	–	–	Prasad et al. (1999)
4.	2	6.5–17.1	1BL, 6AS	–	187 DH lines derived from Courtot × Chinese spring (both bread wheat cvs.)	–	–	Perretant et al. (2000); Sourdille et al. (2003)
5.	9[a]	2.9–7.2	2BL, 7AS	3 (7AS)	106 RILs developed from WL711 (Indian bread wheat cultivar) × PH132 (breeding line)	–	–	Dholakia et al. (2001)
6.	1	6.2	5AL	–	100 RILs developed from WL711 (Indian bread wheat cultivar) × PH132 (breeding line)	–	–	Harjit-Singh et al. (2001)
7.	7	6.5–31.7	4BS, 5AL, 6AS (2 loci), 6BS, 7AS, 7BS	3 (4BS, 6AS, 7AS)	65 RILs developed from Messapia (durum wheat cv.) × MG4343 (wild tetraploid accession)	54–78	−0.11 to −0.75	Blanco et al. (2002)
8.	2		2DS, 7AS	0	114 RILs of ITMI population developed from Opata85 (spring wheat cultivar) × W7984 (synthetic hexaploid wheat)	–	–	Börner et al. (2002)
9.	10	4.2–10.4	1A, 2AS, 3AL, 3BS, 4AS, 4DL, 5BL, 6AL, 7AS, 7DL	4 (2AS, 3AL, 4DL, 7DL)	194 RILs developed from Renan × Récital (two French varieties)	–	−0.19 to −0.66	Groos et al. (2003)

10.	13	0.6–35.8	2AS, 2BL, 2DL, 3DS, 4AL, 6BS, 7AS, 7DS	1 (2DL)	100 RILs developed from WL711 (Indian bread wheat cultivar) × PH132 (breeding line)	–	–	Prasad et al. (2003)
11.	3	32	5B	3 {5B(3)}	133 RICLs for *Triticum turgidum* (L.) var. *dicoccoides* chromosome 5B	57	0.07 to −0.27	Gonzalez-Hernandez et al. (2004)
12.	3	6.2–9.6	3AS, 4D, 7DL	–	165 RILs developed from Renan × Récital (two French varieties)	–	–	Groos et al. 2004
13.	7 +26[b]	8.4–16.6	1AS, 1BL, 1DL, 2AS, 2AL, 2BL, 2DS, 2DL, 3BS, 4AS, 5BL, 5DL, 6DL, 7AL, 7DS	1 (2DS)	Two populations: i) 100 RILs developed from WL711 (Indian bread wheat cultivar) × PH132 (breeding line) and ii) 110 RILs of ITMI population developed from Opata85 (spring wheat cultivar) × W7984 (synthetic hexaploid wheat)	–	–	Kulwal et al. (2005)
14.	3	4.2–21.7	2AS,6AS, 7BL	3 (2AS,6AS, 7BL)	92 backcross inbred lines (BILs)	13–26	−0.28	Blanco et al. (2006)
15.	2	12.6–32.7	4DS, 7BL	–	185 doubled haploid (DH) lines developed from a cross between a Canadian wheat variety "AC karma" and a breeding line 87E03-S2B1	81	−0.043	Huang et al. (2006)
16.	4	15.0–32.0	2AL, 2DS, 5AL, 6DS	–	114 RILs of ITMI population developed from Opata85 (spring wheat cultivar) × W7984 (synthetic hexaploid wheat)	–	–	Nelson et al. (2006)
17.	7	4.5–22.4	3AL, 4AL, 4BL, 5DL, 7BS, 7DS	–	The two advanced backcross populations developed from German winter wheat cultivars and two synthetic wheat accessions (250 lines from Batis × Syn022 and 150 lines from Zentos × Syn086)	–	–	Kunert et al. (2007)

(continued)

Table 1 (continued)

S. No.	No. of QTL	PVE (%)	Chr. (arm)	No. of stable QTL	Plant material	h^2 (%)	Correlation with yield	References
18.	13	–	1B, 2AS, 2AL, 2DS, 3B, 3D, 4B, 5A, 5B, 7D	2 (2AL, 2DS)	222 DH lines developed from Arche × Récital	79–84	–	Laperche et al. (2007)
19.	3	8.6–21.2	3BL, 5AL, 6AS	–	131 RILs derived from Chuan 35,050 × Shannong 483 (both Chinese varieties)	–	–	Sun et al. (2008)
20.	13	–	1B, 2A, 3AS, 3B, 4B, 4D, 5A,5B,7AL, 7D	2 (3AS, 7AL)	160 DH lines developed from kukri × Janz (both Australian cultivars)	69–93	–	Mann et al. 2009
21.	1+2[b]	9.6	4BS, 7BS	0	140 RILs developed from the cross of PDW 233 (Indian durum wheat cultivar) × Bhalegaon 4 (a landrace)	36–55	–	Patil et al. (2009)
22	10		2AL, 2BL, 3BL, 4AL, 5AS, 5BL, 6AS, 6BL, 7AL, 7BS	7 (2AL, 3BL, 4AL, 5AS, 5BL, 6BL, 7AL)	152 RILs from Langdon (durum wheat cultivar) × G18–16 (wild emmer accession)	63	–	Peleg et al. (2009)
23.	1	18–20	4AL	1 (4AL)	190 DH lines from Chara (Australian cultivar) × WW2449 (advanced breeding line)	76–82	–	Raman et al. (2009)
24.	9	–	1A, 1B, 2AS, 2BL, 5B, 6B, 7AL, 7B	3 (2AS, 2BL, 7AL)	185 DH lines from DT695 durum breeding line × Strongfield (durum wheat cultivar)	51–70	−0.45 to −0.51	Supryayogi et al. (2009)
25.	2	8.3–16.8	3AS, 4BS	1 (4BS)	132 RILs from Ning7840 (Chinese hard wheat line) × Clark (soft wheat cultivar)	36–62	0.32	Sun et al. (2010)
26.	3	4.5–16.8	2BS, 5AL, 6DL	3 (2BS, 5AL, 6DL)	139 RILs from MN98550 × MN99394 (both hard red spring wheat lines)	73	0.19 to −0.41	Tsilo et al. (2010)
27	4 + 4 [b]	3.1–8.4	2B, 2D, 3A, 3B, 4A, 5A, 5D, 6D	–	168 DH lines from Huapei 3 × Yumai 57 (both Chinese cultivars)	–	–	Zhao et al. (2010)

28.	22[c]	2.2–5.4	1AS, 1AL, 1BS, 1DL, 2AL, 2BS, 2BL, 2DS, 3BS, 3DL, 4BL, 5BL, 5DL, 6AL, 6BL, 6DL, 7BL	10 {1BS (2), 1DL, 2AL, 2BL, 3DL, 4BL, 6AL, 6BL 7BL}	A core collection consisting of 372 accessions	–	–	Bordes et al. (2011)
29.	10 + 8 [b]	9.3–21.6	1BS, 2AL, 2BS, 2BL, 3BS, 4AL, 5AS, 5BL, 7AS, 7BL.	0	93 RILs from UC1113 (durum breeding line) × Kofa (durum cultivar)	70–85	–	Conti et al. (2011)
30.	4[c]	4.9–11.1	1B, 2D, 3A, 5D	–	207 diverse European elite soft winter wheat lines	87	0.24 to 0.40	Reif et al. (2011)
31.	10	7.8–40.2	1AS, 1AL, 2AS, 2BL, 3BS, 4AL, 4BL, 5AL, 6BS	3 (2AS, 3BS, 4AL)	120 RILs from Svevo × Ciccio (both elite durum wheat cultivars)	47–71	0.01 to −0.68	Blanco et al. (2012)
32.	2	5.3–9.4	1A, 5BL	0	151 F^2 derived F^3 and F^4 lines obtained from Oste-Gata × Massara-1 (durum wheat genotypes)	–	0.29 to −0.35	Golabadi et al. (2012)
33.	9	3.1–9.8	2B, 3A, 4A, 4D, 5B, 7A, 7B,	1 (7AL)	485 RILs from cultivars Weimai 8 × Jimai 20 (both Chinese cultivars)	–	–	Li et al. (2012a)
33.	10	6.3–53.0	1A, 1B, 2A, 2D, 3A, 4B, 5A, 5D, 6B, 7D.	1 (2DS)	229 RILs developed from a cross of two Chinese cultivars Weimai 8 and Yannong 19	–	–	Li et al. (2012a)
34.	8	–	1A, 2D, 4B, 5D, 6A, 6B, 6D, 7B	0	82 introgression lines (BC5F2:F6 families) developed from Am3 × Laizhou953	41	0.13 to 0.25	Li et al. (2012b)
35.	1	16.3–21.9	5BL	1 (5BL)	118 RILs from BR34 (Brazilian soft red spring wheat breeding line) × Grandin (US hard red spring wheat variety)	72	0.28 to 0.35	Simons et al. (2012)
36.	7 +2[d]	4.1–10.90	1D, 2A, 2B, 3B, 4A, 5B, 5D, 6B, 7A	–	302 RILs from Weimai 8 × Luohan 2 (both Chinese cultivars)	–	0.16 to −0.19	Wang et al. (2012)
37.	5 + 2 [b]	1.1–9.3	4B, 4D, 5A, 6A	–	182 RILs from Xiaoyan 54 × Jing 411 (both Chinese wheat cultivars)	–	–	Xu et al. (2012)

(continued)

Table 1 (continued)

S. No.	No. of QTL	PVE (%)	Chr. (arm)	No. of stable QTL	Plant material	h^2 (%)	Correlation with yield	References
38.	14	6.0–25.0	1D, 2AS, 2DS, 3AL, 3BS, 4B, 5A, 5DL, 7BS, 7DS	7 (2AS, 2DS, 3AL, 3BS, 5DL, 7BS, 7DS)	Meta QTL analysis using three related DH populations {Toisondor × Quebon (230 lines), CF9107 × Quebon(316 lines) and Toisondor × CF9107 (140)}	–	−0.64 to 0.11	Bogard et al. (2013)
39.	11	4.3–12.8	1AS, 1BS, 1DL, 2BS, 3BL, 4AS, 5BL, 6AS, 6BL	7 {1BS (2), 1DL, 2BS, 4AS, 5BL, 6BL}	Association mapping panel of 196 accessions of a wheat core collection of old and modern cultivars	–	−0.09 to −0.80	Bordes et al. (2013)
40.	5	5.6–12.3	5BS, 6AL, 6BS, 7BS, 7DL	3 (6AL, 7BS, 7DL)	185 DH lines from CO940610 × Platte (both hard white winter wheats)	68–79	−0.38 to −0.66	El-Feki et al. (2013)
41.	2	–	3BL, 5BL	–	97 RILs from Choteau (spring wheat variety) × Yellowstone (spring version of winter wheat variety)	–	–	Heo and Sherman (2013)
42.	19	0.8–10.5	1B, 1D, 2A, 2B, 2D, 3A, 3B, 4B, 5A, 5B, 5D, 7A	–	Three populations, i) 168 DH lines from Huapei 3 × Yumai 57, ii)256 RILs from Nuomai 1× Gaocheng 8901, and iii) 182 RILs from Shannong 01–35 × Gaocheng 9411	–	–	Deng et al. (2015)
43.	4	–	2B, 2D, 3D, 5A	4	155 RILs from Drysdale × gladius (Australian cultivars)	48–90	–	Maphosa et al. (2015)
44.	13	–	1D, 2A, 2B, 2D, 4A, 4B, 5A, 5B, 5D, 6A, 6B, 6D, 7A	–	180 DH lines from CD87 × Katepwa	–	–	Moore et al. (2015)
45.	11	4.7–16.9	1A, 1B, 2B, 2D, 3D, 4B, 5B, 6B, 7B	3 (5B, 6B 7B)	163 RILs from WCB414 × WCB617 (US hard red spring wheat genotypes)	39	0.37 to −0.59	Echeverry-Solarte et al. (2015)

46.	4	11.5–22	3A, 4BS, 5AL, 5BL	0	127 RILs from Ning7840 (Chinese hard red wheat breeding line) × Clark (US soft winter wheat cultivar)	–	–	Li et al. (2016)
47.	12	7–17	1B, 2D, 3D, 4B, 5A, 5B, 5D, 6A, 7A	–	156 DH lines from RAC875 × Kukri (Australian varieties)	0–68	0.13 to −0.53	Mahjourimajid et al. (2016)
48.	1	18.3–37.9	2B	1 (2BS)	94 DH lines from Kitami 81 × Kachikei 63 (both Japanese elite winter wheat varieties)	–	−0.22 to 0.21	Terasawa et al. (2016)
49.	1	16.2–17.7	1A	1 (1A)	138 DH lines from Berkut × Krichauff	78	–	Tiwari et al. (2016)

PVE phenotypic variation explained by the QTL, *Chr* chromosome, h^2 heritability for GPC, – information not available
[a]Chromosomal location of some QTL was unknown
[b]Number of epistatic QTL
[c]Studies based on association mapping. The number of QTL in case of association mapping is an estimate based on approximate position of associated markers on a particular chromosome. Multiple associated markers located together on a chromosome were considered as one QTL
[d]Conditional QTL for GPC

2012; Suprayogi et al. 2009), 3AS (Groos et al. 2003; Sun et al. 2010), 4AL (Blanco et al. 2012; Prasad et al. 2003; Groos et al. 2003; Raman et al. 2009), 6BS (Joppa et al. 1997; Mesfin et al. 1999; Olmos et al. 2003), 7AS (Börner et al. 2002; Blanco et al. 2002; Suprayogi et al. 2009), and 7BL (Blanco et al. 2006; Suprayogi et al. 2009). However, not all of the major QTL were stable across the environments, meaning they were of limited use for marker-assisted improvement of GPC. Few major QTL with stable effect on GPC across environments have also been identified. Few such QTL and their associated markers are listed in Table 2.

Among all the QTL for GPC, the most important QTL identified so far is *Gpc-B1* (discussed in detail in Sect. 2.2). This QTL was first detected in accession FA-15-3 in a survey of wild populations of tetraploid wheat, *Triticum turgidum* var. *dicoccoides* (Avivi 1978). Later on this accession was used to create a complete set of chromosome substitution lines, each containing a different *dicoccoides* (DIC) chromosome in the background of modern durum wheat (Joppa and Cantrell 1990). Using these chromosome substitution lines, *Gpc-B1* gene was later mapped on chromosome arm 6BS (Joppa et al. 1997; Olmos et al. 2003). This gene explained 66% of the phenotypic variation for GPC. Several breeding programs have now successfully deployed this QTL in their breeding lines (for details, see Sect. 3).

In Japanese winter wheats, Terasawa et al. (2016) identified a major QTL for GPC and FPC (flour protein content) on 2BS over 3 years of testing. The PVE by this QTL ranged from 12.3 to 37.9% across different environments. The lines grouped based on closest flanking microsatellite marker *Xgpw4382* showed differences of about 1.0% in mean protein content. More importantly, the high GPC allele (contributed by Kachikei 63) at this locus was not associated with negative effect on yield or yield-related traits. In durum wheat, Supryayogi et al. (2009) identified two stable QTL, located one each on 2B (*QGpc.usw-B3*) and 7A (*QGpc.usw-A3*). The cultivar Strongfield contributed the alleles for high GPC at those two loci. These QTL were associated with GPC in all the environments in two populations and had no association with kernel weight, but were associated with a small reduction in yield in some environments. A high proportion of lines (58%) with yield and GPC equivalent or better than Strongfield suggests a weak association of these QTL with yield. In another study in durum wheat, Blanco et al. (2012) identified stable QTL for GPC on 2AS, 3BS, and 4AL. Although the QTL on 3BS was also associated with a stable QTL for yield, the other two QTL (2AS and 4AL) had no or a minor negative effect on yield. The desirable alleles for those loci were contributed by the cultivar Svevo.

2.2 *Gpc-B1*: A Major QTL for Grain Protein Content

The *Gpc-B1* remains one of the major QTL conferring stable expression of GPC in wheat. However, the introgression of *Gpc-B1* has shown different effect in different genetic backgrounds (Kumar et al. 2011). Some studies have shown that *Gpc-B1* is associated with consistent improvement in GPC averaging 14 g/kg of grain in tetraploid and hexaploid wheat across environments (Joppa et al. 1997; Mesfin et al.

Table 2 Few important major and stable QTL and their associated markers identified for grain protein content in wheat

Chr.	Associated marker(s)	QTL name	QTL effect (% GPC)	R^2 (%)	Positive alleles	Reference	Association with yield
2AS	*TC82001-Xgwm372c*	–	0.31	19.8	Svevo	Blanco et al. (2012)	No association with yield
2BS	*Xgpw4382*	*QGpc.2B–yume*	0.57	37.9	Kachikei 63	Terasawa et al. (2016)	No association with yield
2BL	*XwPt-0694-Xwmc041*	*QGpc.Usw-B3*	0.26	–	Strongfield	Supryayogi et al. (2009)	No negative effect on kernel weight, but associated with a small reduction in yield in some environments
2DL	*Xgwm1264*	*QGpc.Ccsu-2D*	–	19.6	PH 132	Prasad et al. (1999)	–
	Xcau14.2–ww160.1	*QGpcWY-2D.3*	0.60	12.0	Yannong 19	Li et al. (2012a)	–
3BS	*D_521287-Xgwm389*	–	0.65	40.2	Svevo	Blanco et al. (2012)	Associated with stable negative effect on yield
4AL	*D_376852-Xgwm601*	–	0.31	21.6	Svevo	Blanco et al. (2012)	Minor negative effect on yield in only one environment
4B	*Xgwm368–Xwmc617*	–	0.29	16.8	Clark	Sun et al. (2010)	Associated with test weight and kernel weight

(continued)

Table 2 (continued)

Chr.	Associated marker(s)	QTL name	QTL effect (% GPC)	R^2 (%)	Positive alleles	Reference	Association with yield
6BS	*Xabg387–Xmwg79*	*QGpc.Ndsu-6Bb*	–	66	FA-15-3	Joppa et al. (1997)	–
	Xcdo365–Xucw65	*Gpc-6B1*	1.5	–	FA-15-3	Chee et al. (2001) and Distelfeld et al. (2004)	Weak association with yield
	Xuhw89	*Gpc-6B1*	–	–	FA-15-3	Distelfeld et al. (2006)	–
7A	*Xgwm4-Xbarc108*	*QGpc.Usw-A3*	0.46	–	Strongfield	Supryayogi et al. (2009)	No negative effect on kernel weight, but associated with a small reduction in yield in some environments
	Xgwm473–Xedm16.1	*QGpc.WJ-7A.3*	0.31	9.8%	Weimai 8	Li et al. (2012a)	–

1999; Chee et al. 2001; Olmos et al. 2003). Few recent studies based on marker-assisted introgression of *Gpc-B1* have even reported a gain of up to 37 g/kg in selected lines, compared to the recurrent parent (Vishwakarma et al. 2014, 2016). *Gpc-B1* was cloned using map-based cloning approach (Uauy et al. 2006a). The ancestral wild wheat allele of this gene/QTL encodes a NAC transcription factor (NAM-B1) that accelerates senescence and increases nutrient remobilization from leaves to developing grains. Most of the modern wheat varieties carry a nonfunctional NAM-B1 allele. In addition to *Gpc-B1*, hexaploid wheat has two other functional homoeologous genes, on chromosome 6A (*GPC-A1*) and 6D (*GPC-D1*) and paralogous copies (*GPC-2*) on group 2 chromosomes (Uauy et al. 2006a). Silencing of all homoeologous *GPC-1* and *GPC-2* genes using RNA interference (RNAi) resulted in a 24–30-day delay in senescence and a more than 30% reduction in grain protein, zinc, and iron content compared to the wild-type control plants (Uauy et al. 2006a).

Recent studies have found a wider distribution of wild-type high GPC allele in both tetraploid and hexaploid wheat. A study based on a worldwide collection of 367 bread wheat lines found five accessions carrying the wild-type *Gpc-B1*(NAM-B1), most of which could be traced to a Fennoscandian origin (Hagenblad et al. 2012). Wild type of this gene was more frequent in spring wheat. Screening another collection of 138 spring wheats of northern origin, the same study found that as many as 33% of these accessions, all of Fennoscandian origin, had wild allele (Hagenblad et al. 2012). Similar results were observed in recent study involving Chinese wheat cultivars (Xue-yan et al. 2016). Xue-yan et al. (2016), while studying allelic distribution of *Gpc-B1* in Chinese wheat cultivars observed that this gene is present in 24.3% cultivars and absent in 75.7% cultivars. The *NAM-B1* gene in Chinese wheat cultivars contained a 1-bp insertion in the coding region, compared to the wild-type allele. This insertion caused a frameshift mutation, resulting in a stop codon in the middle of the gene, which rendered it nonfunctional (Xue-yan et al. 2016). The availability of wild-type allele of *Gpc-B1* in different backgrounds could be an excellent resource for introgressing high GPC allele into high yielding elite lines of different regions.

Different studies have reported different results on the effect of *Gpc-B1*. Earlier it was reported that *Gpc-B1* had an independent effect with loose association with yield (Chee et al. 2001). Few recent studies have also suggests that *Gpc-B1* is associated with smaller grain size and reduced yield (Uauy et al. 2006b; Tabbita et al. 2013). Using bread and durum wheat NILs for DIC *Gpc-B1* allele, Brevis and Dubcovsky (2010) observed higher GPC, when the functional *Gpc-B1* allele was present ($P < 0.0001$). However, the DIC *Gpc-B1* allele was associated with decrease in grain weight, although the difference was significant only in hexaploid wheat (2.8%, $P = 0.0004$). The different *Gpc-B1* alleles also did not have any significant effect on grain yield. The authors suggested that the potential negative effect of the *Gpc-B1* DIC allele on yield might be limited by breeding (Brevis and Dubcovsky 2010). In hexaploid NILs with *Gpc*-B1, a significantly lower straw nitrogen (N) concentration at maturity and higher N harvest index suggested that the functional *Gpc-B1* allele probably has more efficient remobilization of N already present in the plant (Brevis and Dubcovsky 2010). It is worth mentioning that this study was limited only to hard spring wheat cultivars adapted to California. Another study involving Swedish varieties observed that the lines with the functional (wild-type) *Gpc-B1* allele reached anthesis earlier and completed senescence faster than the varieties with the nonfunctional allele (Asplund et al. 2013). The study also observed that the functional allele was associated with more ears, lighter grains, and higher yields of P and K. However, contrary to Brevis and Dubcovsky (2010), the Swedish varieties with functional alleles did not have increased grain N concentration or grain N yield.

A recent study using near NILs with and without the *Gpc-B1* allele showed that inclusion of the *Gpc-B1* allele in hard red winter wheat increases GPC (Kuhn et al. 2016). However, this study observed both negative and positive correlation with grain yield and GPC, depending upon the genetic background. Overall, the data showed that lines with *Gpc-B1* allele, showing increased GPC, acceptable grain yield and end-use quality profiles, can be selected (Kuhn et al. 2016). In Argentinean wheat cultivars, Tabbita et al. (2013) observed that the introgression of *Gpc-B1* significantly increases GPC without any significant effect on yield. The lines with *Gpc-B1* showed significant reduction in grain weight and straw nitrogen concentration at maturity and significant increase in nitrogen harvest index. This study suggests that the introgression of high GPC allele of *Gpc-B1* could be a valuable resource for improving GPC in Argentinean wheats without any negative effect on grain yield.

The markers associated with *Gpc-B1* are being currently used in MAS for improvement of GPC in wheat throughout the world (for details, see Sect. 3). Most of the modern tetraploid and hexaploid wheat varieties carry the inactive or nonfunctional allele of *Gpc-B1* gene (Uauy et al. 2006a), meaning that there is a possibility to further improve GPC in both durum and bread wheat, through the introgression of functional copy of *Gpc-B1*. The first step in this direction will be to characterize the germplasm of specific breeding programs for *Gpc-B1* alleles.

2.3 Effect of Environment on Grain Protein Content in Wheat

Knowledge about the relative contributions of genotype (G), environment (E), and genotype × environment interaction (G × E) effects on GPC or any other quantitative trait is important for an effective selection in breeding programs. Numerous studies have now shown that GPC has moderate heritability (Table 1), but at the same time, environmental factors play an important role in phenotypic expression of this trait (for review, see Balyan et al. 2013; Kong et al. 2013). In wheat, the effect of environment on protein content has been interpreted using various statistical procedures. These procedures include (1) analysis of variance (Mariani et al. 1995; Novaro et al. 1997), (2) by comparing the frequency and expression of QTL in different environments (Blanco et al. 2002), and (3) by using specific software to identify QTL × environmental (QE) or QTL× QTL× environmental (QQE) interactions (Kulwal et al. 2005; Zhao et al. 2010; Mahjourimajd et al. 2016).

Some studies found that environment is the most influential variable (Mariani et al. 1995; Novaro et al. 1997; Guttieri et al. 2000), while others found that genotype and G × E are more important for GPC in wheat (Peterson et al. 1992). Panozzo and Eagles (2000) found that environment had the most effect on variability for flour protein, followed by G × E and genotype. In contrast to Panozzo and Eagles (2000), other studies reported genotype had more effect than either environment and/or GE on GPC (Fenn et al. 1994; Hazen et al. 1997; Zhang et al. 2004). Significant location effects for grain and flour protein content under both irrigated and moisture-limited environments were also observed by Souza et al. (2004). The availability of N and micronutrients in the soil also affects GPC and composition in wheat (Hemantaranjan and Garg 1988; Peck et al. 2008; Liu et al. 2015). Increasing the concentrations of N (Mahjourimajd et al. 2016) or Fe and Zn in soils can improve protein content of wheat grains (Hemantaranjan and Garg 1988). Studies have also found that most of Zn applied to wheat was found in the flour protein (Starks and Johnson 1985). Application of Zn influence nitrate reductase (NR) and glutamine synthetase (GS) activity, thus affecting synthesis of GPC (Liu et al. 2015).

In recent times, genomic tools were employed to study the effect of environmental interactions on GPC. Almost all of the QTL mapping studies for GPC observed that only few QTL express across environments (called stable QTL, Table 1). Majority of the QTL for GPC are specific to certain environments. Moreover, stable QTL also showed different levels of expression in different environments suggesting strong genotype × environment interactions (Blanco et al. 2002; Groos et al. 2003; Kulwal et al. 2005). For example, Groos et al. (2003) identified certain regions for GPC which interacted with temperature during grain filling.

In recent years, certain software like QTLMapper (Wang et al. 1999), QTLNetwork (Yang et al. 2008), and IciMapping (Meng et al. 2015) have the necessary statistical methods needed to identify QTL involved in environmental interactions. Kulwal et al. (2005) observed that QE or QQE interactions play a very important role in the control of GPC based on study in two recombinant inbred line (RIL) mapping populations. The QE or QQE interactions involved main effect QTL (M-QTL), epistatic QTL (E-QTL), as well as QTL which have no main effect or epistatic effect, but only significant QE or QQE effect. Overall, the QE and QQE interactions contributed 25.91% and 47.99% of the phenotypic variation for GPC in two populations. The total phenotypic variation explained by all components (main effects, epistasis, and environmental interactions) in the two populations was 35.8% and 61.3%, respectively. This means the environmental interactions explained much higher PV compared to the PV explained by the main effect or the epistatic effects (Kulwal et al. 2005). Similarly, Zhao et al. (2010) identified three QE interactions for GPC in a doubled haploid population derived from two Chinese wheat cultivars Huapei 3 (Hp3) and Yumai 57 (Ym57). However, the total contribution to the PV by these QE interactions was only 9.3%, compared to the 18.3% PV explained by main effect QTL. In another study, Mahjourimajd et al. (2016) studied protein-related traits under different N fertilizations to identify genomic regions associated with protein-related traits and their response to N. This study identified QTL specific to N fertilization. They observed that all of the protein-related QTL at high N were associated with N fertilization and QTL × N interaction.

2.4 Epistatic Interactions for Grain Protein Content in Wheat

Most of the traits of economic importance in wheat, including GPC, are quantitatively inherited and are controlled by complex gene networks consisting not only additive gene actions but epistatic as well as gene × environment interactions. Traditional studies on mode of inheritance suggested that both additive and dominance effects contribute toward genetic variation for GPC (Bains et al. 1972; Bnejdi and El Gazzah 2010; El-Bok et al. 2013; Yao et al. 2014). Some studies suggested a major role of dominance component (Bains et al. 1972), while others suggested a predominant role of additive gene actions (Yao et al. 2014). In recent times, improved statistical and genomic tools have allowed to genetically dissect the role of epistasis and to identify the QTL involved in epistatic interactions for GPC (Kulwal et al. 2005; Patil et al. 2009; Zhao et al. 2010; Conti et al. 2011; Xu et al. 2012). Three different types of epistatic interaction were reported. First type involves the interactions between two main effect QTL (M-QTL), meaning that the individual QTL have significant effect of their own. Second type involves the interaction between an M-QTL and an epistatic QTL (E-QTL). Epistatic QTL do not have significant effect of their own, but when they interact with another QTL, their interaction effect is significant. The third type involves the interaction between two E-QTL.

Kulwal et al. (2005) identified two epistatic interactions in one RIL population and three in another RIL population in bread wheat. In another study using a population developed from two Chinese wheat cultivars Huapei 3 (Hp3) and Yumai 57 (Ym57), Zhao et al. (2010) identified two digenic epistatic interaction for GPC, both involving only E-QTL. The same study also identified five pairs of epistatic interaction between E-QTL, for flour protein content. On the other hand, in another study using a population developed from the Chinese cultivars Xiaoyan 54 and Jing 411, Xu et al. (2012) identified two digenic interaction involving four M-QTL for GPC. In durum wheat cross of "PDW 233" and "Bhalegaon 4," only one pair of epistatic interaction was identified (Patil et al. 2009). The number of M-QTL identified in this study was also one. Another study in durum wheat, using the cross of UC1113 and Kofa, identified five pairs of epistatic interactions for GPC, in addition to ten M-QTL (Conti et al. 2011).

Different studies have reported different levels of contribution by epistatic interaction toward phenotypic variation for GPC. The epistatic interaction in two populations used by Kulwal et al. (2005) contributed a total of 2.68% and 6.04% toward the PV for GPC, compared to 7.24 and 7.22% contributed by M-QTL, in the two populations, respectively. The two epistatic interactions identified by Zhao et al. (2010) explained 24% of the PV for GPC, compared to 18.25% explained by main effects. In the same study, the combined PV for all five-digenic interaction for flour protein content was 10.12%, compared to 31% explained by additive effects (Zhao et al. 2010). In the study by Xu et al. (2012), epistatic interactions contributed 22.3% of PV for GPC, compared to 27.5% PV explained by M-QTL. Individual epistatic interactions contributed up to 0.95–10.8% PV for GPC, in this study. Overall, the studies dissecting the genetics of GPC clearly suggest that epistatic interaction play an important role in the genetic control of protein content and thus should be considered when GPC improvement is targeted in any breeding program.

The studies to identify QTL involved in epistatic interactions were conducted using softwares like QTLMapper (Wang et al. 1999), QTLNetwork (Yang et al. 2008), and IciMapping (Meng et al. 2015). All those software detect only first-order epistatic interactions. However, some studies have also shown a network, which may represent higher-order interactions (Kumar et al. 2007). The metabolic pathways that presumably underlie quantitative traits involve multiple interacting gene products and regulatory loci, which could generate higher-order epistatic interactions (McMullen et al. 1998). This suggested that higher-order interactions might also contribute to a part of the substantial variation that remained unexplained in most of the studies of complex traits like GPC. Due to the complexity of higher-order interactions, most of the software dealing with epistasis includes only first-order or two-locus interaction. It may be noted that higher-order interactions require too many parameters in the genetic model, which would be difficult to estimate properly except in extremely large populations (Xu and Crouch 2008). Future studies using populations of large sample size and suitable statistical methods may be able to provide a better picture of gene networks for complex traits like GPC.

2.5 Relationship Between GPC and Grain Yield

Grain protein content and grain yield are two most important traits which determine the economic value of wheat. However, most of the studies suggest a negative relationship between these two traits, not only in wheat but also in most of the cereals (Table 1; Mesfin et al. 2000; Oury et al. 2003; Oury and Godin 2007; Blanco et al. 2012; Wang et al. 2012; Bogard et al. 2013). This relationship hampers simultaneous improvement of both GPC and grain yield, and because of it, most of the earlier high-yielding cultivars generally had lower GPC values (Simmonds 1995). Different studies have shown a wide range of values for correlation between GPC and yield-related traits (Table 1). In one study, multiple year and location data showed a strong negative correlation with yield when environmental effects were reduced in comparison to genotype effects (Oury et al. 2003). Another study based on two RIL populations showed smaller (−0.17, −0.22) but negative phenotypic correlation between these two traits (Mesfin et al. 2000). Positive correlation values between GPC and yield-related traits have also been reported in wheat (Table 1; Sun et al. 2010; Tsilo et al. 2010; Reif et al. 2011; Golabadi et al. 2012; Li et al. 2012a, b; Simons et al. 2012; Echeverry-Solarte et al. 2015). Using several segregating populations, Kibite and Evans (1984) observed that GPC and yield had a highly significant phenotypic correlation, but mostly nonsignificant genotypic correlations. It means that the negative relationship between grain yield and GPC was not caused by genetic factors. It is probably caused mainly by environmental factors, source sink relationship, and dilution of protein by nonprotein compounds (Kibite and Evans 1984). Identification of some genotypes showing simultaneous increase in both GPC and yield in few studies further confirms that the negative relationship between GPC and grain yield is not entirely genetic (Sears 1998; Oury et al. 2003; Clarke et al. 2005). Mesfin et al. (2000) while studying two RIL populations observed that the highest-yielding line in both populations was a low GPC line, but among the ten highest-yielding lines, there were five to six high GPC lines in both the populations. Therefore, this study suggested that it might be possible to combine high GPC with acceptable level of yield in wheat. It was also suggested in the past that if a sufficient number of sources of genes for increasing the GPC are used in a breeding program, it is possible to increase both GPC and yield (Sears 1998).

Using molecular markers, few studies also attempted to understand the relationship between GPC and yield (or yield contributing traits) at genetic level and to identify GPC loci which have no association with loci controlling yield (Groos et al. 2003; Suprayogi et al. 2009; Tsilo et al. 2010; Blanco et al. 2012; Wang et al. 2012; Bogard et al. 2013). In hexaploid wheat, Groos et al. (2003) identified four stable loci for GPC, one each on chromosomes 2A, 3A, 4D, and 7D and each explaining about 10% of the phenotypic variation. Among them, the loci on chromosome 3A and 7D shared some common genomic region with a QTL for thousand-kernel weight and grain yield, respectively. The other two QTL located on 2A and 4D were associated only with GPC. Similarly, out of the two QTL identified for GPC by Huang et al. (2006), in a hexaploid wheat population developed from Canadian wheat variety "AC Karma" and a breeding line 87E03-S2B1, only one QTL identified on 4D was co-localized with a QTL for thousand grain weight. The other QTL identified on chromosome 7B was independently associated with GPC only (Huang et al. 2006); none of these two QTL were associated with grain yield in this population.

In a RIL population developed from two hard red spring wheat breeding lines (MN98550 and MN99394) adapted to the Upper Midwest region of the United States, Tsilo et al. (2010) identified three stable QTL for GPC, which were located on 2B, 5A, and 6D. Although, as expected, a negative phenotypic correlation between GPC and grain yield was observed, QTL analysis showed that only the 2B genomic region was negatively associated with grain yield. The other two QTL for GPC located on 5A and 6D were independent of grain yield, grain weight, and grain size (Tsilo et al. 2010).

Wang et al. (2012) used conditional QTL mapping to dissect the genetic interrelationship between GPC and yield traits in hexaploid wheat. In this study seven additive QTL for GPC were detected using unconditional QTL mapping. However, three of the seven QTL did not show significant effects when GPC was conditioned on grain number/spike. Similar results were observed when GPC was conditioned on thousand grain weight, grain weight per spike, spike number per row, and grain yield per plot, indicating a strong genetic association between GPC and yield traits. However, conditional QTL mapping analysis identified two additional QTL for GPC, located one each on chromosomes 2B and 4A, which were independent from yield. Most importantly, this study found one "stable" QTL on 7A which exhibited similar effects when GPC was conditioned on four yield-related traits, indicating that this QTL is not affected by yield. This QTL accounted for about 0.27% of the additive effect of GPC (Wang et al. 2012). Such QTL could be important for improving GPC through MAS, without effecting yield.

In a meta-analysis study based on three interconnected populations of bread wheat, QTL for GPC and GY with negative association were identified on chromosomes 2A, 2D, 3B, 7B, and 7D (Bogard et al. 2013). However, the same study identified two QTL for GPC, one each on chromosome 3A and 5D, which had no association with GY. The QTL on 3A was identified in a 10 cM confidence interval with SSR marker *Xgwm666* being closest to the QTL peak. Comparison of map location of this QTL with earlier studies suggests that this region have been found to be associated with GPC and dough strength (Groos et al. 2003; Groos et al. 2004; Reif et al. 2011), low molecular weight glutenin, total gliadin content (Charmet et al. 2005), bread-making quality (Groos et al. 2007), and flour protein content (Zhao et al. 2010). The other independent QTL for GPC was located on long arm of 5D, about 1 cM away from DArT marker *wPt-1999* and about 6 cM away from SSR marker *gpw5174*. This region has also been reported to carry a QTL for glutamine synthetase (GS) activity under high nitrogen conditions in another mapping population (Fontaine et al. 2009). In addition to those two independent QTL for GPC, the study by Bogard et al. (2013) also identified another stable QTL for GPC on 2A, which was associated with reduced yield in only some environments. Although the QTL was associated with GPC in seven environments, the negative association between GPC and GY was observed in only three environments, suggesting that probably it contributes moderately to the negative GPC-GY relationship and might be useful for improving GPC (Bogard et al. 2013). In another study in bread wheat conducted under different nitrogen conditions, Bordes et al. (2013) identified 11 regions including the genes *SPA-D*, *Glu-D1*, *FdGOGAT-B*, and *DeK1-A* associated with GPC. Seven of these regions were associated with GY.

In a recent study, Terasawa et al. (2016) used a double haploid population derived from a cross between hard red winter wheat variety "Yumechikara" (known for a high protein content) and soft red winter wheat "Kitahonami" (low protein content) to detect QTL associated with GPC. A novel single major QTL (*QGpc.2B–yume*) mapped between markers *Xgwm37* and *Xwmc179* on chromosome 2BS was identified. This QTL did not show any negative effect on yield-related traits. High-protein haplotype of this QTL was observed in major North American hard red winter wheat varieties. Independence from yield loss makes this an excellent target of breeding programs to increase GPC. The two markers can assist the incorporation of *QGpc.2B–yume* into adapted productive cultivars in many countries to enhance the levels of the GPC without compromising overall yield. In another recent study, a team in Australia identified several QTL that were associated with GPC and yield in both negative and positive direction. Genomic regions on chromosomes 1B, 2D, 3A, 5A, 5B, and 5D were potential candidates for the improvement of GPC without compromising the overall yield of cultivars carrying those genes. Some loci on chromosomes 1A, 2A, 3B, 3D, 4A, 4B, 5A, 6A, 7A, 7B, and 7D with possible negative correlation between grain yield and GPC were also identified (Mahjourimajd et al. 2016).

In a durum wheat, a double haploid (DH) population developed from DT695 × Strongfield identified nine QTL for GPC in six environments (Suprayogi et al. 2009). Four of these genomic regions were also associated with grain yield, while three of these QTL were associated with thousand-kernel weight in some of the environments. This study identified three stable QTL (identified in at least half of the environments). One QTL located on 2A (*QGpc.usw-A2*) was associated with GPC in three environments, another QTL (*QGpc.usw-B3*) was detected on 2B in four environments, and the third stable QTL associated with all the six environments was located on 7A (*QGpc.usw-A3*). None of these stable QTL were associated with thousand-kernel weight in any environment. However, Strongfield alleles at *QGpc.usw-B3* and *QGpc.usw-A3* were associated with reduced grain yield, but each only in a single environment (Suprayogi et al. 2009). The most stable QTL *QGpc.usw-A3* was also validated in another population, where it was associated with GPC in all the five environments studied, but was associated with yield in only two of the environments. Despite this negative association with yield in few environments, a high proportion of lines (58%) have high GPC equivalent to Strongfield coupled with yield equal to or greater than Strongfield in those environments suggesting that this QTL is only weakly associated with grain yield. The *QGpc.usw-A3* could be the same QTL which was earlier reported in hexaploid wheat, as both the QTL are linked to SSR marker *Xbarc108* (Turner et al. 2004). The results showed that the selection of Strongfield alleles using *Xbarc108* resulted in an increase of 0.4–1.0% GPC (Suprayogi et al. 2009). The *QGpc.usw-A3* and the SSR marker *Xbarc108* could serve as an excellent genomic resource for selecting/introgressing high GPC in durum wheat, at least in crosses involving Strongfield. Another study in durum wheat, conducted using adjusted values of GPC in relation to several grain yield components, identified four QTL for high GPC which have no effect on grain yield (Blanco et al. 2012). Six other QTL for increased GPC identified in the same study were, however, found to be associated with decrease in values of yield traits. The four independent QTL for GPC always showed significant effects and were located on 1AL, 2AS, 2BL, and 4AL. The comparison of the map locations of these independent QTL found in "Svevo" × "Ciccio" RIL population with QTL found in other genetic materials, however, shows that some of those QTL were associated with reduced grain yield in some environments, thus indicating the influence of genotypes and environmental factors on the effect of the GPC QTL. For example, the independent GPC QTL detected on the proximal region of chromosome arm 2AS in Svevo × Ciccio RIL population (Blanco et al. 2012) was in the same location where QTL were earlier reported in durum backcross line 3BIL-85 (Blanco et al. 2006) and in a Canadian durum line DT695 (Suprayogi et al. 2009). However, the QTL allele for higher GPC from 3BIL-85 (Blanco et al. 2006) and DT695 (Suprayogi et al. 2009) was associated with reduced grain yield. Similarly, the independent GPC QTL detected on chromosome arm 2BL in Svevo × Ciccio RIL population (Blanco et al. 2012) was also found in DT695 × Strongfield population (Suprayogi et al. 2009), but it was also associated with reduced grain yield. The independent QTL for GPC detected on chromosome arm 4AL by Blanco et al. (2012) was in a similar position to a QTL found in three earlier studies (Groos et al. 2003; Prasad et al. 2003; Raman et al. (2009). However, interestingly, these QTL were not associated with QTL for reduced grain yield component in any of the four mapping populations. No QTL for GPC was reported earlier on 1AL, where Blanco et al. (2006) reported an independent QTL for GPC.

Overall, the results of molecular genetic dissection are also in agreement with the results of correlation analysis, which suggested a negative association between GPC and yield traits. At present, it is, however, not possible to distinguish whether the association is due to true pleiotropic effect or it is due to tight linkage, because of the low resolution of the maps used in QTL analysis studies. However, both traditional and molecular marker-based studies suggest that the negative relationship between grain yield and protein content is not entirely genetic. Molecular genetic dissection of the relationship between GPC and yield traits in multiple populations has identified several genomic regions that are only associated with GPC, without any adverse effect on yield. In view of this, one of the strategies to improve both GPC and GY simultaneously should be to identify independent QTL/genes for these two traits that have no negative association with each other. Such genes should be incorporated in breeding programs, through MAS, to improve GPC without decreasing GY. The primary targets for this purpose should be the stable QTL identified in multiple environments and/or multiple populations.

3 Marker-Assisted Selection for Grain Protein Content

Marker-assisted selection continues to gain acceptance as a breeding tool in crop improvement. In the last few years, many genes/quantitative trait loci (QTL) have been deployed in different plant breeding programs with the help of molecular markers (Gupta et al. 2010a, b; Kumar et al. 2010, 2011; Miedaner and Korzun 2012; Randhawa et al. 2013; Vishwakarma et al. 2014, 2016; Mishra et al. 2015). Marker-assisted selection strategies are being used for introgression of genes, marker-assisted backcrossing, marker-assisted recurrent selection, gene pyramiding, and early generation selection for genes/QTL. Higher genetic gains could be achieved through marker-assisted selection for traits like GPC which are controlled by a complex gene network and are significantly affected by environment. Marker-assisted selection is not affected by environmental factors and thus could be more effective in selecting genotypes with high GPC, in early generations. More importantly, it can precisely introgress yield-independent GPC QTL, thus avoiding any linkage drag associated with yield loss. For GPC, a study carried out to compare the effectiveness of using MAS and phenotypic selection observed that the overall selection efficiency was comparable for both methods (Davies et al. 2006). The same study observed that because GPC is controlled by many genes with smaller effects, the phenotypic selection was found to be little more effective in few environments (Davies et al. 2006). This study suggested that MAS was a more sustainable and effective approach for the improvement of GPC if the target is the introgression of major QTL/genes. However, in recent years, the availability of dense genetic maps and tools like genomic selection has made it possible to introgress significant loci for trait of interest from the whole genome (Battenfield et al. 2016). Genomic selection is being applied in wheat, and studies indicate that the use of genomic selection tool is powerful for selection of GPC in wheat (Battenfield et al. 2016).

As of today, the successful application of genomic tools for improvement in GPC has been limited to the introgression of major gene *Gpc*-B1. The detailed procedure for the introgression of *GPC-B1* in durum and bread wheat cultivars using MAS can be found on MASWheat web (http://maswheat.ucdavis.edu/protocols/HGPC/index.htm) or elsewhere (Kumar et al. 2011; Tyagi et al. 2014; Vishwakarma et al. 2014, 2016). These studies involving the introgression of *Gpc-B1* showed that it has different effects in different recipient genotypes and in different environments (Kumar et al. 2011). In an attempt to improve GPC in Indian wheat genotypes, Kumar et al. (2011) introgressed *Gpc-B1* from cultivar "Yecora Rojo" into ten wheat genotypes using MAS. After multilocation evaluation of 124 BC_3F_5/F_6 progenies with *Gpc-B1*, seven progenies with significantly higher GPC (14.83–17.85%) than their recipient parental genotypes and having no significant yield loss were obtained. In another study, same group used MAS to pyramid *Gpc-B1* and seven other QTL/genes, in the background of a popular wheat cultivar "PBW343" (Tyagi et al. 2014). Four lines with different desirable QTL/genes in the background of PBW343 were used to pyramid all eight QTL/genes. The high GPC lines developed by Kumar et al. (2011) are being used by a number of Indian institutes as the donor genotype for introgression of the high GPC into other high-yielding Indian bread wheat varieties (Vishwakarma et al. 2014, 2016; Mishra et al. 2015).

Two successful examples of improvement of GPC through MAS were provided by Vishwakarma et al. (2014, 2016). In cultivar "HUW468", the functional allele of *Gpc-B1* was introgressed from Glu269, while in cultivar "HUW234", high GPC allele was introgressed from Glu133. In both the studies, the SSR marker *Xucw108* was used for foreground selection, while 86 SSRs from across the whole genome were used for background selection. The selected lines with functional *Gpc-B1* allele had 26–34% higher GPC and no yield loss when compared to the recurrent cultivars (Vishwakarma et al. 2014, 2016; Mishra et al. 2015). The BC_2F_3 lines selected in the background of HUW468 had 13–17.2% GPC compared to 10% in HUW468 and 14.3% in the donor parent. The selected progenies also showed significant increase in Fe and Zn (Vishwakarma et al. 2014). Similarly, selected BC_2F_4 families in the background of HUW234 had 13–14.8% GPC, compared to 11.1% for HUW234. Iron and Zn contents of the selected families ranged from 40.3 to 49.3 ppm and 41.4 to 55.7 ppm, compared to 38 ppm and 28 ppm for HUW234, respectively. Majority of the selected lines also out-yielded the recurrent parent by up to 8.9% (Vishwakarma et al. 2016).

As mentioned above, MAS has been successfully used to introgress a major gene for GPC into many elite varieties. Most of the selected lines with improved GPC are being evaluated in extensive multilocation trials, and hopefully some of them will be released as varieties in near future. However, one successful example of a cultivar developed using MAS for GPC is the durum wheat variety "Desert King-High Protein" (DK-HP). Desert King-High Protein was developed by introgressing

Gpc-B1 into high-yielding variety "Desert King," at the University of California wheat breeding program (http://www.triticeaecap.org/t-cap-variety-and-germ-plasm-releases_2011_2013/). The newly developed variety DK-HP with functional *Gpc-B1* has significantly higher GPC (8–10% increase), compared to the original variety Desert King. The introgression of *Gpc-B1* also accelerates the maturity 5–8 days in DK-HP. However, as a result of shorter grain filling duration, DK-HP has a 3–4% average reduction in yield compared to the original Desert King. However, this decrease is compensated by an increase in GPC. Another variety which used MAS for *Gpc-B1* is "Egan," a hard red spring wheat released by Montana State University (http://www.triticeaecap.org/tcap-variety-and-germ-plasm-releases-2014-2016/). In this variety, MAS was used to pyramid gene for resistance to the orange wheat blossom midge, high grain protein content, stripe rust resistance, and gluten strength.

Another success story of application of MAS for GPC improvement is available in Canadian wheat breeding programs (Randhawa et al. 2013). Markers linked to *Gpc-B1* are being routinely used to introgress/select this gene in both hexaploid and durum wheats. Marker-assisted selection was used to introgress/select high GPC allele of *Gpc-B1* in several of the released wheat cultivars (DePauw et al. 2005; Humphreys et al. 2010; Randhawa et al. 2013). Moreover, many wheat breeding lines developed through MAS for high *Gpc-B1* allele are currently in pre-variety registration trials (Randhawa et al. 2013).

It is important to mention that the whole process of introgression of high GPC allele for *Gpc-B1* can be completed in 2–3 years (five or six crop cycles) (Kumar et al. 2011; Vishwakarma et al. 2014, 2016). The examples mentioned above clearly show that, through a combination of MAS and phenotypic selection, improvement in GPC can be successfully accomplished in a much shorter duration compared to when relying on only conventional breeding approaches. In a breeding program, markers could be used in early generations to reject lines with low GPC. However, in the later generations, a combination of MAS and phenotypic analysis may help improve the effectiveness of selection for GPC.

4 Conclusions

Grain protein content is a complex quantitative trait with intermediate heritability and strong genotype by environment interaction. Genetic dissection studies have shown that GPC in wheat is controlled by multiple QTL/genes. To date, more than 325 main effect QTL and 42 epistatic QTL located on all 21 wheat chromosomes have been reported for GPC. Studies have also shown that environmental (QTL × E or QTL × QTL × E) and epistatic interactions (QTL × QTL) explain significant amount of phenotypic variation for GPC and thus should be studied in more details using larger mapping populations. Most of the identified QTL for GPC have smaller effects and are expressed under specific environments. However, few major and stable QTL for GPC have also been reported. The most important QTL for GPC (named *Gpc-B1*) is located on 6BS. This QTL has been successfully deployed in several breeding programs in different parts of the world and has shown significant improvement (up to 37 g/kg) in GPC, in both tetraploid and hexaploid wheat. This QTL/gene encodes a NAC transcription factor (NAM-B1) that accelerates senescence and increases nutrient remobilization from leaves to developing grains. Most of the modern wheat varieties carry a nonfunctional NAM-B1 allele, thus providing an opportunity to introgress wild-type allele of *Gpc-B1* to elite wheat lines for further improvement in GPC. Of course, the first step in this direction will be to characterize the germplasm of specific breeding programs for *Gpc-B1* alleles. In addition to *Gpc-B1*, several other stable QTL and associated markers with significant effect on GPC have been reported and could serve as excellent resource for improvement of GPC through MAS.

Several studies attempting to understand the relationship between GPC and yield traits have observed mostly negative correlation between GPC and yield traits. However, many studies have also reported weak negative association or even positive association between GPC and yield traits. Similarly, studies based on molecular genetic dissection have identified many co-localized QTL for GPC and yield with opposite effects. Importantly, those studies have also identified several independent QTL for GPC with no adverse effect on yield-related traits. In view of this, one of the strategies to improve both GPC and grain yield simultaneously should be to identify independent QTL for these two traits and incorporate them in breeding programs, through MAS, to improve GPC without significantly affecting grain yield.

References

Asplund L, Bergkvist G, Leino MW, Westerbergh A, Weih M (2013) Swedish spring wheat varieties with the rare high grain protein allele of *NAM-B1* differ in leaf senescence and grain mineral content. PLoS One 8(3):e59704. https://doi.org/10.1371/journal.pone.0059704

Avivi L (1978) High protein content in wild tetraploid *Triticum dicoccoides* Korn. In: Ramanujam S (ed) Proceedings of the 5th international wheat genetics symposium. Indian Society of Genetics and Plant Breeding (ISGPB), New Delhi, India, pp 372–380

Bains KS, Gill KS, Sehgal KL (1972) Epistatic bias in additive and dominance variation and genetic associations between two quality traits in bread wheat. Can J Genet Cytol 14:49–55

Balyan HS, Gupta PK, Kumar S, Dhariwal R et al (2013) Genetic improvement of grain protein content and other health-related constituents of wheat grain. Plant Breed. https://doi.org/10.1111/pbr.12047

Battenfield SD, Guzmán C, Gaynor RC, Singh RP et al (2016) Genomic selection for processing and end-use quality traits in the CIMMYT spring bread wheat breeding program. Plant Genome 9(2)

Blanco A, De Giovanni C, Laddomada B, Sciancalepore A et al (1996) Quantitative trait loci influencing grain protein content in tetraploid wheats. Plant Breed 115:310–316

Blanco A, Pasqualone A, Troccoli A, Di Fonzo N, Simeone R (2002) Detection of grain protein content QTLs across environments in tetraploid wheats. Plant Mol Biol 48:615–623

Blanco A, Simeone R, Gadaleta A (2006) Detection of QTLs for grain protein content in durum wheat. Theor Appl Genet 112:1195–1204

Blanco A, Mangini G, Giancaspro A, Giove S et al (2012) Relationship between grain protein content and grain yield components through quantitative trait locus analyses in a recombinant inbred line population derived from two elite durum wheat cultivars. Mol Breed 30:79–92

Bnejdi F, El Gazzah M (2010) Epistasis and genotype-by-environment interaction of grain protein content in durum wheat. Genet Mol Biol 33:125–130

Bogard M, Allard V, Martre P, Heumez E et al (2013) Identifying wheat genomic regions for improving grain protein concentration independently of grain yield using multiple inter-related populations. Mol Breed 31:1380–3743

Bordes J, Ravel C, Le Gouis J, Charmet G, Balfourier F (2011) Use of global wheat core collection for association analysis of flour and dough quality traits. J Cereal Sci 54:137–147

Bordes J, Ravel C, Jaubertie JP, Duperrier B et al (2013) Genomic regions associated with the nitrogen limitation response revealed in a global wheat core collection. Theor Appl Genet 126:805–822

Börner A, Schumann E, Fürste A, Cöster H, Leithold B, Röder MS, Weber WE (2002) Mapping of quantitative trait loci for agronomic important characters in hexaploid wheat (*Triticum aestivum* L.) Theor Appl Genet 105:921–936

Brevis JC, Dubcovsky J (2010) Effects of the chromosome region including the grain protein content locus *Gpc-B1* on wheat grain and protein yield. Crop Sci 50:93–104

Charmet G, Robert N, Branlard G, Linossier L, Martre P, Triboi E (2005) Genetic analysis of dry matter and nitrogen accumulation and protein composition in wheat kernels. Theor Appl Genet 111:540–550

Chee PW, Elias EM, Anderson JA, Kianian SF (2001) Evaluation of a high grain protein QTL from *Triticum turgidum* L. var. *dicoccoides* in an adapted durum wheat background. Crop Sci 41:295–301

Clarke JM, McCaig TN, DePauw RM, Knox RE et al (2005) Strongfield durum wheat. Can J Plant Sci 85:651–654

Conti V, Roncallo PF, Beaufort V, Cervigni GL et al (2011) Mapping of main and epistatic effect QTLs associated to grain protein and gluten strength using a RIL population of durum wheat. J Appl Genet 52:287–298

Cox MC, Qualset CO, Rains DW (1985) Genetic variation for nitrogen assimilation and translocation in wheat. III. Nitrogen translocation in relation to grain yield and protein. Crop Sci 26:737–740

Davies J, Berzonsky WA, Leach GD (2006) A comparison of marker-assisted and phenotypic selection for high grain protein content in spring wheat. Euphytica 152:117–134

Deng Z, Hu S, Chen F, Li W, Chen J, Sun C, Zhang Y, Wang S, Song X, Tian J (2015) Genetic dissection of interaction between wheat protein and starch using three mapping populations. Mol Breed 35:1–9. https://doi.org/10.1007/s11032-015-0216-6

DePauw RM, Townley-Smith TF, Humphreys G, Knox RE, Clarke FR, Clarke JM (2005) Lillian hard red spring wheat. Can J Plant Sci 85:397–401

Dholakia BB, Ammiraju JSS, Sandra DK, Singh H et al (2001) Molecular marker analysis of protein content using PCR-based marker in wheat. Biochem Genet 39:325–338

Distelfeld A, Uauy C, Olmos S, Schlatter AR, Dubcovsky J, Fahima T (2004) Microcolinearity between a 2-cM region encompassing the grain protein content locus *Gpc-6B1* on wheat chromosome 6B and a 350-kb region on rice chromosome 2. Funct Integr Genomic 4(1):59–66

Distelfeld A, Uauy C, Fahima T, Dubcovsky J (2006) Physical map of the wheat high-grain protein content gene *Gpc-B1* and development of a high-throughput molecular marker. New Phytol 169:753–763

Echeverry-Solarte M, Kumar A, Kianian S, Simsek S et al (2015) New QTL alleles for quality-related traits in spring wheat revealed by RIL population derived from supernumerary × non-supernumerary spikelet genotypes. Theor Appl Genet 128:893–912. https://doi.org/10.1007/s00122-015-2478-0

El-Bok S, Bnejdi F, El-Gazzah M (2013) Evidence of cytoplasmic and epistatic effects in inheritance of grain protein content in durum wheat. J Food Agri Environ 11:804–806

El-Feki WM, Byrne PF, Reid SD, Lapitan NLV, Haley SD (2013) Quantitative trait locus mapping for end-use quality traits in hard winter wheat under contrasting soil moisture levels. Crop Sci 53:1953–1967

FAO (2014) FAOSTAT http://faostat3.fao.org/faostat-gateway/go/to/home/E
Fenn D, Lukow OM, Bushuk W, DePauw RM (1994) Milling and baking quality of 1BL/1RS translocation wheats. I. Effects of genotype and environment. Cereal Chem 71:189–195
Fontaine J, Ravel C, Pageau K, Heumez E, Dubois F, Hirel B, Le Gouis J (2009) A quantitative genetic study for elucidating the contribution of glutamine synthetase, glutamate dehydrogenase and other nitrogen-related physiological traits to the agronomic performance of common wheat. Theor Appl Genet 119:645–662
Gillies SA, Futardo A, Henry RJ (2012) Gene expression in the developing aleurone and starchy endosperm of wheat. Plant Biotechnol J 10:668–679. https://doi.org/10.1111/j.1467-7652.2012.00705.x
Golabadi M, Arzani A, Mirmohammadi Maibody SAM (2012) Identification of microsatellite markers associated with grain protein content in durum wheat grown under drought stress at terminal growth stages. Cereal Res Commun 40:215–224
Gonzalez-Hernandez JL, Elias EM, Kianian SF (2004) Mapping genes for grain protein concentration and grain yield on chromosome 5B of *Triticum turgidum* (L.) var. dicoccoides. Euphytica 139:217–225
Groos C, Robert N, Bervas E, Charmet G (2003) Genetic analysis of grain protein content, grain yield and thousand–kernel weight in bread wheat. Theor Appl Genet 106:1032–1040
Groos C, Bervas E, Charmet G (2004) Genetic analysis of grain protein content, grain hardness and dough rheology in a hard × hard bread wheat progeny. J Cereal Sci 40:93–100
Groos C, Bervas E, Chanliaud E, Charmet G (2007) Genetic analysis of bread-making quality scores in bread wheat using a recombinant inbred line population. Theor Appl Genet 115:313–323
Gupta PK, Kumar J, Mir RR, Kumar A (2010a) Marker-assisted selection as a component of conventional plant breeding. Plant Breed Rev 33:145–217
Gupta PK, Langridge P, Mir RR (2010b) Marker-assisted wheat breeding: present status and future possibilities. Mol Breed 26:145–161
Guttieri MJ, Ahmad R, Stark JC, Souza E (2000) End use quality of six hard red spring wheat cultivars at different irrigation levels. Crop Sci 40:631–635
Hagenblad J, Asplund L, Balfourier F, Ravel C, Leino MW (2012) Strong presence of the high grain protein content allele of NAM-B1 in Fennoscandian wheat. Theor Appl Genet 125:1677–1686
Harjit-Singh PM, Varshney RK, Roy JK, Balyan HS, Dhaliwal HS, Gupta PK (2001) STMS markers for grain protein content and their validation using near-isogenic lines in bread wheat. Plant Breed 120:273–278
Hazen SP, Ng PKW, Ward RW (1997) Variation in grain functional quality for soft winter wheat. Crop Sci 37:1086–1093
Hemantaranjan A, Garg OK (1988) Iron and zinc fertilization with reference to the grain quality of *Triticum aestivum* L. J Plant Nutr 11:1439–1450
Heo H, Sherman J (2013) Identification of QTL for grain protein content and grain hardness from winter wheat for genetic improvement of spring wheat. Plant Breed Biotech 1(4):347–353
Huang XQ, Cloutier S, Lycar L, Radovanovic N et al (2006) Molecular detection of QTLs for agronomic and quality traits in a doubled haploid population derived from two Canadian wheats (*Triticum aestivum* L.) Theor Appl Genet 113:753–766
Humphreys DG, Townley-Smith TF, Lukow OM et al (2010) Burnside extra strong hard red spring wheat. Can J Plant Sci 90(1):79–84
Joppa LR, Cantrell RG (1990) Chromosomal location of genes for grain protein content of wild tetraploid wheat. Crop Sci 30:1059–1064
Joppa LR, Du C, Hart GE, Hareland GA (1997) Mapping gene(s)for grain protein in tetraploid wheat (*Triticum turgidum* L.) using a population of recombinant inbred chromosome lines. Crop Sci 37:1586–1589
Kibite S, Evans LE (1984) Causes of negative correlations between grain yield and grain protein concentration in common wheat. Euphytica 33:801. https://doi.org/10.1007/BF00021906

Kong LG, Si JS, Zhang B, Feng B, Li SD, Wang FH (2013) Environmental modification of wheat grain protein accumulation and associated processing quality: a case study of China. Aust J Crop Sci 7(2):173–181

Kuhn JC, Stubbs TL, Carter AH (2016) Effect of the *Gpc-B1* Allele in hard red winter wheat in the US Pacific Northwest. Crop Sci 56:1009–1017

Kulwal P, Kumar N, Kumar A, Balyan HS, Gupta PK (2005) Gene networks in hexaploid wheat: interacting quantitative trait loci for grain protein content. Funct Integr Genomic 5:254–259

Kumar N, Kulwal PL, Balyan HS, Gupta PK (2007) QTL mapping for yield and yield contributing traits in two mapping populations of bread wheat. Mol Breed 19:163–177

Kumar J, Mir RR, Kumar N, Kumar A et al (2010) Marker-assisted selection for pre-harvest sprouting tolerance and leaf rust resistance in bread wheat. Plant Breed 129:617–621

Kumar J, Jaiswal V, Kumar A, Kumar N et al (2011) Introgression of a major gene for high grain protein content in some Indian bread wheat cultivars. Field Crops Res 123:226–233

Kunert A, Naz AA, Dedeck O, Pillen K, Léon J (2007) AB-QTL analysis in winter wheat: I. Synthetic hexaploid wheat (*T. turgidum* ssp. dicoccoides · *T. tauschii)* as a source of favourable alleles for milling and baking quality traits. Theor Appl Genet 115:683–695

Laperche A, Brancourt-Hulmel M, Heumez E, Gardet O et al (2007) Using genotype × nitrogen interaction variables to evaluate the QTL involved in wheat tolerance to nitrogen constraints. Theor Appl Genet 115:399–415

Li J, Cui F, Ding A-M, Zhao C-H et al (2012a) QTL detection of seven quality traits in wheat using two related recombinant inbred line populations. Euphytica 183:207–226

Li Y, Zhou R, Wang J, Liao X, Brandland G, Jia J (2012b) Novel and favorable QTL allele clusters for end-use quality reveled by introgression lines derived from synthetic wheat. Mol Breed 29:627–643

Li C, Bai G, Chao S, Carver B, Wang Z (2016) Single nucleotide polymorphisms linked to quantitative trait loci for grain quality traits in wheat. Crop J 4:1–11

Liu HE, Wang QY, Rengel Z, Zhao P (2015) Zinc fertilization alters flour protein composition of winter wheat genotypes varying in gluten content. Plant Soil Environ 61:195–200. https://doi.org/10.17221/817/2014-PSE

Mahjourimajd S, Taylor J, Rengel Z, Khabaz-Saberi H et al (2016) The genetic control of grain protein content under variable nitrogen supply in an Australian wheat mapping population. PLoS One 11(7):e0159371. https://doi.org/10.1371/journal.pone.0159371

Mann G, Diffey S, Cullis B, Azanza F et al (2009) Genetic control of wheat quality: interactions between chromosomal regions determining protein content and composition, dough rheology, and sponge and dough baking properties. Theor Appl Genet 118:1519–1537

Maphosa L, Langridge P, Taylor H, Emebiri LC, Mather DE (2015) Genetic control of grain protein, dough rheology traits and loaf traits in a bread wheat population grown in three environments. J Cereal Sci 64:147–152. https://doi.org/10.1016/j.jcs.2015.05.010

Mariani BM, D'Egidio MG, Novaro P (1995) Durum wheat quality evaluation: influence of genotype and environment. Cereal Chem 72:194–197

McMullen MD, Byrne PF, Snook ME, Wiseman BR et al (1998) Quantitative trait loci and metabolic pathways. Proc Natl Acad Sci U S A 95:1996–2000

Meng L, Li H, Zhang L, Wang J (2015) QTL IciMapping: integrated software for genetic linkage map construction and quantitative trait locus mapping in bi-parental populations. Crop J 3:265–279

Mesfin A, Frohberg RC, Anderson JA (1999) RFLP markers associated with high grain protein from *Triticum turgidum* L. var. *dicoccoides* introgressed into hard red spring wheat. Crop Sci 39:508–513

Mesfin A, Frohberg RC, Khan K, Olson TC (2000) Increased grain protein content and its association with agronomic and end-use quality in two hard red spring wheat populations derived from *Triticum turgidum* L. var. *dicoccoides*. Euphytica 116:237–242

Miedaner T, Korzun V (2012) Marker-assisted selection for disease resistance in wheat and barley breeding. Phytopathology 102:560–566

Mishra VK, Gupta PK, Arun B, Chand R et al (2015) Introgression of a gene for high grain protein content (*Gpc-B1*) into two leading cultivars of wheat in Eastern Gangetic Plains of India through marker assisted backcross breeding. J Plant Breed Crop Sci 7(8):292–300

Moore CM, Richards RA, Rebetzke GJ (2015) Phenotypic variation and QTL analysis for oil content and protein concentration in bread wheat (*Triticum aestivum* L.) Euphytica 204:371–382

Nelson JC, Andreescu C, Breseghello F, Finney PL et al (2006) Quantitative trait locus analysis of wheat quality traits. Euphytica 149:145–159

Novaro P, D'Egidio MG, Bacci L, Mariani BM (1997) Genotype and environment: their effect on some durum wheat quality characteristic. J Genetics Breed 51:247–252

Olmos S, Distelfeld A, Chicaiza O, Schlatter AR (2003) Precise mapping of a locus affecting grain protein content in durum wheat. Theor Appl Genet 107:1243–1251

Oury FX, Godin C (2007) Yield and grain protein concentration in bread wheat: how to use the negative relationship between the two characters to identify favourable genotypes? Euphytica 157:45–57

Oury FX, Berard P, Brancourt-Hulmel M, Depatureaux C et al (2003) Yield and grain protein concentration in bread wheat: a review and a study of multi-annual data from a French breeding program. J Genet Breed 57:59–68

Panozzo JF, Eagles HA (2000) Cultivar and environmental effects on quality characters in wheat. 2. Protein. Aust J Agric Res 51:629–636. https://doi.org/10.1071/AR99137

Patil RM, Oak MD, Tamhankar SA, Rao VS (2009) Molecular mapping of QTLs for gluten strength as measured by sedimentation volume and mixograph in durum wheat (*Triticum turgidum* L. ssp. *durum*). J Cereal Sci 49:378–386

Peck AW, McDonald GK, Graham RD (2008) Zinc nutrition influences the protein composition of flour in bread wheat (*Triticum aestivum* L.) J Cereal Sci 47:266–274

Peleg Z, Cakmak I, Osturk L, Yazici A et al (2009) Quantitative trait loci conferring grain mineral nutrient concentrations in durum wheat × wild emmer wheat RIL population. Theor Appl Genet 119:353–369

Perretant MR, Cadalen T, Charmet G, Sourdille P et al (2000) QTL analysis of bread-making quality in wheat using a doubled haploid population. Theor Appl Genet 100:1167–1175

Peterson CJ, Graybosch RA, Baenziger PS, Grombacher AW (1992) Genotype and environment effects on quality characteristics of hard red winter wheat. Crop Sci 32:98–103

Prasad M, Varshney RK, Kumar A, Balyan HS et al (1999) A microsatellite marker associated with a QTL for grain protein content on chromosome arm 2DL of bread wheat. Theor Appl Genet 99:341–345

Prasad M, Kumar N, Kulwal P, Röder MS, Balyan H, Dhaliwal H, Gupta P (2003) QTL analysis for grain protein content using SSR markers and validation studies using NILs in bread wheat. Theor Appl Genet 106:659–667

Raman R, Allen H, Diffey S, Raman H, Martin P, McKelvie K (2009) Localization of quantitative trait loci for quality attributes in a doubled haploid population of wheat (*Triticum aestivum* L.) Genome 52:701–715

Randhawa HS, Asif M, Pozniak C, Clarke JM, Graf RJ, Fox SL, Humphreys DG, Knox RE, DePauw RM, Singh AK, Cuthbert RD, Hucl P, Spaner D (2013) Application of molecular markers to wheat breeding in Canada. Plant Breed 132:458–471

Reif JC, Gowda M, Maurer HP, Longin CFH et al (2011) Association mapping for quality traits in soft winter wheat. Theor Appl Genet 122:961–970

Sears RG (1998) Improving grain protein concentration and grain yield in USA hard winter wheat. In: Fowler DB, Geddes WE, Johnston AM, Preston KR (eds) Wheat protein production and marketing. Printcrafters Inc, Winnipeg

Shewry PR (2009) Improving the protein content and quality of temperate cereals: wheat, barley and rye. Impacts of agriculture on human health and nutrition-Volume II 118

Simmonds NW (1995) The relation between yield and protein in cereal grain. J Sci Food Agric 67:309–315
Simons K, Anderson JA, Mergoum M, Faris JD et al (2012) Genetic mapping analysis of bread-making quality traits in spring wheat. Crop Sci 52:2182–2197
Sourdille P, Cadalen T, Guyomarc'h H, Snape J et al (2003) An update of the Courtot × Chinese Spring intervarietal molecular marker linkage map for the QTL detection of agronomic traits in wheat. Theor Appl Genet 106:530–538
Souza EJ, Martin JM, Guttieri MJ, O'Brien KM et al (2004) Influence of genotype, environment, and nitrogen management on spring wheat quality. Crop Sci 44:425–432
Starks TL, Johnson PE (1985) Techniques for intrinsically labeling wheat with ^{65}Zn. J Agricult Food Chem 33:691–698
Sun HY, JH L, Fan YD, Zhao Y et al (2008) Quantitative trait loci (QTLs) for quality traits related to protein and starch in wheat. Prog Nat Sci 18:825–831
Sun X, Marza F, Ma H, Carver BF, Bai G (2010) Mapping quantitative trait loci for quality factors in an inter-class cross US and Chinese wheat. Theor Appl Genet 120:1041–1051
Suprayogi Y, Pozniak CJ, Clarke FR, Clarke JM et al (2009) Identification and validation of quantitative trait loci for grain protein concentration in adapted Canadian durum wheat populations. Theor Appl Genet 119(3):437–448
Tabbita F, Lewis S, Vouilloz JP, Ortega MA et al (2013) Effects of the *Gpc-B1* locus on high grain protein content introgressed into Argentinean wheat germplasm. Plant Breed 132:48–52
Terasawa Y, Ito M, Tabiki T, Nagasawa K, Hatta K, Nishio Z (2016) Mapping of a major QTL associated with protein content on chromosome 2B in hard red winter wheat (*Triticum aestivum* L.) Breed Sci 66(4):471–480
Tiwari C, Wallwork H, Arun B, Mishra VK et al (2016) Molecular mapping of quantitative trait loci for zinc, iron and protein content in the grains of hexaploid wheat. Euphytica 207:563–570
Tsilo TJ, Hareland GA, Simsek S, Chao SM, Anderson JA (2010) Genome mapping of kernel characteristics in hard red spring wheat breeding lines. Theor Appl Genet 121:717–730
Turner AS, Bradburne RP, Fish L, Snape JW (2004) New quantitative trait loci influencing grain texture and protein content in bread wheat. J Cereal Sci 40:51–60
Tyagi S, Mir RR, Chhuneja P, Ramesh B, Balyan HS, Gupta PK (2014) Marker-assisted pyramiding of eight QTLs/genes for seven different traits in common wheat (*Triticum aestivum* L.) Mol Breed 34:167–175
Uauy C, Distelfeld A, Fahima T, Blechl A, Dubcovsky J (2006a) A NAC gene regulating senescence improves grain protein, zinc, and iron content in wheat. Science 314:1298–1301
Uauy CA, Brevis JC, Dubcovsky J (2006b) The high grain protein content gene *Gpc-B1* accelerates senescence and has pleiotropic effects on protein content in wheat. J Exp Bot 57:2785–2794
Vishwakarma MK, Mishra VK, Gupta PK, Yadav PS, Kumar H, Joshi AK (2014) Introgression of the high grain protein gene *Gpc-B1* in an elite wheat variety of Indo-Gangetic Plains through marker assisted backcross breeding. Curr Plant Biol 1:60–67
Vishwakarma MK, Arun B, Mishra VK et al (2016) Marker-assisted improvement of grain protein content and grain weight in Indian bread wheat. Euphytica 208:313–321. https://doi.org/10.1007/s10681-015-1598-6
Wang DL, Zhu J, Li ZK, Paterson AH (1999) Mapping QTLs with epistatic effects and QTL × environment interactions by mixed linear model approaches. Theor Appl Genet 99:1255–1264
Wang L, Cui F, Wang J, Jun L et al (2012) Conditional QTL mapping of protein content in wheat with respect to grain yield and its components. J Genet 91:303–312
Xu Y, Crouch JH (2008) Marker-assisted selection in plant breeding: from publications to practice. Crop Sci 48:391–407
Xu YF, An DG, Liu DC, Zhang AM, HX X, Li B (2012) Molecular mapping of QTLs for grain zinc, iron and protein concentration of wheat across two environments. Field Crops Res 138:57–62
Xue-yan C, Guo-qi S, Shu-juan Z, Yu-lian L et al (2016) The allelic distribution and variation analysis of the *NAM-B1* gene in Chinese wheat cultivars. J Integr Agr 15:60345–60347

Yang J, CC H, Hu H, Yu R et al (2008) QTLNetwork: mapping and visualizing genetic architecture of complex traits in experimental populations. Bioinformatics 24:721–723

Yao J, Ma H, Yang X, Zhou M, Yang D (2014) Genetic analysis of the grain protein content in soft red winter wheat (*Triticum aestivum* L.) Turkish J Field Crops 19(2):246–251

Zhang Y, He ZH, Ye GY, Zhang AM, Van Ginkel M (2004) Effect of environment and genotype on bread-making quality of spring-sown spring wheat cultivars in China. Euphytica 139:75–83. doi:https://doi.org/10.1007/s10681-004-2131-5

Zhao L, Zhang KP, Liu B, Deng Z, HL Q, Tian JC (2010) A comparison of grain protein content QTLs and flour protein content QTLs across environments in cultivated wheat. Euphytica 174:325–335

Statistical Data Analysis Tools: Software Prospects for Crop Productivity

Sunil Kumar, A. S. Panwar, Sudhir Kumar, M. Shamim, and Dushyant Mishra

Abstract

In agriculture research, the role of statistical analysis is quite predominant. Knowledge of statistics play a pivotal role for recommendation of specific farming system to be adopted, choice of tree/crop species, crop spacing, and other components of a farming system. This leads to the problem of choosing a tool for data analysis and subsequent activities that require statistical content. Spread sheet is one the oldest and command tool used to explain the statistical peculiarities in the research data. Yet, Excel is not a full-fledged statistical tool. Besides this basic tool, other tools are also available based on different languages to perform statistical functions. However, these language-based packages need extensive learning and training. The user must know about programming before they can be put into good use. Of course, several commercial statistics packages like SAS, SPSS, SYSTAT, STATA, and MINITAB, etc. with graphical user interface (GUI) allow command line input and programming for analysis. However, such softwares are expensive and in many cases beyond the means of the individual users. But still a number of programs have statistical functionality; however before deciding upon a statistical software, one should be clear about the requirements of statistical software and it applicability with the data available.

Keywords

Statistical · Program · Analysis · Farming system

S. Kumar (✉) · A. S. Panwar · M. Shamim · D. Mishra
Indian Institute of Farming Systems Research (IIFSR), Modipuram, Meerut, Uttar Pradesh, India

S. Kumar
Division of Plant Physiology, ICAR-Indian Agricultural Research Institute, New Delhi, India

R. S. Sengar, A. Singh (eds.), *Eco-friendly Agro-biological Techniques for Enhancing Crop Productivity*, https://doi.org/10.1007/978-981-10-6934-5_12

1 Introduction

Statistical data analysis software is a viable tool for solving problems in all aspects of agriculture. A statistical data model is a simplified representation of a system for studying about system its behavior using model. Calibration and validation are the two most important processes before giving any outcome of the model for the study area or the systems studied. Calibration or parameterization of model is needed for adjusting certain model parameters or relationships to make the model work for study sites. Validation is the comparison between the results of model simulations with observations from crops that were not used for the calibration. Application part of the model fulfills the requirement for which it was developed. Statistical models are being very much used for climatic variation-based change, forecasting (both weather and yield), yield gap analyses, yield trend analysis, devising agronomic management strategies, and extrapolation to other locations. Several software models have also been used to estimate national food production and help government for policy decisions.

Statistical software data analysis is an important method to explain the effect of different scenarios of climate and management options on agriculture, more specifically on crop productivity and its interaction among themselves. Modeling is very effective for designing and analysis of the system. Further, the applications of statistical modeling for crops, weather, soil, climate, and horticulture can show the path to the agriculture researcher and rural farming community to replace some of the traditional techniques.

The first question that needs to be answered is what peculiarities should a statistical software package have to make data analysis possible. There are so many software packages available with different functionalities to address the problem in data analysis. SPSS, SAS, SYSTAT, STATA, and MINITAB are the few known names in the statistical community for data analysis. These professional tools have great capabilities. Consequently, these professional packages are very expensive, sometimes not easy to use, and most of the time due to financial constraint remain out of reach for most research and teaching departments. In addition, most departments will not have these packages. The fact is that most statistical functions needed in agriculture research are in the areas of statistical designs of experiments, statistical inference, multivariate analysis, sampling theory, etc. There are some free tools available for the need-based data analysis.

The second requirement is feasibility in software operation. The learning curve should not take too much time. If a software package is not user friendly in operation and learning, then majority of users will try to look for other options. Ease of use for statistical software generally means that entry of data should be simple; with as many as file formats of input data, the analysis needs to be simple with a number of all required functions. Along with these the software must have GUI which is simple and easy to understand, and software should have good capabilities to bring the results in presentable forms, i.e., graphs and tables.

The third requirement is inbuilt support with the software like tutorials and help files. If there are such functions like probability distributions, statistical plots, and various exercises incorporated into the package, then data analysis and interpretation becomes more easy. Mostly common statistical functions and procedures have to be supported so that most routine analysis can be performed with ease. Linear progression, bi-plots, chi-square univariate and multivariate analysis, scatter plots, and ANOVA analysis should have the minimum capabilities and should be available within the software.

Finally, Windows operating systems being the most common OS, we have to look for packages compatible for Windows. Needless to say, many excellent tools for Mac OS or Linux operating systems are available, but being a user-friendly interface with Windows, these are not widely used in statistical community.

Here, we will describe important capabilities and properties of commonly used software packages.

2 Main Software Packages

2.1 Vista

Vista is a complete application for statistical analysis. It also has the capabilities to work with excel. Some plug-ins are inbuilt, but for specific analysis like multivariate analysis, homogeneity analysis, lognormal analysis, and multilevel analysis, they need to be downloaded. They are freely available at the site www.visualstats.org. Vista which is a free and open source package and can be installed on Windows, Unix, and Mac. This software gives the freedom to the users to develop own script to define certain functions to analysis-specific data. It has specific function which helps us to understand the processing details inside. It can execute LISP, C++, and FORTRAN programs. In terms of cross application use, it can be used to read SAS- and Excel-based data. In fact, the current software still needs more functionality to be added to get the analysis more smoothly.

2.2 AM

This package was developed by American Institute of Research. Though it's not open source, it is freely available to download for noncommercial purposes from the website address http://am.air.org.

AM has many good features for statistical analysis. A few are as follows:

(a) AM has capabilities to offer drag and drop interface for the users. Integrated help is always a boon for the users.
(b) In terms of analysis capabilities, AM offers functions to analyze including models estimated via marginal maximum likelihood (MML); besides this, it also has the option to work with assessment and survey data; MML procedures recognize that tests can't pinpoint proficiency, but instead define a probability

distribution over the proficiency scale. AM works alike the NAEP program which has the options to analyze the plausible values.

(c) AM packages include not only assessment analysis but also have functionalities to work for the non-assessment-based data also.

(d) AM has inbuilt capabilities to calculate and show the standard errors. All procedures are available with a Taylor series approximation for the standard errors and also provide the options to use replication techniques.

(e) This software is freely available, and the developers have provided the right to share this software with any one.

One of the most important property of this software is the option to provide working environment for the cross platform. Almost all the file format used to save data can be used here with a few exceptions. Most command file format like SPSS and SAS can be used with ease. It allows univariate and multivariate analysis. Though this software has some limitations in terms of the graphic capabilities and for developing own script for analysis. In terms of the help files for operation, they are easy to use and can be very helpful in this initial handling of the software.

2.3 WinIdams

WinIdams is also a very useful software for analysis. It was developed for data analysis by UNESCO. With integrated functions for analysis, this software provides the option for the users to use own developed programs. With capabilities to analyze interactive data, it also has option to present the data, i.e., it has functions to plot the data. This software has the capabilities to work the most common statistical analysis like ANOVA, frequency tables, etc. This software can be downloaded freely from the website http://www.unesco.org/idams.

2.4 IRRISTAT

IRRI has given us so many rice varieties, but a few knows that they also developed a statistical software for field-based data analysis. This software was developed mainly to analyze rice-based data grown in the field. IRRISTAT is the software developed by IRRI. It has the capabilities to perform ANOVA, regression, and several other statistical procedures. Besides these this is also capable to perform univariate and multivariate analysis. User interface is very easy and handy and can be used with ease to get the data analyzed. In terms of graphic option, it is somewhat scary, but still for general use, it can be used. This software is freely available and can be downloaded from the IRRI website.

2.5 Openstat

It is also a free and open source software and can be downloaded from the statpages website. Most of the command statistical analysis can be performed with this software, but this software has many limitations as it does not give the users the freedom to make own scripts or program to analyze data. Besides these limitations, it is also very specific for the file format; all format files can be used here. So mostly CSV files can be used here with ease.

2.6 Instat

Instat+ is a free software for single user, but for commercial and bulk user, it requires license to be purchased. It has many capabilities for data analysis, but still in terms of represented data in presentable form, it has limited capabilities. It is generally used in class to explain the data analysis by the teachers. It has the functions to analyze most of the statistical functions like ANOVA and regression; besides these it has the capabilities to work with multivariate analysis. This software has certain limitations as nonlinear regression can be done with this software. User-based functions can be used with this software, and programming options are not available with this software.

2.7 SAS

(a) SAS is the ultimate name for statistical analysis.
(b) It has the capabilities to access database via ODBC.
(c) Data can be put in several different locations including mainframes.
(d) It has the capabilities to run across the different platforms.
(e) Being a complete package, SAS has a wider role starting from research data analysis to the warehouse and other related analysis.
(f) There are a number of highly experienced, dedicated SAS programmers around the world.
(g) Almost all the analysis can be done through SAS in this world.
(h) It is expensive (unless one gets an academic discount). Base SAS is quite limited, but for SAS/STAT and other modules, the prices add up.
(i) SAS requires a lot of support, by someone who knows computers very well and knows SAS very well. It is a pain to learn. This is partially mitigated by the fact that SAS is very modular. Basically, there is the data step, where SAS data sets are made, manipulated, and saved. Then there is a long list of procedures (called PROCs), which can be learned one at a time, for the most part.
(j) The graphics in SAS are difficult to a certain extent. They came out with SAS/Graph, which has (theoretically) vast capability, but is almost impossible to use.

However, SAS being the most comprehensive software can be used in all operating systems. It also has the capabilities to work with all file formats. Although all the functions are integrated, but still any functions user in need, it has given the programming functions and script can also be developed to get the desired analysis. It has to be purchased and license has to be renewed every year.

2.8 SPSS

(a) SPSS is owned by the same company, which owns some other statistical package copyright such as SYSTAT. This experience of statistical programming makes it more advanced and user friendly.
(b) SPSS owes a lot to the days when the manual was one of the best introductory books around on statistics.
(c) It has inbuilt functions so nonprogrammer finds some ease to work with this package.
(d) Slower performance
(e) With wide application it is more used in professional field.
(f) Training is available everywhere.
(g) It seems that SPSS is not the software of choice in industries that need data analyst. But SPSS has its audience too. It depends on our area of expertise.

SPSS is a nice program for doing social research. SAS is a wonderful tool for doing data mining. They both have their place. It has been observed that in smaller databases, such as market research, SPSS is a far better tool for some quick and painless analysis. When you need to get some numbers flying around and are doing transformations like a maniac, you better know how to do some SAS. In terms of creating easily exportable, attractive output SPSS blows SAS away. It is not even close. SAS/Graph is well, not so good. The reason SAS is more valuable is because SPSS cannot really handle large databases very well. If there is a need to manipulate thousands and thousands of cases, SAS has the power and reliability to get you there. It also exists in a UNIX/mainframe environment; SPSS does not.

2.9 STATA

(a) It is interactive, very fast performance. Analysis can be done iteratively or through programming (called ado files). STATA loads the entire data set into RAM (which makes it much faster than SPSS in the normal case) which means the performance degrades considerably when data set size exceeds available memory, whether this matter depends on the exact size of the data sets and availability of RAM within the system.
(b) It is resource hungry and it needs large memory to work. For better speed and performance of the software, the system has to be fast.
(c) It offers “net courses” and less expensive training accomplished via e-mail.

(d) Analysis tools and support are available, both from the STATA company and from other users on the Statalist.
(e) STATA does not play as nicely with other programs as SAS does. Also, STATA does not have many of the modules available in SAS.
(f) STATA can handle extremely large data sets, both in terms of number of cases and variables.
(g) Updates are easy to install, and there are many very good user-written plug-ins available.
(h) STATA's documentation is superior to those available from SAS.
(i) STATA is much more competitively priced for the entire package rather than getting pieces of it.

STATA has many advantages from a statistics point of view. They have excellent support for complex samples (cluster, stratified, weighted) and offer lots of statistics especially for limited dependent variables. SPSS is ahead of STATA in terms of the friendly interface, and it is easier to do data management with SPSS. If one is looking for a single program to handle big data sets and do relatively simple analyses, SPSS is certainly up to the job.

2.10 MINITAB

MINITAB is very quick to learn. More than 400 textbook supplement references make it easy to use MINITAB in academic courses. There are a lot of macros written for MINITAB. MINITAB statistical software is available for PC and Macintosh microcomputer systems. MINITAB is also available for mainframes, minicomputers, and workstations including VAX and other DEC platforms, Sun, IBM, Prime, Data General, Hewlett-Packard, and others.

MINITAB is much more suited to statistical analysis than Excel, but definitely has some significant disadvantages compared to SAS or SPSS. It has a good interface and decent graphics, a reasonable macro facility, and decent support.

2.11 SYSTAT

SYSTAT is a statistical graphics software package, developed by Leland Wilkinson in the University of Illinois at Chicago.

SYSTAT offers solution for almost all statistical data with the advanced algorithms. General linear model and regression models fitting which are not possible with most of the software can be done with ease. Even for fitting model, SYSTAT is the first choice for most of the users. Besides these other statistical analyses like ANOVA, frequency table can be done with ease. The software comes with proper documentation and help, so that a user with the help of these help files can perform data analysis. User interface is also very attractive. All the menu buttons are well placed and can be used with the help of the image-based help files. Representation of the data with SYSTAT is also very nice. Graphs and table options are also available which can help in presenting the data in nice layout. This software is not freely available. It has to be purchased and license can be obtained online which can be obtained one time for complete use.

3 Conclusions

Several software packages have been described along with their capabilities, compatibilities, usefulness, and limitations. We use several statistical packages for the same data analysis. Each of the packages has core statistical capabilities which can be used, but every package has its own pros and cons of use features for the different types of analysis. Most of the software makers try to make GUI capabilities more user friendly. Among the open/free access softwares, Vista and WinIdams are the most common softwares which provide the most functionality. Better GUI helps in research/teaching statistics and analysis, not spending a lot of time and effort in learning the software. Vista and WinIdams both software packages have good graphic abilities. WinIdams is an effort of UNESCO and it might in the future be developed faster. The open access and free software have some nice capabilities, but it is questionable how they can be used in more advanced procedures. All the statistical packages yield almost the same results (number of decimal places or some decimal points may change at best) although it follows different algorithm in each cases. In addition, some more information is being provided in each package compared to others. Thus, according to need, users should choose their packages carefully.

References

http://am.air.org/. Retrieved December 12, 2015
http://www.irri.org/science/software/irristat.asp. Retrieved December 12, 2015
http://www.reading.ac.uk/ssc. Retrieved December 12, 2015
http://www.statpages.org. Retrieved December 12, 2015
http://www.unesco.org/idams. Retrieved December 12, 2015
http://www.visualstats.org/. Retrieved December 12, 2015
http://support.sas.com/training/tutorial/global.html. Retrieved May 28, 2017
https://systatsoftware.com/. Retrieved May 28, 2017
https://www.ibm.com/analytics/us/en/technology/spss/. Retrieved May 28, 2017

MIX
Papier aus verantwortungsvollen Quellen
Paper from responsible sources
FSC® C105338

If you have any concerns about our products,
you can contact us on
ProductSafety@springernature.com

In case Publisher is established outside the EU,
the EU authorized representative is:
Springer Nature Customer Service Center GmbH
Europaplatz 3, 69115 Heidelberg, Germany

Printed by Libri Plureos GmbH
in Hamburg, Germany